ACPL ITEM DISCARDED

*620 .112 N27A 7041124
NEATHERY, RAYMOND F.
APPLIED STRENGTH OF
MATERIALS

**DO NOT REMOVE
CARDS FROM POCKET**

ALLEN COUNTY PUBLIC LIBRARY

FORT WAYNE, INDIANA 46802

You may return this book to any agency, branch,
or bookmobile of the Allen County Public Library

Applied Strength of Materials

Applied Strength of Materials

Raymond F. Neathery
School of Technology
Oklahoma State University

JOHN WILEY & SONS
NEW YORK   CHICHESTER   BRISBANE   TORONTO   SINGAPORE

Copyright © 1982, by John Wiley & Sons, Inc.

All rights reserved. Published simultaneously in Canada.

Reproduction or translation of any part of
this work beyond that permitted by Sections
107 and 108 of the 1976 United States Copyright
Act without the permission of the copyright
owner is unlawful. Requests for permission
or further information should be addressed to
the Permissions Department, John Wiley & Sons.

Library of Congress Cataloging in Publication Data:

Neathery, Raymond F. (Raymond Franklin), 1939–
   Applied strength of materials.

  Includes index.
  1. Strength of materials.   I. Title.
TA405.N39   620.1'12   81-14732
ISBN 0-471-07991-X   AACR2

7041124

To Kay, Raymond, Paige, and Neal

Preface

In recent years, numerous textbooks have appeared to meet the needs of the growing discipline of engineering technology. In mechanics and other areas, many books have used only basic mathematics and have omitted the use of calculus. Today, however, many schools—and students—are ready for texts that have a more sophisticated mathematical approach. This book is an attempt to fill that need.

Applied Strength of Materials is intended primarily for students who have a knowledge of technical calculus. Throughout the text, the understanding and application of the derivative and the integral are emphasized. Since these two concepts can be used to characterize much of the technical working world, they are too important for today's engineering technologists to ignore. To facilitate the students' understanding, a careful review of these concepts is included in Chapter 1.

Throughout the text there is also an attempt to show the student that mathematics is our trustworthy friend and not an adversary to be avoided. I recognize that although calculus is very important in developing and understanding the concepts of strength of materials, in applications it is rarely used. Thus, a student without a calculus background will not have difficulty in the routine application of most of the material. By emphasizing Chapter 1, such students should be fairly comfortable with the development of the material as well.

The text also is based on an understanding of the principles of statics, which are necessary for an understanding of most of the equations developed and used in strength of materials. The student is encouraged to understand stress and deflection phenomena, to know what equations mean, and to recognize when they may be used. Chapter 2 presents an abbreviated statics course, while Chapters 4 and 6 present additional statics topics. This material will serve as a review and reference for students already familiar with statics, or it can be used to introduce the subject. If the statics material is used as an introduction, it must be given substantial attention, since the remainder of the book depends on an understanding of statics topics.

Engineering technology has often been characterized as applied engineering, since it does address the service, operation, and production end of the engineering employment spectrum. Therefore, any book written for the engineering technologist must consistently emphasize the application and practical importance of the material. I have tried to do this not by adding a chapter on applications, but by emphasizing the practical nature of the material on every page. This aspect of the book will make it a valuable reference to those in industry as well.

I wish to acknowledge the following reviewers who have made this a better book with their constructive suggestions: Donald E. Breyer, California State Polytechnic University, Pomona; T. M. Brittain, The University of Akron; Stanley M. Brodsky, New York City Technical College; Donald S. Bunk, Dutchess Community College; Eugene F. Keutzer, Illinois Valley Community College; Dan M. Parker, Southern Technical Institute; John O. Pautz, Middlesex County College; Richard S. Rossignol, Central Virginia Community College. In addition, my friend Charles Kroell, of General Motors Research Laboratories, made numerous helpful suggestions. And, finally, Fran Maurizzi has surely spent as much time typing as I have writing. I am grateful to all.

R. F. Neathery

Contents

1. **INTRODUCTION** ... 1
   1.1 Baseball, Hotdogs, Apple Pie, and Chevrolet ... 1
   1.2 Dimensional Algebra ... 3
   1.3 Significant Figures ... 6
   1.4 Unity on Units? ... 9
   1.5 The Derivative in a Changing World ... 12
   1.6 Summing Up Integration ... 14
   Problems ... 19

2. **STATICS** ... 22
   2.1 Newton's Laws ... 22
   2.2 The Parallelogram Law for the Addition of Forces ... 23
   2.3 Moments of Force ... 26
   2.4 Free Body Diagrams ... 27
   2.5 Equilibrium ... 30
   2.6 Two and Three Force Members ... 33
   2.7 Summary ... 39
   Problems ... 40

3. **STRESS AND STRAIN** ... 46
   3.1 Normal Stress ... 46
   3.2 Shear Stress ... 49
   3.3 Normal Strain ... 53
   3.4 Hooke's Law ... 55
   3.5 Characteristics of Other Materials ... 61
   3.6 Poisson's Ratio ... 63
   3.7 Thermal Strain ... 66
   3.8 Shear Stress and Strain ... 68
   3.9 Allowable Stress ... 69
   3.10 Summary ... 73
   Problems ... 74

4. DISTRIBUTED LOADS AND PROPERTIES OF AREAS — 84
- 4.1 Distributed Load and Center of Gravity — 85
- 4.2 Centroids — 87
- 4.3 Centroids of Composite Areas — 89
- 4.4 Second Moment of Area or Moment of Inertia — 92
- 4.5 Parallel Axis Theorem — 94
- 4.6 Moment of Inertia of Composite Areas — 97
- 4.7 Radius of Gyration — 102
- 4.8 Polar Moment of Inertia — 103
- 4.9 Summary — 105
- Problems — 106

5. TORSIONAL LOADS — 111
- 5.1 Power Transmission — 111
- 5.2 Shear Stress — 117
- 5.3 Longitudinal Shear Stress — 121
- 5.4 Hollow Shafts — 123
- 5.5 Angular Twist — 126
- 5.6 Shaft Couplings — 132
- 5.7 Summary — 134
- Problems — 135

6. INTERNAL FORCES — 140
- 6.1 Axial Loads — 141
- 6.2 Beams — 143
- 6.3 Internal Shear and Bending Moment — 145
- 6.4 Beam Graphs — 151
- 6.5 Relation between Load, Shear and Bending Moment — 154
- 6.6 General Procedure for Beam Diagrams — 157
- 6.7 Summary — 164
- Problems — 165

7. STRESS AND STRAIN FROM BENDING LOADS — 168
- 7.1 Stress and Strain Distribution — 169
- 7.2 Flexure Formula — 173
- 7.3 Beam Design Using the Flexure Formula — 180
- 7.4 Summary — 183
- Problems — 184

8. BEAM DEFLECTION — 189
- 8.1 Deflection in Design — 189
- 8.2 Relation between Bending Moment and Curvature — 190
- 8.3 Load Rate to Shear to Bending Moment to Slope to Deflection — 193
- 8.4 Deflection by Multiple Integration — 195
- 8.5 Deflection by Graphical Integration — 199
- 8.6 Deflection Formulas — 202
- 8.7 Deflection by Superposition — 204

	8.8	The Moment-Area Method	206
	8.9	Summary	211
		Problems	212
9.	**SHEARING STRESSES IN BEAMS**	219
	9.1	The Shear Stress Formula	219
	9.2	Shear Stress Distribution	224
	9.3	Maximum Shear Stress for Common Cross Sections	229
	9.4	Shear Flow	230
	9.5	Summary	232
		Problems	232
10.	**PUTTING IT ALL TOGETHER: COMPOUND STRESSES**	238
	10.1	Nonaxial Loads	240
	10.2	A Special Case: Concrete and Masonry Structures	245
	10.3	Loads Not Parallel to the Axis	247
	10.4	Superposition of Shearing Stresses	252
	10.5	Coiled Springs	260
	10.6	Summary	264
		Problems	264
11.	**PLANE STRESS ANALYSIS**	271
	11.1	Stress on an Inclined Plane	271
	11.2	General Stress Formulas	275
	11.3	Principal Stresses	279
	11.4	Maximum Shear Stress	284
	11.5	Mohr's Circle	287
	11.6	Application of Mohr's Circle to Failure Analysis	294
	11.7	Summary	297
		Problems	298
12.	**COLUMNS**	304
	12.1	Buckling	304
	12.2	Euler's Equation	305
	12.3	Using Euler's Equation	308
	12.4	End Conditions for Columns	311
	12.5	Short Columns and Long Columns	316
	12.6	Idealized Design of Columns	319
	12.7	Design Formulas	320
	12.8	Summary	321
		Problems	322
13.	**MISCELLANEOUS APPLICATIONS**	327
	13.1	Cylindrical Pressure Vessels	327
	13.2	Spherical Pressure Vessels	330
	13.3	Bolted and Riveted Connections	332
	13.4	Weldments	335

13.5	Stress Concentration	338
13.6	Fatigue	342
13.7	Summary	344
	Problems	345

14. **STATICALLY INDETERMINATE MEMBERS** — 351

14.1	Axially Loaded Members	352
14.2	Beams	358
14.3	Beams by Superposition	359
14.4	Summary	362
	Problems	362

APPENDIX TABLES — 370

ANSWERS TO PROBLEMS — 399

INDEX — 417

1

Introduction

1.1 Baseball, hotdogs, apple pie and Chevrolet
1.2 Dimensional algebra
1.3 Significant figures
1.4 Unity on units?
1.5 The derivative in a changing world
1.6 Summing up integration

1.1 BASEBALL, HOTDOGS, APPLE PIE, AND CHEVROLET

Once upon a time there were three little pigs. ...Oh well, you know the story of these poor, misguided creatures. Of course, they were excellent and hard-working craftsmen. (Not unlike your everyday average technology student.) And, my, such efficiency has seldom been seen. But, alas, they had failed to recognize the intricacies of statically indeterminate structures and temperamentally indeterminate wolves. This text can be of great assistance in designing wolf-resistant structures. It won't be much help with the wolf. (Similarities between the wolf and your instructor are purely coincidental.)

The topic at hand, strength of materials, is fundamental to the design of houses—straw, hay, or brick—airplanes, spaceships, pool rooms, football stadia, beer cans, elevators, dirt bikes, lawn mowers, ferris wheels, freeways, toilet booths, vending machines, skyscrapers, tug boats, nuclear missiles, nuclear reactors, canal locks, casket handles, voting machines, snuff cans, bar stools, batting helmets, artificial limbs, seat belts, speedometer cables, paper clips, electric power cables, hospital beds, socket wrenches, screwdrivers, paint scrapers, eyeglass frames, and one or two other items. It is an applied subject!

In the Bible in the tenth chapter of Ecclesiastes it is recorded of King

Solomon, "Through indolence the rafters sag, and through slackness the house leaks." Thus we find that the wisest of men from antiquity appreciated hard work *and* the deflection of beams. It has always been so! The successful student in this subject will likewise come to appreciate hard work and the deflection of beams.

There are several points which we will emphasize in this introductory chapter. The first is that this text is written for the student. (This statement is commonly used in the introduction to textbooks. It often is the last thing you understand.) That is a bit of a problem, because although you, the student, buy the book, your instructor selects it. So, you'll forgive me if from time to time I include a paragraph or two for your instructor's benefit. In writing for the student I'll use a personal style. That is not the usual style for technical writing. However, I have found that the student who benefits most from a text is the one who reads it. I want to make that pleasant—even enjoyable. I'll try to lighten the load with a touch of humor.

You should understand from the outset that, although this subject is extremely important, your passing or failing will do little to change the course of the world. Pass or fail, God will still be in His heaven. Most successful people in the world live rewarding lives without having ever heard of it. Why, some of my best friends didn't even know Poisson had a ratio! So it is important—but not the meaning of life.

Second, we must recognize the subjects' name. We call it *strength of materials* because that is traditional, but we will be concerned with strengths in only an incidental way. It is not a very descriptive name. Mechanics of materials is a better title, and the one I prefer. In engineering schools it might be called statics of deformable bodies or even continuum mechanics.

Most students find the subject very demanding. You can take it one of two ways—seriously or again. It will lean heavily on the mathematics and statics you have learned previously. Weaknesses in these subjects will cause severe problems. Substantial deficiencies will be very hard to overcome and should be remedied before proceeding with this study. I have taught this subject many times and have discussed it at length with others who have also. The three most common difficulties students seem to have are (1) algebra, (2) trigonometry, and (3) statics. All three of these are prerequisites for this course. A distant recognition of these topics is inadequate. You must have a working command of them.

In addition to the above, this book requires an understanding of the basics of differential and integral calculus. Few problems will actually require the manipulation of calculus equations. However, since calculus is universally required in engineering technology programs and since the differential and integral are so powerful in explaining and developing the concepts of the subject, it seems negligent not to use them. A brief review of calculus is given at the end of this chapter. This will be sufficient to allow the uninitiated to read the text and handle practically all the problems (most of the problems do not require calculus). There will be a few problems which the uninitiated will find excessively difficult.

This will be the second course in mechanics for most students. As in the previous course, statics, a proper approach to problems is very important. An orderly approach to problem solution cannot be overemphasized. Significant problems cannot be worked three to a page. Nor can they often be worked without error. Hence the objective of the neat and orderly format is not only to make it easier to arrive at the correct answer, but to make it easier to find the errors that you are sure to make. An appropriate format will be emphasized throughout the book. An "engineer's pad" or some similar form of grid paper will be very helpful.

Chapter 2 gives a review of statics. The principles of statics are easy to understand, but mastering their use is another matter. Those who are uninitiated to statics should study Chapter 2 very carefully and in detail. If sufficient time is devoted to this brief study of statics, an adequate comprehension can be had for the purposes of this course. However, a separate course is recommended for most students. The material of Chapters 4 and 6 is covered in some texts on statics but not universally. Therefore, it is presented in this text as though it were new material, although some will treat it as review.

1.2 DIMENSIONAL ALGEBRA

Dimensional algebra is an important area to be emphasized now and throughout this text. Mastery of this concept will be useful to the student in all courses based on physical principles and throughout his professional life. A primary application is in the conversion from one unit system to another. Students of this generation are likely to be heavily involved in this activity as we convert to the metric-based International System (SI—French, you know). There is an elaborate and very useful theory called dimensional analysis associated with the principle put forth here. We do not go into that in this book but emphasize what we call *dimensional algebra*. The necessary tools for using dimensional algebra are simply the basics of algebra we already know.

When we talk about the dimensions of an equation, we are using a general term for the units. If we say something is 12 ft long, the units are feet and the dimensions are length. If we say a car is moving at 60 mph, the units are miles per hour and the dimensions are length per unit time. Other units of velocity are feet per second, meters per second, and furlongs per fortnight. These different units all have the same dimensions—length per unit time. With these definitions of dimensions and units, what we are going to do here might be better described as unit algebra. However, we will stick to the phrase *dimensional algebra*.

The principle of dimensional algebra is that if terms are physically related by an equation, they must be dimensionally equivalent as well. That is, all physical equations must be dimensionally homogeneous. For example, if a body moves at constant velocity, its displacement is its velocity multiplied by the elapsed time, or

$$s = vt \qquad (1.1)$$

4 INTRODUCTION

This equation says that if we travel at 60 mph for 2 hr then we will have traveled 120 mi, found by

$$s = \left(60\,\frac{\text{mi}}{\text{hr}}\right) 2\text{ hr} = 60(2)\left(\frac{\text{mi}}{\cancel{\text{hr}}}\right)(\cancel{\text{hr}})$$

$$s = 120\text{ mi}$$

Note that in arriving at this answer we do the algebra on the numbers *and* the units. The algebra of the units determines the units for distance traveled. We may examine units dimensionally as follows:

$$s \stackrel{D}{=} \frac{\text{mi}}{\cancel{\text{hr}}}(\cancel{\text{hr}}) \stackrel{D}{=} \text{miles}$$

The equal sign with a D over its means "s is equal dimensionally to." Other equations are illustrated below.

$$a = \frac{v}{t} \stackrel{D}{=} \frac{\text{ft}}{\text{s}} \cdot \frac{1}{\text{s}} \stackrel{D}{=} \frac{\text{ft}}{\text{s}^2}$$

$$s = v_1 t + \tfrac{1}{2} a t^2$$

$$s \stackrel{D}{=} \left(\frac{\text{ft}}{\cancel{\text{s}}}\right)\cancel{\text{s}} + \frac{\text{ft}}{(\cancel{\text{s}})^2}(\cancel{\text{s}})^2$$

$$s \stackrel{D}{=} \text{ft} + \text{ft} \stackrel{D}{=} \text{ft}$$

The above equation illustrates another use of dimensional algebra. It calls for adding $v_1 t$ to $\tfrac{1}{2} a t^2$. The units for $v_1 t$ and $\tfrac{1}{2} a t^2$ must be the same; otherwise meaningful addition cannot occur. In other words, we cannot add apples to oranges. (Actually we can, and if we include bananas and melon balls it makes an excellent fruit salad which is quite tasty and nourishing—but meaningless physically.) This point also allows us to check the accuracy of equations about which we have doubt. If in the above we were not sure whether the last term were $\tfrac{1}{2} a t^2$ or $\tfrac{1}{2} a t$, the dimensions would dictate $\tfrac{1}{2} a t^2$. On the other hand, the $\tfrac{1}{2}$ is a dimensionless constant, and this analysis is blind to it.

There is a class of equations that is confounding to dimensional algebra. These equations are empirical and design equations which have constants in them that carry dimensions. Constants in basic physical equations never carry dimensions. An example of a design equation is

$$H = \frac{s_s N D^3}{321{,}000}$$

This equation gives the power in horsepower that can be carried by a shaft turning at N revolutions per minute of D diameter, in inches, and of an allowable shear stress, in pounds per square inch. This equation is dimensionally consistent only when the 321,000 carries with it the units of pound-revolution per horsepower-minute. Without these units being written with the constant—and they usually are not—the equation has only limited application. It is further handicapped by the fact that it is only valid when s_s, N, and D are entered with the proper units. It is, however, quite valuable if we are routinely making this calculation, because it relieves us of repeatedly making the same units conversion.

Equations of this type can be generally recognized by their uneven constants. Naturally occurring constants which are usually dimensionless are integers, π, e, and the like.

The last usage of dimensional algebra that we consider is unit conversion and unit inconsistency. Returning to Eq. 1.1, we ask how far can we go at 60 mph in 30 s. We write

$$s = vt = 60 \frac{\text{mi}}{\text{hr}} (30) \text{s}$$

$$s = 1800 \frac{\text{mi-s}}{\text{hr}}$$

The above answer is physically valid! However, it is not conventional, nor does it communicate to most people the distance traveled. Hence we transform it to more conventional and understandable units, as follows:

$$s = 1800 \frac{\text{mi-s}}{\text{hr}} \times \left(\frac{1 \text{ hr}}{60 \text{ min}} \right) \left(\frac{1 \text{ min}}{60 \text{ s}} \right)$$

$$s = \frac{1800}{(60)(60)} \text{mi} = \mathbf{0.500 \text{ mi}}$$

Notice that unit conversion is achieved by multiplying by factors that are physically equal to one. They are written as fractions in which the numerator and denominator each express the same quantity, but in different units. The fractions selected are those that will lead to the desired result through cancellation. Examples are

$$1 = \frac{1 \text{ hr}}{60 \text{ min}}$$

and

$$1 = \frac{60 \text{ s}}{\text{min}}$$

This is a preferred method of unit conversion which always works and eliminates confusion over conversion factors.

We conclude that there is a threefold usage of dimensional algebra.

1. It checks the dimensional accuracy of the equation. All equations must be dimensionally consistent.
2. It checks the dimensional accuracy of the execution of a solution. Carrying through the algebra of the units must result in units that are appropriate for the variable being evaluated.
3. It allows efficient and accurate conversion from one set of units to another—achieved by multiplying terms by 1 until the desired units are acquired.

One final note is in order concerning the units for angles. As shown in Fig. 1.1 the angle θ is defined as the arc length s divided by the radius r.

$$\theta = \frac{s}{r} \tag{1.2}$$

In this equation, if s is equal to r, the radius, the angle defined is one radian. By dimensional algebra we see

$$\theta \underset{=}{\text{D}} \frac{\text{ft}}{\text{ft}} = \text{dimensionless}$$

Thus we conclude that an angle (being the ratio of two lengths) is, in fact, dimensionless when measured in radians. Any other unit of angular measure will not be dimensionless. Basic equations involving angles will invariably call for their measurement in radians. It is not surprising that the SI calls for angular measurement in radians.

1.3 SIGNIFICANT FIGURES

The electronic calculator has revolutionized technical education. Calculations are now made much more rapidly and accurately than before. For instance, with unbelievable ease my handy-dandy SR-56 tells me that $10^{2.1} \sin 28° = 59.10296805$. Fantastic! And as fast as I can punch it in. Amazing! The only problem is—in the words of Sportin' Life—"It ain't necessarily so!" In fact, it is most likely not so.

The difficulty comes with the calculator. Although it is extremely fast and

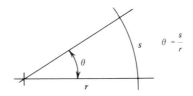

FIG. 1.1 Angle definition.

almost intelligent, it is—well—naive. That is, it believes whatever we tell it. So when we say we want $10^{2.1}$ it thinks we mean $10^{2.1000000}$—and so forth. The truth is we know very few numbers that precisely, and in engineering technology it is a rare event when that kind of precision is needed. In any case, the naive calculator assumes the information we give it is exact, and it gives us answers according to the best of its ability—which is also limited. The information given by the numbers 2.1 and 28 was given to two significant figures. The answer cannot be more precise. Were our calculator more intelligent it would have said 59—two significant figures at best.

Before proceeding with this discussion, let's define what we mean by significant figures. The number of significant figures indicates the precision with which a quantity is known. We cannot measure any physical quantity *exactly*—all measurements are approximations. We can, of course, count exactly, but as the number grows larger we often resort to less precise estimates. For the remainder of this discussion we will exclude exact counts.

Let's say we measure a piece of steel with a ruler and find it to be 2.0 in. long. We then measure it with a vernier caliper and find it to be 2.01 in. long. Lastly, we measure it with a micrometer and find it to be 2.008 in. Each successive instrument allows for a more precise estimate of the length of the piece proceeding from two to three and then to four significant figures. The difference in these three numbers is the precision with which we indicate the value. None is exact.

To determine the number of significant figures move left to right on the number. Start counting at the first nonzero and continue to the last nonzero. If the last nonzero is to the right of the decimal, keep counting until the last zero. This gives the number of significant figures. For example:

$$\underset{\text{xxxx}}{308.2} \rightarrow 4 \text{ significant figures}$$

$$\underset{\text{xxxxx}}{0.0012300} \rightarrow 5 \text{ significant figures}$$

$$\underset{\text{xxx}}{418,000} \rightarrow 3 \text{ significant figures}$$

Note that lead zeros do not count, middle zeros do, and trailing zeros do not unless they are to the right of the decimal. The last case is actually ambiguous. It is not clear if there are three, four, five, or six significant figures, though if it is written as 418,000.0, there are clearly seven. Scientific notation eliminates this ambiguity. With it we write

$$4.18 \times 10^5 \text{ or}$$

$$4.180 \times 10^5 \text{ or}$$

8 INTRODUCTION

$$4.1800 \times 10^5 \text{ or}$$

$$4.18000 \times 10^5$$

according to out intent.

Because neither the computer nor our delightfully pleasant typist likes to use superscripts, we have taken to writing "computereeze." We will often write

$$4.18 \times 10^5 \quad \text{as } 4.18E5$$

in the manner of an IBM 360 or even my hand-held calculator. I think you'll find it convenient, too.

Let us consider now the number 2. By it we mean some number between 1.5 and 2.5. With only 2 we absolutely cannot distinguish between 1.7 and 2.2. We represent this as

$$2 = \begin{cases} 1.5 \\ 2.5 \end{cases}$$

In a similar manner

$$3 = \begin{cases} 2.5 \\ 3.5 \end{cases}$$

Next consider the product 2×3. We evaluate this and its extremes

$$1.5 \times 2.5 = 3.75$$

$$2 \times 3 \quad = 6$$

$$2.5 \times 3.5 = 8.75$$

We see that 2×3 really includes any value from 3.75 to 8.75—a very wide range. We say that the answer is

$$6 = \begin{cases} 5.5 \\ 6.5 \end{cases} \simeq \begin{cases} 3.75 \\ 8.75 \end{cases}$$

As can be seen from the above, when we say $2 \times 3 = 6$ we are limiting the answer to a range of possible values that is a small portion (20%) of the actual range. This cannot be avoided without writing the number with tolerances (which is sometimes necessary). We should, however, point out the absurdity of saying

$$2 \times 3 = 6.0 \begin{cases} 5.95 \\ 6.05 \end{cases}$$

which includes only 2% of the possible values.

It will be helpful for comparison to consider the product 2.00×3.00 and its extremes

$$\begin{matrix} 1.995 \times 2.995 = 5.975 \\ 2.00 \ \times 3.00 \ = 6.00 \\ 2.005 \times 3.005 = 6.025 \end{matrix} = \begin{cases} 5.995 \\ 6.005 \end{cases}$$

Again we see that by preserving the original number of significant figures, three, we indicate a range of answers that is approximately 20% of the range of possible answers. Greater precision (6.000) is clearly ludicrous (look it up). Since 6.00 is overly precise, why not drop to 6.0? That gives

$$6.0 = \begin{cases} 5.95 \\ 6.05 \end{cases}$$

This range is twice as large as the actual range 5.975 to 6.025. We are faced with the dilemma of either overspecifying the answer as 6.00 or underspecifying the answer as 6.0. The absurd option of 6.000 and finer tuned numbers is clearly not acceptable. We opt to use 6.00, because it does not enlarge the original uncertainty of the numbers. Additional operations will continually degrade the precision of the numbers.

In most calculations it is not necessary or desirable to do this type of error analysis. We can get by just fine by applying a simple rule:

The number of significant figures in an answer can be no more than the smallest number of significant figures in any of the factors forming the answer.

This is simplified further in our application, since most of the factors we will deal with can be conveniently determined to three significant figures. Unless we indicate otherwise, we will assume all given values are accurate to three significant figures. Therefore, all derived answers will be good to three significant figures.

1.4 UNITY ON UNITS?

In the United States engineering calculations have traditionally been made using English units, especially in the mechanical and civil fields. Undoubtedly this system will be around for years to come. Most technical literature available today continues to be in English units in these fields. Nonetheless the switch is on, and the switch is to metric, or more correctly Système International d' Unités (SI). This switch is already complete in some industries, while meeting significant resistance in others. There is no doubt that the switch is expensive—and inconvenient—so continued resistance by industry is understandable and predictable. However, the greatest resistance appears to be in the public sector.

Now, the focus on problems is not always the sharpest in the public sector. Political organizations have already been formed to oppose the adoption of the

SI. Some antagonists see the SI as a communist plot, others only as part of a takeover by the international oil cartel. Still others believe it results in sterility to left-handed people. I, personally, have access to a suppressed report showing that Canadian rats kept in metric cages develop intestinal cancer 37.3% more frequently than rats kept in cages with English measurements. (This paper was to be presented at the 1979 Annual Meeting of the Flat Earth Society but was stricken from the program due to pressure from the International Committee for a True Two-by-Four.)

In spite of the pressure and reluctance, the cold, hard reality is that today's technologist will function with a dual measurement system for much of his professional life. Thus in this text we present both. The real crunch comes when one goes from one system to the other.

The methods of unit transformation presented in the previous article on dimensional algebra are totally adequate for transforming between different English units and English and SI units. Because our purpose here is to gain understanding and skill in the use of the concepts of strength of materials, we are reluctant to confound the issue and burden the student further with the transformation of units. Hence we generally keep them separate, which makes it a whole lot easier. Also, our reluctance to tackle the problem of transformation will not make it go away. And, sooner or later, it's going to get you!

Having rejected all responsibility for your ability in transforming units, I will now demonstrate (out of the goodness of my heart) how the principles of dimensional algebra can be applied. Units which are important in our study are given in Table 1.1, and conversion factors are given in Table A1.1.

EXAMPLE PROBLEM 1.1

An automobile is tested for crash-protection capability by crashing it into a wall at 30 mph. What is this speed in meters per second?

Solution

All conversions are accomplished by multiplying the number by 1, physically. Multiplying by one will not change the true value. Thus selecting factors from Table A1.1 and using other basic information, we have

$$\frac{30 \text{ mi}}{\text{h}} \cdot \frac{5280 \text{ ft}}{\text{mi}} \cdot \frac{\text{h}}{60 \text{ min}} \cdot \frac{\text{min}}{60 \text{ s}} = 44.0 \frac{\text{ft}}{\text{s}}$$

We have multiplied by 1, three times, doing the indicated algebra on the unit as well as the conversion factors. We continue

$$\frac{44.0 \text{ ft}}{\text{s}} \cdot \frac{0.3048 \text{ m}}{\text{ft}} = 13.4 \frac{\text{m}}{\text{s}}$$

Table 1.1 Metric and English units

Dimension	SI (symbol)	Common English (symbol)
Length	meter (m)	foot (ft) or inch (in.)
Mass	kilogram (kg)	$\left(\dfrac{\text{lb-s}^2}{\text{ft}}\right)$ or slug
Time	second (s)	second (s)
Force	newton (N) or $\left(\dfrac{\text{kg}\cdot\text{m}}{\text{s}^2}\right)$	pound (lb) or kilopound (kip)
Stress or pressure	pascal (Pa) or $\left(\dfrac{\text{N}}{\text{m}^2}\right)$	$\left(\dfrac{\text{lb}}{\text{in.}^2}\right)$ or (psi)
Energy or work	joule (J) or (N·m)	ft-lb —
Angle	radian (rad)	degree (°)
Power	watt (W) or $\left(\dfrac{\text{N}\cdot\text{m}}{\text{s}}\right)$ or $\left(\dfrac{\text{J}}{\text{s}}\right)$	$\left(\dfrac{\text{ft-lb}}{\text{s}}\right)$ or horsepower (hp) or watts (W), etc.
Area	(m²)	(in.²), etc. —
Volume	(m³)	(in.³), etc. —
Velocity	$\left(\dfrac{\text{m}}{\text{s}}\right)$	$\left(\dfrac{\text{ft}}{\text{s}}\right)$
Angular velocity	$\left(\dfrac{\text{rad}}{\text{s}}\right)$	revolution per minute (rpm)
Density	$\left(\dfrac{\text{kg}}{\text{m}^3}\right)$	$\left(\dfrac{\text{lb-s}^2}{\text{ft}^4}\right)$
Acceleration	$\left(\dfrac{\text{m}}{\text{s}^2}\right)$	$\left(\dfrac{\text{ft}}{\text{s}^2}\right)$
Moment of force	(N·m)	(lb-ft)

Automobile speeds would more commonly be expressed in kilometers per hour. Thus with a bit of backtracking

$$\frac{13.4\text{ m}}{\cancel{\text{s}}} \cdot \frac{60\,\cancel{\text{s}}}{\cancel{\text{min}}} \cdot \frac{60\,\cancel{\text{min}}}{\text{h}} = 48{,}300\ \frac{\text{m}}{\text{h}}$$

Or adjusting the decimal $48.3\ \dfrac{\text{km}}{\text{h}}$

Of course, this conversion could also have been made by dividing by 1. In fact, we might interpret some of the calculations as being division, for example, 60 min/hr. Whether multiplying or dividing, I find it convenient to work with a single division bar and will frequently do so. It minimizes the opportunity for error in the units algebra.

1.5 THE DERIVATIVE IN A CHANGING WORLD

There are two principal ideas in calculus—the derivative and the integral. Both of these concepts can be best understood by their geometric interpretations. Figure 1.2 shows a plot of the function $y=2(x)^{1/2}$. In the language of the mathematician—or as Charlie Brown would say, for those who speak algebra—we can express this in the general form as $y=f(x)$. What this says is that y is a function of x, and it represents the above function, or for that matter, most any other function. Now let's suppose we wish to know the slope of the curve at the point (1,2). It escapes me for the moment just why we want to know the slope at this point, but please bear with me. We construct a line through the point and tangent to the curve at the point (Fig. 1.2). The slope of this line is the slope of the curve at this point, and it is equal to the tangent of the angle θ, which is the angle the line makes with a horizontal. This slope is approximated by the value $\Delta y/\Delta x$.

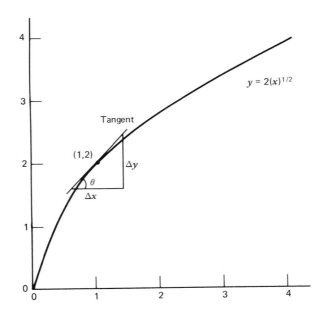

FIG. 1.2

This approximation becomes more and more accurate as Δy and Δx become smaller. This idea is expressed mathematically as:

$$\lim_{\Delta x \to 0} \frac{\Delta y}{\Delta x} = \frac{dy}{dx}$$

This expression, dy/dx, is known as the derivative. Geometrically it is the slope of the curve at the point in question. Physically it has many other interpretations. Rules for taking the derivative are given in calculus texts, and a summation of these is given in Table A1.2. From this table we take Eq. (6) which reduces to

$$\frac{d}{dx}(x^n) = nx^{n-1}$$

Applying this to the equation

$$y = 2(x)^{1/2}$$

we get

$$\frac{d}{dx}(2x^{1/2}) = 2(\tfrac{1}{2})x^{-1/2}$$

and thus

$$\frac{dy}{dx} = \frac{1}{x^{1/2}} = y'$$

y' (say "y prime") is a shorthand notation for the first derivative. Now evaluating this for the point $(1, 2)$ gives

$$\frac{dy}{dx} = \frac{1}{(1)^{1/2}} = 1.00$$

$$\tan \theta = \frac{dy}{dx} = 1.00$$

and

$$\theta = 45° = \frac{\pi}{4} \text{ radians}$$

The angle θ is the angle the tangent makes with the horizontal. We have specifically found the slope at the point $(1, 2)$. However, in doing so, we also found the equation of the derivative in general form, namely:

$$y' = \frac{1}{x^{1/2}}$$

This equation is the equation of the slope of the curve $y=2x^{1/2}$ for all values of x. To find the slope at any point, one need only substitute the appropriate value of x into equation $y'=1/\sqrt{x}$. This relatively insignificant problem illustrates the power of using the appropriate mathematical tool to solve an entire class of problems in a single step.

The above application is a geometric one and is illustrative of many cases in which geometry can be applied directly or by analogy to physical problems. To illustrate the idea further, consider an object in free-fall motion, say a gravity drop-type pile driver. It has been determined that the motion of the pile driver can be described by the equation

$$s = 16.1\, t^2$$

where s is the displacement of the driver (in feet) at any time t (in seconds). We wish to know its velocity after it has fallen for one-half second. The velocity can be defined as the first derivative of displacement with respect to time.

$$v = \frac{ds}{dt}$$

Again using Eq. (6) from Table A1.2 we have

$$v = \frac{ds}{dt} = 2(16.1)t = 32.2t$$

Substituting in $t=0.5$ s we have:

$$v = 32.2\,(.5) = 16.1 \text{ ft/s}$$

We have found the velocity to be 16.1 ft/s after free falling for 0.5 s. Notice that in the process we have again solved a more general problem and have found an equation for velocity as a function of time.

1.6 SUMMING UP ON INTEGRATION

Integration is essentially a summation process. If we wish to know the total area bounded by the curve in Fig. 1.3 we find it by summing up the small areas, ΔA, shown thus

$$A = \Delta A_1 + \Delta A_2 + A_3 + \cdots + \Delta A_n$$
$$= \Sigma \Delta A_i$$

Where Σ is the capital Greek letter sigma (English s) and stands for summation. The subscript i takes on the values of 1 to n as the summation is executed. If we now let the ΔAs become smaller, then the ΔAs at the border such as ΔA_n fit the

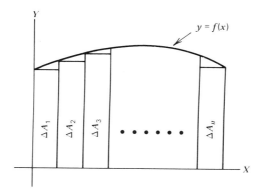

FIG. 1.3 Area under the curve $y=f(x)$.

border better, and we approximate the total area more closely. In fact, if we could let the ΔAs become small enough, the total would approach the exact area to whatever degree of precision we desired. Mathematically we can let ΔA decrease without bound, and if we do, the total area is exactly equal to the summation. We express this as

$$A = \lim_{\Delta A \to 0} (\Sigma \Delta A_i) = \int dA$$

This geometric interpretation of the integral is a very important one. You should remember that integration is a summation process. Note the symbol for summation, Σ, is the Greek equivalent to a capital S while the integral sign, $\int$, is an old English script S. Both are indicating summation. We use the symbol dA for the infinitesimal limit of ΔA. This is known as the differential of A. The integrals for several common expressions are given in Table A1.3.

There are, of course, other interpretations of the integral as well as applications. We will see some of these as we go through this text. An additional important one can be observed if we consider the differential equation

$$\frac{dy}{dx} = x$$

This equation is solved by separating the variables

$$dy = x\, dx$$

and integrating using Eq. (4) from Table A1.3

$$y = \int x\, dx$$

$$y = \frac{x^2}{2} + C$$

16 INTRODUCTION

The C is for a constant of integration. It is called an *arbitrary constant*, as it identifies a specific curve out of a family of curves. In physical problems it is not "arbitrary" at all but has a specific value based on the initial or boundary conditions of the particular problem. If we differentiate the above equation we get

$$\frac{dy}{dx} = x$$

which is what we started with.

From the above we may observe that the integral is the inverse of the differential and vice versa. Because of this, the integral is often called the antiderivative.

Table A1.3 lists some common integral forms. Note the similarity between these forms and the differential forms in Table A1.2. The table does not include the arbitrary constant, which must be added when the integration is indefinite. The forms are used without the arbitrary constant for definite integration (or integration with limits). Indefinite integration (with an arbitrary constant) is frequently used to find a functional relation between variables. Definite integration is more often used to find the specific numerical answer to a problem, although these roles are reversible. (Examples of definite integration are given in the following problems.)

EXAMPLE PROBLEM 1.2

For Fig. 1.4 find the area bounded on the top and bottom by the curve

$$y = 2 + 2x$$

and the X axis; and on the left and right by the Y axis and the line $x = 6$.

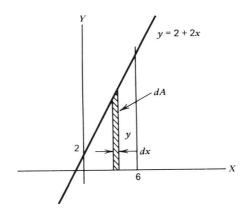

FIG. 1.4

Solution

A rectangular element of area is identified

$$dA = y\,dx$$

The total area is the summation of all the elements of area between $x=0$ and $x=6$. Mathematically we say

$$A = \int dA = \int_0^6 y\,dx$$

Before this equation can be evaluated we must know y as a function of x. This is the equation of the bounding curve, namely

$$y = 2 + 2x$$

By substitution,

$$A = \int_0^6 (2 + 2x)\,dx$$

Because integration is a summation process, we may split this into parts

$$A = \int_0^6 2\,dx + \int_0^6 2x\,dx$$

These integrals can be evaluated by Eqs. (2), (3), and (4) from Table A1.3. Thus

$$A = 2\int_0^6 dx + 2\int_0^6 x\,dx$$

$$= 2x\Big|_0^6 + 2\left(\frac{x^2}{2}\right)\Big|_0^6$$

$$= 2[(6) - (0)] + [(6)^2 - (0)^2]$$

$$A = 48$$

We have worked this problem without units. Dimensionally the area will be length squared.

EXAMPLE PROBLEM 1.3

The deceleration of a car during a crash is as shown in Fig. 1.5. This deceleration pulse is approximated by the function:

$$a = -644 \sin\left(\frac{\pi}{0.12}\right)t$$

18 INTRODUCTION

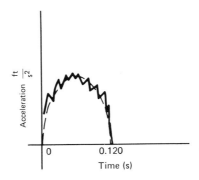

FIG. 1.5

where t is given in seconds, the angle is evaluated in radians, and a will be in ft/s^2.

The change in velocity of the car during the crash will be the area under the acceleration curve or the integral

$$\Delta v = \int a \, dt$$

Find the change in velocity.

Solution

$$\Delta v = \int a \, dt$$

$$\Delta v = -644 \int_0^{0.12} \sin\left(\frac{\pi}{.12}\right) t \, dt$$

where the constant has been factored outside the integral and the limits of integration are set from 0 to 0.12 s. Equation (7), Table A1.3 is

$$\int \sin x \, dx = -\cos x$$

To make our problem fit this form, we observe

$$x = \frac{\pi}{.12} t$$

So

$$dx = d\left(\frac{\pi t}{.12}\right) = \frac{\pi \, dt}{.12}$$

Thus we rewrite our problem as

$$-644 \left(\frac{.12}{\pi}\right) \int_0^{0.12} \sin \frac{\pi}{.12} t \left(\frac{\pi \, dt}{.12}\right)$$

And on integration we get

$$\Delta v = \frac{-644(.12)}{\pi}\left[-\cos\frac{\pi t}{.12}\right]_0^{.120}$$

$$\Delta v = +\frac{644}{\pi}(.12)\left[\cos\frac{\pi}{.12}(.12) - \cos\frac{\pi}{.12}(0)\right]$$

$$\Delta v = +24.6[\cos\pi - \cos 0]$$

The cosine is evaluated for the angle in radians.

$$\Delta v = 24.6[(-1)-(+1)]$$

$$\Delta v = -49.2\frac{\text{ft}}{\text{s}} = -33.5 \text{ mph}$$

PROBLEMS

1.1 Using dimensional algebra determine an acceptable set of units for the indicated variable in each of the following equations:

(a) $S_s = \dfrac{Tc}{J}$ for c

(b) $S_s = \dfrac{VQ}{Ib}$ for V

(c) $\dfrac{P}{A} = \dfrac{\pi^2 E}{(L/k)^2}$ for k

(d) $S = Ee$ for E

(e) $\delta = \dfrac{PL^3}{3EI}$ for I

Table of units for problems 1.1 and 1.2

S, S_s, E	$\dfrac{\text{lb}}{\text{in.}^2}$
T, M	in.-lb
J, I	in.^{4}
$b, d, c, L, \ell, k, \delta$	in.
A	in.^{2}
Q	in.^{3}
π, e	Dimensionless
w	lb/ft
P	lb

1.2 Which of the following equations are dimensionally valid?

(a) $S = \dfrac{Mc}{I}$

(b) $M = \dfrac{w\ell^3}{8}$

(c) $S = \dfrac{P\ell^2}{I}$

(d) $I = \bar{I} + Ad^2$

1.3 Convert the following values to the units indicated:

(a) 42.3 ft-lb to in.-lb

(b) 1250 $\dfrac{\text{lb}}{\text{ft}^2}$ to $\dfrac{\text{lb}}{\text{in.}^2}$

(c) 86.0 hp-days to ft-lb

(d) 928 $\dfrac{\text{lb-ft}^2}{\text{in.}^4}$ to $\dfrac{\text{lb}}{\text{in.}^2}$

(e) 650 $\dfrac{\text{revolutions}}{\text{min}}$ to $\dfrac{\text{rad}}{\text{s}}$

1.4 Write each of the following numbers to four, three, two, and one significant figures.

(a) 866,723

(b) 942.06

(c) 0.00012345

(d) Exactly 2

(e) Exactly 0.02

(f) π(3.14159)

1.5 Do Problem 1.3, observing the rules for three significant figures.

1.6 Compute the range of values covered by each of the following products; compare that with the range represented when the answer is given to the same number of significant figures as are contained in the original factors:

(a) 4×5

(b) 4.0×5.0

(c) 4.00×5.00

1.7 Convert each of the following terms to the indicated units:

(a) 30 psi to Pa

(b) 215 ft^3 to m^3

(c) 22 $\dfrac{\text{ft}}{\text{s}}$ to $\dfrac{\text{km}}{\text{h}}$

(d) 225 $\dfrac{\text{lb}}{\text{ft}}$ to $\dfrac{\text{N}}{\text{m}}$

(e) 20,000 $\dfrac{\text{ft-lb}}{\text{min}}$ to W

1.8 Differentiate each of the following expressions with respect to x.
(a) $y = 6x^3$
(b) $y = 8x^2 + 2x + 3$
(c) $y = 2\sin(2x)$
(d) $y = 3 \ln x$
(e) $y = x \sin\left(\dfrac{x}{2}\right)$

1.9 Find the slope of the following curves at the indicated values of x.
(a) $y = 4x - 3$ at 2, 4, and 6
(b) $y = 6x^2 + 4$ at 2, 4, and 6
(c) $y = \sin 2x$ at 0, 45°, and 90°
(d) $y = x$ at 0, 4, and 9
(e) $y = 3 \ln x$ at 1, 4, and 9

1.10 Integrate the following expressions:
(a) $\int 2x^3 \, dx$
(b) $\int \left(\dfrac{x^2}{3} + 2x + 4\right) dx$
(c) $\int \sin 2x \, dx$
(d) $\int \sqrt{3x} \, dx$
(e) $\int 4\cos 3\theta \, d\theta$

1.11 Evaluate each of the integrals in Problem 1.10 for the indicated limits:
(a) 0 to 6
(b) 2 to 4
(c) 0 to 20°
(d) 2 to 8
(e) 0 to 45°

1.12 Find the area bounded by the x and y axes, the vertical line $x = 8$, and the curve
(a) $y = 4 + 10x$
(b) $y = 6 + 8x^2$
(c) $y = x^2 + 2x + 2$

1.13 An automobile is accelerated at 12 m/s². If its initial velocity is zero, what will its speed be after 5 s? How far will it have traveled?

Hint: $v = \int a \, dt$, $s = \int v \, dt$

1.14 An automobile crashes into a rigid barrier and comes to rest. It is decelerated at a constant 20 g (1 g = 32.2 ft/s²) for 200 ms. What was its initial velocity?

Hint: $v = \int a \, dt$

2

Statics

2.1 Newton's laws
2.2 The parallelogram law for the addition of forces
2.3 Moments of force
2.4 Free body diagrams
2.5 Equilibrium
2.6 Two and three force members
2.7 Summary

2.1 NEWTON'S LAWS

The field of mechanics is based on Newton's three laws of motion. These may be stated as:

FIRST LAW. If the sum of the external forces acting on a body at rest is zero, the body will remain at rest (or continue on with constant velocity, if in motion).

$$\Sigma F = 0 \qquad (2.1)$$

SECOND LAW. If the sum of the external forces acting on a body is not zero, the body will have an acceleration proportional to and in the direction of the unbalanced force.

$$\Sigma F = Ma \qquad (2.2)$$

THIRD LAW. If a body exerts a force on a second body, the second body will exert an equal but opposite force on the first body.

These principles are of such a fundamental nature that the student should commit them to memory if they are not already at his command. Statics is based on the first and third laws. The second is the basis of dynamics. In addition to Newton's laws there are just a few principles that are essential in the study of statics. Among these are the parallelogram law for the addition of forces, the concept of a moment of force, the application of Newton's laws to moments of force, and the free body diagram. They are discussed in the following sections.

2.2 THE PARALLELOGRAM LAW FOR THE ADDITION OF FORCES

Forces may be characterized as vectors—that is, they behave like vectors. Vector addition (and, therefore, force addition) is according to the parallelogram law. Two vectors (forces), **A** and **B**, may be added as shown in Fig. 2.1. (The boldface **A** and **B** denote vector as opposed to scalar quantities.) The vector addition shown in Fig. 2.1 can be done graphically or mathematically. Both methods are useful. The sum of **A** and **B** is known as the resultant and is designated as **A + B**. It is the diagonal of the parallelogram formed by **A** and **B** when all three, **A**, **B**, and **A + B**, have a common origin. The inverse of the parallelogram law of addition is also true; that is, the vector **A + B** can be resolved into the component vectors **A** and **B**. Again, this may be done graphically or mathematically.

A particular set of components is most commonly used. These are called rectangular components. The vector **C** in Fig. 2.2 is resolved into rectangular components graphically. Note that the vector **C** = **C**$_x$ + **C**$_y$. Denoting the angle between **C**$_x$ and **C** as θ results in the following relations:

$$C_x = C \cos \theta \tag{2.3}$$

and

$$C_y = C \sin \theta \tag{2.4}$$

to mathematically resolve the vector **C** into its components. Note the omission of the boldface **C**, **C**$_x$, and **C**$_y$ in these two equations, meaning we are referring to the associated scalar quantity or magnitude. When adding vectors mathematically, it

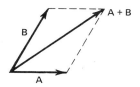

FIG. 2.1

FIG. 2.2

most often is desirable to first break them into rectangular components and then add the components algebraically.

If

$$C = A + B \quad \text{vector addition} \qquad (2.5)$$

then

$$C_x = A_x + B_x \quad \text{scalar addition} \qquad (2.6)$$

$$C_y = A_y + B_y \quad \text{scalar addition} \qquad (2.7)$$

and

$$C = C_x + C_y \quad \text{vector addition} \qquad (2.8)$$

EXAMPLE PROBLEM 2.1

Add the vectors A and B shown in Fig. 2.3a.

Solution

Graphically

The parallelogram is constructed with the diagonal $A + B$. The diagonal is scaled to show $A + B = 156$ lb at an angle of 56.3° with the horizontal.

Analytically

A and B are resolved into components and added

$$(A+B)_x = A_x + B_x = 100 \cos 30 + 0 = 86.6 \text{ lb}$$

$$(A+B)_y = A_y + B_y = 100 \sin 30 + 80 = 130 \text{ lb}$$

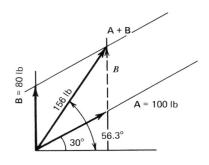

FIG. 2.3a

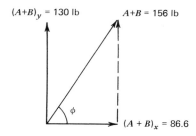

FIG. 2.3b

The components are resolved into the vector $A+B$ as shown in Fig. 2.3b.

$$\tan\phi = \frac{130}{86.6} = 1.501$$

$$\phi = 56.3°$$

$$\sin\phi = \frac{(A+B)_y}{(A+B)}$$

$$(A+B) = \frac{(A+B)_y}{\sin\phi} = \frac{130}{\sin 56.3}$$

$$A+B = 156 \text{ lb} \qquad \text{at } 56.3°$$

Alternate Analytical Solution

Sketch the force polygon as in Fig. 2.3c and note the angle opposite R is 120°. Then by the law of cosine

$$c^2 = a^2 + b^2 - 2ab\cos C$$

$$R^2 = (100)^2 + (80)^2 - 2(100)(80)\cos 120°$$

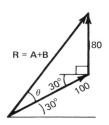

FIG. 2.3c

$$= 10{,}000 + 6400 - (-8000)$$

$$= 24{,}400$$

$$R = 156 \text{ lb.}$$

Then by the law of sines

$$\frac{\sin \theta}{80} = \frac{\sin 120°}{156}$$

$$\sin \theta = \frac{80}{156} \sin 120° = 0.444$$

$$\theta = 26.3°$$

When added to 30° this gives

$$R = A + B = 156 \text{ lb} \qquad \text{at } 56.3°$$

2.3 MOMENTS OF FORCE

A moment of a force about a point is defined as the magnitude of the force times the perpendicular distance between the point and force, Fig. 2.4.

$$M = d \times F \qquad (2.9)$$

As is illustrated in this figure, the moment of force of F has a tendency to rotate about point A. Note particularly these things:

1. There must be a reference point, point A in this case.
2. The distance is the perpendicular distance.
3. The force has a tendency to rotate about the reference point.
4. Rotation in a counterclockwise direction is called positive; clockwise is negative.
5. Moments from individual forces may be summed, yielding a resultant moment.

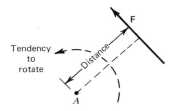

FIG. 2.4 Moment of force.

6. The unit will be a force times a distance, usually ft-lb or in.-lb. In the SI the standard unit is N·m.

EXAMPLE PROBLEM 2.2

Find the moment of the force system about point A in Fig. 2.5.

Solution

We sum the moments of force about point A, noting forces which tend to produce a counterclockwise rotation about point A as positive and those tending to a clockwise rotation as negative.

$$M_A = \Sigma M = -6(150) - 5(100) + 8(150) + (0)250$$

$$= -1400 + 1200$$

$$M_A = -200 \text{ ft-lb}$$

The negative sign means the resultant moment of force is in the clockwise direction. Thus we write:

$$M_A = \mathbf{200 \text{ ft-lb}} \curvearrowright$$

2.4 FREE BODY DIAGRAMS

An essential concept in most mechanics courses is that of the free body diagram (sometimes abbreviated FBD). (It may occur to you that a great number of things have been called "essential." Unfortunately, that is the case!) The free body diagram is, rather logically, the diagram of a free body. To arrive at a free body

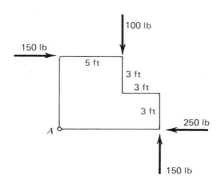

FIG. 2.5

diagram a body of interest is defined. Its boundary may be real or imaginary—it doesn't matter. A sketch is then made of the body. All external forces acting on the body are then drawn. This diagram constitutes a free body diagram. Fig. 2.6a shows a frame holding a 100-lb weight. A free body diagram of the entire structure is shown in Fig. 2.6b. The externally applied load of 100 lb is shown at E. To make the body free we cut the support at A. Since this is a fixed support, three components of reaction R_x, R_y, and M are required. A reaction is required for each type of motion that is prevented by the support. The fixed support resists motion horizontally—giving R_x; vertically—giving R_y; and rotationally—giving M. The only way motion can be resisted is by a force (or moment of force). Figure 2.7 indicates several supports, the motions they prevent, and the reactions they consequently produce. The student should affirm in his or her own mind that the supports prevent motion in the direction of the reactions and no other.

We also note that in general we will not know the sense of the unknown reactions R_x, R_y, and M at this point. Not to worry—guess! It we guess right, we will get positive answers; if wrong, negative answers.

We can also draw free body diagrams of any of the parts, as in Fig. 2.6c where we show the member BD. Note that we show the external load of 100 lb at

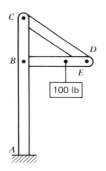

FIG. 2.6a

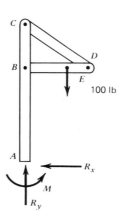

FIG. 2.6b

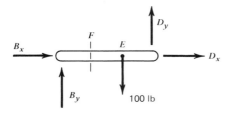

FIG. 2.6c

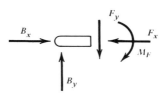

FIG. 2.6d

FREE BODY DIAGRAMS   29

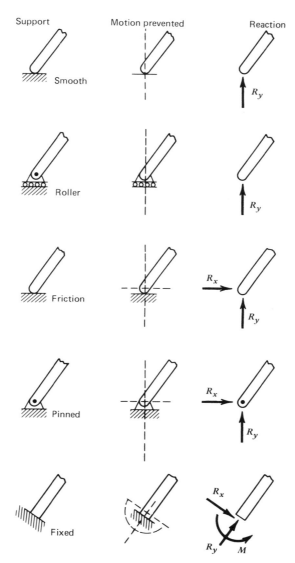

FIG. 2.7 Supports and their reactions in free body diagrams.

E and the reactions at B and D. These reactions, B and D, are external to member BD and are, therefore, shown in Fig. 2.6c. They are internal to the entire structure and, therefore, are not shown in Fig. 2.6b.

We can carry our free body diagram concept even further by sectioning member BD at the location F, shown in Fig. 2.6c. Figure 2.6d is the resulting FBD. Since BD is a rigid member, when it is cut we have three components, as in the case of a rigid support. Note that the reactions, F_x, F_y, and M are external to the free body diagram of Fig. 2.6d and are shown accordingly. However, they are internal to the member BD and are not shown in the FBD of Fig. 2.6c.

The first step in the solution of any statics problem is an accurate and complete free body diagram. Given such a free body diagram, equations satisfying Newton's first law are written in accordance with the free body diagram. The rest is algebra.

2.5 EQUILIBRIUM

Earlier we reviewed Newton's first law. It says that if a body is at rest (or at constant velocity) the sum of the external forces acting on it are zero. We call this condition *equilibrium*. For two-dimensional problems we express it by these three equations:

$$\Sigma F_x = 0 \qquad (2.10)$$

$$\Sigma F_y = 0 \qquad (2.11)$$

$$\Sigma M = 0 \qquad (2.12)$$

Equation 2.12 is a statement of the first law as it applies to moments of force. This equation may be written with respect to any reference point. In fact, it must be satisfied for every point in space.

Equipped with a free body diagram and these three equations, we are ready to attack statics problems. We should recall from algebra that if we have three equations we can solve for three unknowns. Thus if we have more than three unknowns in a planar (two-dimensional) problem, we cannot solve the problem by the method of statics alone. Such problems are called statically indeterminate.

Additional equations may be written by applying Eq. 2.12 to additional points. However, this does not allow us to handle additional unknowns, that is, more than three. Rather, these equations can be used instead of 2.10 or 2.11 when they are more convenient. They can also be used to check results, but in no case do they result in additional independent equations beyond the original three.

In some problems all the forces intersect at a single point. This is called a *concurrent force problem*. In such a problem Eq. 2.12 yields the result that zero equals zero. (I have suspected this for some time!) Such a result is called trivial; that is, it yields no useful information. That leaves us with only two equations, 2.10 and 2.11, and thus in concurrent force problems we can only handle two unknowns.

Equations 2.10 and 2.11 may also be solved graphically. If we draw the force vectors to scale laying them successively tip to tail, they will produce a force polygon that must close. This method is illustrated in Example Problem 2.6. In the case of concurrent forces, the force polygon is sufficient for solution. For nonconcurrent force systems the force polygon still must close, but this alone does not assure equilibrium. The closing of the force polygon assures that Eq. 2.10 and 2.11 are satisfied, but it says nothing about Eq. 2.12.

EXAMPLE PROBLEM 2.3

Find the external loads at A and B for Fig. 2.8a

Solution

Draw a free body diagram of the entire structure and resolve A into components as in Fig. 2.8b. We are now ready to apply Eqs. 2.10, 2.11, and 2.12.

If we arbitrarily apply them, the worst that can happen is that we have three equations in three unknowns. Following well-established procedures, the

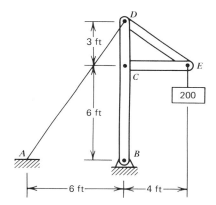

FIG. 2.8a

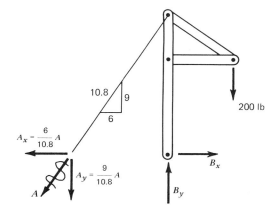

FIG. 2.8b

algebra will yield a solution. But "that ain't exactly no picnic." We can reduce the work substantially if we are more clever in writing equations, especially moment equations. We note, for instance, that the unknowns B_x, B_y and the component A_x of the unknown A all pass through point B. If we sum moments about B, these terms will not appear in the equation (their moment arms being zero). This leaves only A_y, and the equation may be solved quickly and directly for it. Thus

$$\Sigma M_B = 0 = 6 A_y - 4(200)$$

$$A_y = \frac{4(200)}{6} = \mathbf{133 \ lb} \downarrow$$

By proportion

$$\frac{A_x}{6} = \frac{A_y}{9}$$

$$A_x = \frac{6}{9} A_y = \frac{6}{9} \cdot \left(\frac{4(200)}{6} \right)$$

$$A_x = \frac{800}{9} = \mathbf{88.9 \ lb} \leftarrow$$

Resolving the components A_x and A_y gives

$$A = \mathbf{160 \ lb}$$

We now need to find B_x and B_y. Applying the principle used in obtaining our first equation, we would sum the moments about point D yielding B_x or about point A yielding B_y. That is fine, and some will prefer these steps. In taking these steps we are replacing Eqs. 2.10 and 2.11 with additional versions of 2.12, which is legitimate. However, in this case even simpler equations may be obtained by writing 2.10 and 2.11 directly. These equations are always preferred when they contain a single unknown (because they will be simpler). Thus

$$\Sigma F_x = 0 = -88.9 + B_x$$

$$\mathbf{B_x = 88.9 \ lb}$$

and

$$\Sigma F_y = 0 = -133 + B_y - 200$$

$$B_y = 200 + 133 = \mathbf{333 \ lb}$$

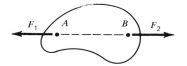

FIG. 2.9 Two force body.

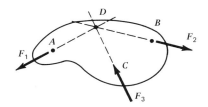

FIG. 2.10 Three force body.

2.6 TWO AND THREE FORCE MEMBERS

If a body has only two forces acting on it, Fig. 2.9, and it is in equilibrium, it can be shown (by applying Newton's first law) that the two forces must be colinear along the line connecting the two points of application. They must also be equal in magnitude but opposite in direction. Such a body is rather obviously called *der bodkinmitzweiblitzken in der unter vey*, but is better known by the obscure English term, *two force body*. Recognizing two force bodies is never necessary, but it can simplify analysis problems, as illustrated in Example Problem 2.4.

Figure 2.10 is of a three force body. Analysis of it, based on Newton's first law, leads to the conclusion that all three forces must pass through a common point. Therefore, these are called concurrent forces, meaning they come together at a point. Recognition of three force bodies or members can also simplify the analysis of some statics problems, as is illustrated in Example Problem 2.5.

The terms two force and three force are unnecessarily limiting. If we have, say, three forces at point A in Fig. 2.9, these three forces can be resolved into a single force at A. If we carry this argument to point B as well, we would have *six* forces on this two force body. Nonetheless, the principle applies. We see, then, that what is at issue in two and three force members is the number of points at which forces are applied and not necessarily the number of forces.

EXAMPLE PROBLEM 2.4

For Fig. 2.11a find the reactions at A and B.

Solution

We start by drawing a free-body diagram (Fig. 2.11b). We cut the pinned connection at A so we need two components of force, A_x and A_y. Similarly, we cut the pin at B, giving the two components B_x and B_y. We notice that there are four unknowns, A_x, A_y, B_x, and B_y. From Newton's first law we have three equations, and from algebra we know that to solve for four unknowns requires four equations, not three. Numerous expressions have been developed over the years to describe this situation. One of the more polite phrases is "up a creek without a paddle."

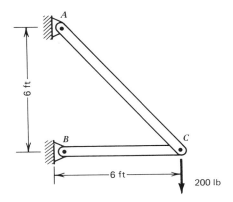

FIG. 2.11a

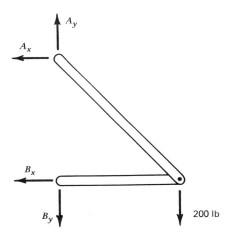

FIG. 2.11b Free body diagram.

A "paddle" may be appropriated by recognizing members AC and BC to be two force members, that is, forces come into these members at only two locations. This recognition gives us the *directions* of the forces A and B, providing two of the original four unknowns, as shown in the second free body diagram (Fig. 2.11c). Resolving these forces into components (Fig. 2.11d), we may write the equations

$$\Sigma F_y = 0 = A \sin 45° - 200$$

$$A = \frac{200}{\sin 45} = \mathbf{283\ lb}$$

TWO AND THREE FORCE MEMBERS

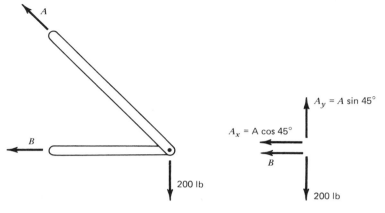

FIG. 2.11c Free body diagram recognizing two force members.

FIG. 2.11d

$$\Sigma F_x = 0 = -A\cos 45° - B$$

$$B = -A\cos 45° = -283\cos 45°$$

$$\boldsymbol{B = -200 \text{ lb}}$$

The minus sign tells us we guessed wrong in selecting B to the left in the free body diagram. It is actually to the right.

EXAMPLE PROBLEM 2.5

Find the loads at joints C and D in Example Problem 2.3 (Fig. 2.8a).

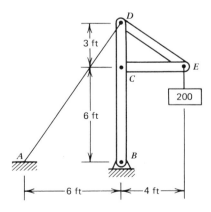

FIG. 2.8a

36 STATICS

Solution

Draw a free body diagram of member CE observing that it is a two-force member, and therefore, the force C is in the direction CE. We also resolve DE into its components DE_x and DE_y (Fig. 2.12a).

$$\Sigma F_y = 0 = DE_y - 200$$

$$DE_y = 200 \text{ lb}$$

By proportion

$$\frac{DE_x}{4} = \frac{200}{3}$$

$$DE_x = \tfrac{4}{3}(200) = 267 \text{ lb}$$

and

$$\Sigma F_x = 0 = C - DE_x = C - 267 \text{ lb}$$

$$C = \mathbf{267} \text{ lb}$$

Draw a free body diagram of member BCD (Fig. 2.12b).

Note that the force at joint C is known. It exerts 267 lb to the right on member CE. According to Newton's third law (action-reaction), it will exert 267 to the left on member BD. Summing the forces horizontally we have

$$\Sigma F_x = 0 = D_x - 267 + 88.9$$

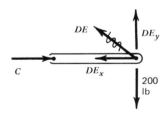

FIG. 2.12a

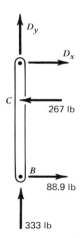

FIG. 2.12b

TWO AND THREE FORCE MEMBERS 37

$$D_x = 267 - 88.9 = \mathbf{178 \ lb}$$

And vertically

$$\Sigma F_y = 0 = D_y + 333$$

$$D_y = \mathbf{-333 \ lb} \downarrow$$

EXAMPLE PROBLEM 2.6

Find the reactions at A and B for the structure shown in Fig. 2.13a.

Solution

Member AC is a cable, and the force on it must be along it. If it were a rigid link, the same would be true, because it is a two force member. We draw a free body diagram of the structure $BDCE$ (Fig. 2.13b).

To solve this problem, sum moments about B, solving for AC, or better, sum moment about C, solving for B_x. Then sum forces horizontally and vertically, solving for the remaining unknowns.

However, we can take a different approach by recognizing the frame to be a three force member; that is, forces are applied only at B, C, and E. This requires the three forces to be concurrent—come together at a common point. This point, labeled F, is found by extending the lines of action for the 600-lb vertical force and the force AC, as shown in Fig. 2.13b.

This can be done analytically but is much simpler graphically. The force at B must also pass through F. Its direction is determined by the line BF as shown. The force polygon is then drawn by laying out the 600-lb vertical force to scale (Fig. 2.13c). The vectors AC and B are drawn by constructing them parallel to the figure above. We then scale their values, yielding

$$AC = \mathbf{700 \ lb}$$

$$B = \mathbf{1280 \ lb}$$

Plotting the last figure to a larger scale will yield more accurate results.

EXAMPLE PROBLEM 2.7

Find the internal load in member DE in Example Problem 2.6.

Solution

This time we will isolate joint E. We do this by passing cutting planes through members CE and DE. In general, when we cut a rigid member we

38 STATICS

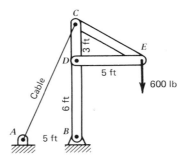

FIG. 2.13*a*

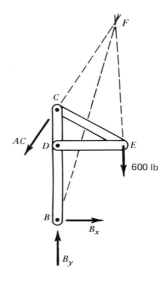

FIG. 2.13*b*

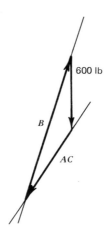

FIG. 2.13*c*

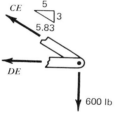

FIG. 2.14a

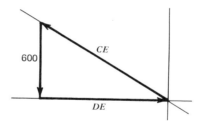

FIG. 2.14b

also get a shear force and bending moment, as shown in Fig. 2.6d. In this case both members CE and DE are two force members, so the forces are along the members, as shown in the free body diagram (Fig. 2.14a).

Given the free body diagram, the results may be obtained several ways. We have solved it graphically in this instance by laying the 600-lb force out to scale and drawing lines in the directions of DE and CE, Fig. 2.14b. Their intersection completes the force triangle. We can then scale DE to be 1000 lb to the right, that is, compression and not tension, as shown in the free body diagram. Furthermore,

$$CE = 1170 \text{ lb tension}$$

2.7 SUMMARY

Newton's first and third laws form the basis for statics. For two-dimensional problems the first law may be expressed by three equations

$$\Sigma F_x = 0$$
$$\Sigma F_y = 0$$
$$\Sigma M = 0$$

For a body to be in equilibrium (at rest, for our purposes), these three equations must be satisfied. Alternately, if these three equations are satisfied, the body is in equilibrium.

The moment of a force is the tendency of a force to rotate a body about a point. It is determined by

$$M = F \cdot d$$

To solve statics problems an accurate and complete free body diagram must be drawn. This is done by the following steps:

40   STATICS

1. Sketch the selected body.
2. Draw all external forces acting on the body.
3. Cut the body free from each support, replacing the support with the appropriate reaction force or moment. (If the support being cut is a two force member, the reaction will be along the member.)

If the number of unknown forces and moments does not exceed the number of independently available equations, from above, the statics problems may be solved. At most we can handle three unknowns on a single free body diagram. Solutions may be graphical, analytical, or a combination of the two.

PROBLEMS

2.1 Resolve the force P into horizontal and vertical components for the following values of P and θ for Fig. 2.15.

	P	θ
(a)	100 lb	30°
(b)	100 lb	60°
(c)	100 lb	75°
(d)	50 kN	0.3 rad
(e)	50 kN	0.6 rad
(f)	50 kN	1.0 rad

2.2 Resolve the force P into horizontal and vertical components for the following values of P and θ for Fig. 2.16. (Note: 1 kip is 1000 lb.)

	P	θ
(a)	250 kips	20°
(b)	250 kips	40°
(c)	250 kips	80°
(d)	1800 N	0.40 rad
(e)	1800 N	0.80 rad
(f)	1800 N	1.20 rad

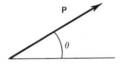

FIG. 2.15   Problems 2.1, 2.3, 2.4.

FIG. 2.16   Problem 2.2.

2.3 Resolve the force **P** into horizontal and vertical components for the following values of **P** and θ for Fig. 2.15.

	P	θ
(a)	18 tons	$+35°$
(b)	18 tons	$-35°$
(c)	18 tons	$-125°$
(d)	18 tons	$+125°$
(e)	18 tons	$215°$
(f)	18 tons	$-215°$

2.4 Resolve the force **P** into horizontal and vertical components for the following values of **P** and θ for Fig. 2.15.

	P	θ
(a)	25 kN	$\frac{\pi}{4}$ rad
(b)	25 kN	$-\frac{\pi}{4}$ rad
(c)	25 kN	$\frac{3\pi}{4}$ rad
(d)	25 kN	$\frac{-3\pi}{4}$ rad
(e)	25 kN	$\frac{5\pi}{4}$ rad
(f)	25 kN	$\frac{5\pi}{4}$ rad

2.5 Find the resultant force (that means magnitude *and* direction, without our saying so) from the following components:

	R_x	R_y
(a)	30 lb	40 lb
(b)	30 lb	-40 lb
(c)	-50 lb	-120 lb
(d)	-50 lb	$+120$ lb
(e)	120 N	50 N
(f)	-120 N	-50 N
(g)	20 kips	-30 kips
(h)	30 kips	20 kips
(i)	400 kN	600 kN
(j)	-400 kN	600 kN
(k)	-400 kN	-600 kN
(l)	400 kN	-600 kN

2.6 Find the resultant force from the following components graphically:

	A	B
(a)	50 lb at 30°	100 lb at 75°
(b)	100 lb at 30°	50 lb at 75°
(c)	800 lb at 45°	800 lb at 135°
(d)	60 kips at 90°	80 kips at $-135°$
(e)	300 N at $-20°$	400 N at $-80°$
(f)	10 kN at $-\frac{\pi}{4}$ rad	25 kN at $+\frac{\pi}{4}$ rad
(g)	25 kN at $\frac{3\pi}{4}$ rad	10 kN at $\frac{\pi}{6}$ rad

2.7 Find the resultant forces in Problem 2.6 by sketching the force polygon and using the laws of sines and cosines.

2.8 Find the resultant forces in Problem 2.6 by summing components.

2.9 Find the moment of the force system in Fig. 2.17 shown about point:
(a) A
(b) B
(c) C
(d) D
(e) E
(f) F
(g) G
(h) H

2.10 Find the moment of the force system in Fig. 2.18 about point:
(a) A
(b) B
(c) C

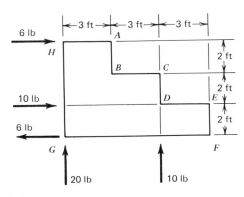

FIG. 2.17 Problem 2.9.

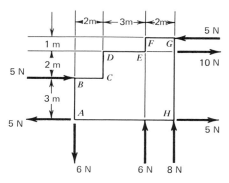

FIG. 2.18   Problem 2.10.

(d) D
(e) E
(f) F
(g) G
(h) H

2.11 For Fig. 2.19 find the reactions at A and C when $a=4$ in., $b=6$ in., $c=4$ in. and

(a) $P=100$ lb, $Q=0$
(b) $P=0$, $Q=100$ lb
(c) $P=100$ lb, $Q=100$ lb
(d) $P=250$ lb, $Q=150$ lb

2.12 For Fig. 2.19 find the reactions at A and C when: $a=3$ m, $b=5$ m, $c=4$ m and

(a) $P=50$ kN, $Q=0$
(b) $P=0$, $Q=50$ kN

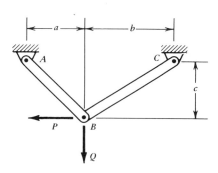

FIG. 2.19   Problems 2.11, 2.12.

44 STATICS

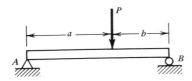

FIG. 2.20 Problem 2.13.

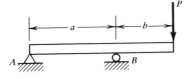

FIG. 2.21 Problem 2.14.

(c) $P=50$ kN, $Q=50$ kN
(d) $P=80$ kN, $Q=100$ kN

2.13 For Fig. 2.20 find the reactions at A and B when:

	a	b	P
(a)	6 in.	4 in.	80 lb
(b)	5 in.	3 in.	120 lb
(c)	80 mm	40 mm	20.0 N
(d)	2 m	1 m	40 kN

2.14 For Fig. 2.21 find the reactions at A and B when:

	a	b	P
(a)	6 in.	4 in.	80 lb
(b)	5 in.	3 in.	120 lb
(c)	80 mm	40 mm	200 N
(d)	2 m	1 m	40 kN

2.15 Find the reactions at A and D for Fig. 2.22.
2.16 Find the loads at the joints A, B, and D for Fig. 2.23.
2.17 Find the reactions at B and E for Fig. 2.24.

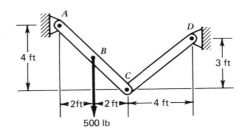

FIG. 2.22 Problem 2.15.

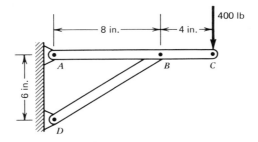

FIG. 2.23 Problem 2.16.

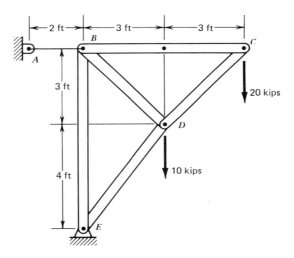

FIG. 2.24 Problem 2.17.

3

Stress and Strain

3.1 Normal stress
3.2 Shear stress
3.3 Normal strain
3.4 Hooke's law
3.5 Characteristics of other materials
3.6 Poisson's ratio
3.7 Thermal strain
3.8 Shear stress and strain
3.9 Allowable stress
3.10 Summary

3.1 NORMAL STRESS

Stress is introduced by re-examining Example Problem 2.7 of the previous chapter. In this problem a section was passed through member *CE* in order to solve a statics problem, namely, finding the load in the member. At the same time, unknowingly, the first step was taken in determining the stress in the member. The force found in member *CE* was the load necessary to maintain equilibrium. The force was found by passing a cutting plane through the member and is, therefore, an *internal force* or internal load. This is the first step in any stress analysis problem—to find the internal load. The second step is to find the stress produced by that load, and that is the principal subject of this book, but the first and necessary step is always to find the load causing the stress.

Now, given the load in the member *CE*, the question is raised, "How is that load carried?" For the moment it is assumed that it is uniformly carried, as shown

NORMAL STRESS 47

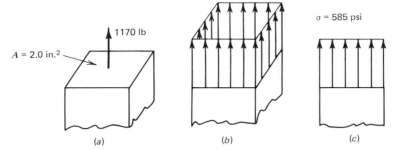

FIG. 3.1a-c Normal stress.

in Figure 3.1. If the load is equally shared by all the 2 in.² of cross-sectional area, then the stress in the member is the load divided by the area, or

$$\text{Stress} = \frac{\text{load}}{\text{area}} = \frac{P}{A} = \frac{1170 \text{ lb}}{2 \text{ in.}^2}$$

$$\sigma = \frac{585 \text{ lb}}{\text{in.}^2} = 585 \text{ psi}$$

There are several things to be observed regarding the above exercise. The first is the symbol for stress, σ. This is the Greek letter sigma; the equivalent to the English s. Some texts use the English s, but σ is much more common, and so we will use it. Becoming used to it now will make later work more convenient. In structural design it is common to use a small f for stress.

The second point is the equation we now label

$$\sigma = \frac{P}{A} \qquad (3.1)$$

This equation, fundamental to the subject at hand, should be learned by the student, and normally it will be, from repeated use. In obtaining the equation it was *assumed* the stress was uniform, that is, equally distributed. It turns out that was a very good assumption, which is approximately true in a very large number of cases. Even when it is patently untrue, the design stress is frequently based on the average stress, so Eq. 3.1 will have very wide application.

The direction of this stress should also be noted. It is perpendicular or normal to the surface over which it acts and is, therefore, called a *normal stress*. Normal means perpendicular to the surface. In addition to these two properties (direction and magnitude) a third characteristic is its distribution—uniform in this case. It is often convenient to sketch the stress distribution. When it is not sketched the student should always visualize a mental picture. A three-dimensional representation of this stress distribution is shown in Fig. 3.1b. The more common two-dimensional representation is shown in Fig. 3.1c.

48 STRESS AND STRAIN

The sense of the stress is also important. It cannot be determined from the sign of the force vector. It depends instead on the action of the stress on the body. If the stress is tending to stretch the body or pull it apart, it is called *tension*. Normally, stress-producing tension is considered positive. If the stress is compressing or squashing the body, it is called *compression* and carries a negative sign.

The last aspect considered here is that of the units for stress. These follow from Eq. 3.1

$$\sigma = \frac{P}{A} \triangleq \frac{\text{force}}{\text{area}} = \frac{\text{lb}}{\text{in.}^2} = \text{psi}$$

These are the most common units for stress in the English system. Students who have studied topics in fluid mechanics will recognize these units as those for pressure. This follows from the fact that pressure is nothing more than normal stress. Because of the properties of many engineering materials, it is common to think in terms of thousands of pounds per square inch or kilopounds per square inch. This is abbreviated as kpsi or simply ksi. (It is also common to abbreviate kilopounds force as kips. Say "kips" and spell out "k-s-i.") There are, of course, other units, and any unit of force divided by any unit of area is legitimate.

In the SI the unit of force is the newton and the unit of area is the meter squared. Thus the unit of stress is the newton per meter squared. This is a derived unit in the SI and carries the name pascal, abbreviated Pa,

$$1 \text{ Pa} = \frac{1 \text{ N}}{\text{m}^2}$$

A newton is a bit more than a fifth of a pound, and a square meter is a relatively large area. The pascal is, therefore, a relatively small unit of stress. More precisely

$$1 \text{ Pa} = 1.450 E - 4 \text{ psi}$$

or

$$1 \text{ psi} = 6.895 E 3 \text{ Pa}$$

Because this unit is so small, it is commonly used with the multiplying prefixes

mega (MPa = 1,000,000 Pa)   and   giga (GPa = 1,000,000,000 Pa)

when referring to stresses and other properties of engineering interest.

A special type of normal stress is called a bearing stress. It occurs when two objects come into contact or bear on one another. The stress that a table leg exerts on the floor is an example. Of course, bearing stresses are always compressive. Otherwise they are just normal "Normal stresses..." normally.

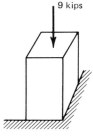

FIG. 3.2a

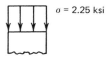

FIG. 3.2b

EXAMPLE PROBLEM 3.1

A short column (Fig. 3.2) has a 2 in.×2 in. cross section. It is used to carry a 900-lb load. Find the normal stress in the member.
Note: "kips" is a commonly used abbreviation for kilopound or 1000 pounds.

Solution

$$\sigma = \frac{P}{A} = \frac{9 \text{ kips}}{2 \text{ in.} \times 2 \text{ in.}}$$

$$\sigma = \frac{9}{(2)(2)} \frac{\text{kips}}{\text{in.}^2} = \mathbf{2.25 \text{ ksi compression}}$$

The stress distribution is uniform. The two-dimensional sketch is Fig. 3.2b. Note that the stress is perpendicular (or normal) to the surface it acts over.

EXAMPLE PROBLEM 3.2

If, in the above problem, the applied load is 15 kN and the cross section is 2 cm×3 cm, find the bearing stress of the block on the table it sits on.

Solution

$$\sigma = \frac{P}{A} = \frac{15 \text{ kN}}{(.02 \text{ m})(.03 \text{ m})} = 25000 \frac{\text{kN}}{\text{m}^2}$$

$$\sigma = 25000 \text{ kPa} = \mathbf{25.0 \text{ MPa}}$$

3.2 SHEAR STRESS

In Example Problem 3.1 just discussed, the applied force was normal, that is, perpendicular to the cross section. Figure 3.3a represents a section in which the

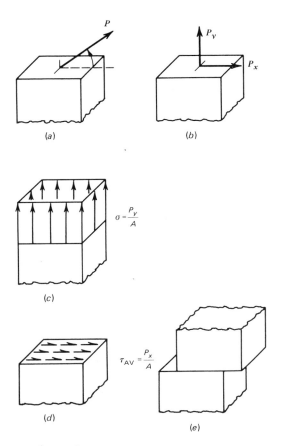

FIG. 3.3a–e Average shear stress.

internal load is not normal. In Fig. 3.3b this force **P**, a vector, has been resolved into a normal component P_y and a tangential component P_x. The normal component P_y can be related to a normal stress, as in Fig. 3.3c. Thus $\sigma = P_y/A$ gives the average normal stress which approximates the true situation very closely. The effect of tangential component P_x is to shear the member, as indicated in Fig. 3.3e. An average shear stress may be calculated

$$\tau_{AV} = \frac{P_x}{A} \tag{3.2}$$

This equation, however, differs considerably from the true stress situation. Nonetheless, for many practical reasons Eq. 3.2 is widely used in many engineering applications. The subscript AV is used to indicate that an *average*, and not the true, stress is being calculated.

SHEAR STRESS

The Greek latter tau (equivalent to t) is most frequently used for shear stress, although s_s is not uncommon. Since this is also a load divided by an area, it will also have the units of psi, ksi, Pa, MPa, and so on.

In the previous section it was noted that the magnitude, direction, and distribution of the stress represented by Eq. 3.1 were important. That is also true for the shear stress. The magnitude is, of course, given by Eq. 3.2. The direction is parallel to the surface, tending to shear, and this is called *shear stress*. The stress distribution is assumed to be uniform, as shown in Fig. 3.3d.

A common application of Eq. 3.2 is in the analysis of bolted or riveted connections. The following example problems illustrate its application.

EXAMPLE PROBLEM 3.3

A 400-lb load is carried by a 1.00-in. diameter rivet as shown in Fig. 3.4. Find the average shear stress in the rivet.

Solution

A free body diagram is drawn (Fig. 3.4b) and the load is determined

$$\Sigma F_x = 0$$

$$400 - V = 0$$

$$V = 400 \text{ lb}$$

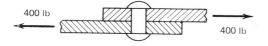

FIG. 3.4a

FIG. 3.4b

FIG. 3.4c

52   STRESS AND STRAIN

The shear load is 400 lb. Now calculate the average shear stress

$$\tau_{AV} = \frac{P}{A} = \frac{P}{\dfrac{\pi D^2}{4}} = \frac{4P}{\pi D^2}$$

Again, we usually use a single division bar

$$\tau_{AV} = \frac{400 \text{ lb }(4)}{\pi (1 \text{ in.})^2}$$

$$\tau_{AV} = 509 \text{ psi}$$

The stress distribution is uniform, as shown in Fig. 3.4c.

EXAMPLE PROBLEM 3.4

A 1-kN load is carried by a 2-cm-diameter rivet, as shown in Fig. 3.5. Find the average shear stress in the rivet.

Solution

A free body diagram of the top plate is drawn (Fig. 3.5b) and the shear load is determined.

$$\Sigma F_x = 0$$

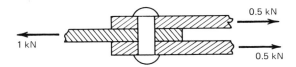

FIG. 3.5a

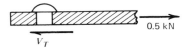

FIG. 3.5b

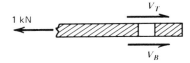

FIG. 3.5c

$$0.5 - V_T = 0$$

$$V_T = 0.500 \text{ kN}$$

A free body diagram of the middle plate is then drawn, Fig. 3.5c

$$\Sigma F_x = 0$$

$$-1.0 + V_T + V_B = 0$$

$$V_B = 1.0 - 0.5 = 0.500 \text{ kN}$$

Thus we see the shear load on either side of the middle plate is 0.500 kN. This condition is known as double shear, and in it the rivet carries only one-half of the total load. In single shear it carries the total load. To calculate the stress we have

$$\tau_{AV} = \frac{P}{A} = \frac{P}{\frac{\pi D^2}{4}} = \frac{4P}{\pi D^2} = \frac{4(0.500 \text{ kN})}{\pi (0.02 \text{ m})^2} = 1590 \text{ kPa}$$

$$\tau_{AV} = 1.59 \text{ MPa}$$

3.3 NORMAL STRAIN

Consider now a member of uniform cross section whose length is L, Fig. 3.6. An axial tensile load P is applied to the member, and it stretches in the direction of the applied load. This elongation, labeled δ (Greek delta—English d) is called

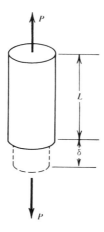

FIG. 3.6 Deformation under axial load.

total strain or deformation. Unit strain is then defined as the deformation per unit length or

$$\text{unit strain} = \frac{\text{deformation}}{\text{original length}}$$

or symbolically

$$\varepsilon = \frac{\delta}{L} \qquad (3.3)$$

The symbol for unit strain is the Greek epsilon (equivalent to the English e). Most often we refer to unit strain simply as strain, and that will be our practice throughout this text.

Note the units for strain:

$$\varepsilon = \frac{\delta}{L} \underset{=}{\text{D}} \frac{\text{in.}}{\text{in.}}$$

So the units are inch per inch, which really is dimensionless. When working with metals the strains that are allowed are usually very small. In Example Problem 3.5, a strain of 0.00165 in./in. will be found. Because the magnitude of the number is so small, it is convenient to write it as 1650 (10^{-6} in./in.). The value inside the parentheses is called microstrain (micro represents the metric prefix 10^{-6}). The Greek letter mu (μ) is used to represent it, so we write

$$\varepsilon = 1650 \, \mu$$

This terminology is common in the experimental stress analysis field. Of course, since strain is dimensionless, it makes absolutely no difference which unit system we are in.

EXAMPLE PROBLEM 3.5

A test piece (Fig. 3.7) has a diameter of 0.505 in. and a gage length of 2.00 in. (Gage length is the active length of the test piece.) Under a tensile load of 10,000 lb it elongates 0.00330 in. Find the stress and strain in the member.

Solution

From Eq. 3.1

$$\sigma = \frac{P}{A}$$

$$A = \frac{\pi d^2}{4} = 0.200 \text{ in.}^2$$

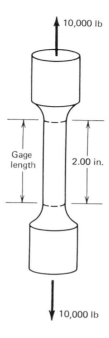

FIG. 3.7

and

$$\sigma = \frac{P}{A} = \frac{P}{0.200} = 5P = 5(10,000) = \mathbf{50{,}000 \text{ psi}}$$

The 0.505-in. diameter test piece gives an area of 0.200 in.² and the stress in psi may be calculated as 5 times the load in pounds. Because of this convenience, the 0.505-in. diameter is a standard for tensile tests. From Eq. 3.3

$$\varepsilon = \frac{\delta}{L} = \frac{0.00330 \text{ in.}}{2 \text{ in.}} = \mathbf{0.00165} \; \frac{\text{in.}}{\text{in.}}$$

Since strain is dimensionless we may leave the units off. We can also write the answer as 1650 μ meaning microstrain.

3.4 HOOKE'S LAW

In 1676 Robert Hooke postulated a principle which now bears his name. He arrived at this principle from studying springs. The original, which was in Latin, can be roughly translated as "the force varies with deflection." Today we are more likely to say that "stress is proportional to strain." We write it as

56   STRESS AND STRAIN

$$\sigma = E\varepsilon \qquad (3.4)$$

In this equation E is the constant of proportionality. E turns out to be a property of the material. It is called the modulus of elasticity or Young's modulus. It can be determined from the experiment described in Example Problem 3.5 as follows. Solving for E in Eq. 3.4, we have

$$E = \frac{\sigma}{\varepsilon}$$

Substituting from the example problem

$$E = \frac{50{,}000 \text{ lb}}{\text{in.}^2 \left(0.00165 \frac{\text{in.}}{\text{in.}} \right)} = 30.3 \times 10^6 \frac{\text{lb}}{\text{in.}^2}$$

$E = 30.3 \times 10^6$ psi

This is a very large number. Numbers of this magnitude are common for metals. Note that the units are the same as those for stress—psi. This can be clearly seen in the algebraic manipulation above, where E is equal to the stress—psi—divided by the strain—dimensionless; thus the units for E are the same as for stress. Values of the modulus of elasticity for various metals are given in Table A3.1.

In the SI we have for this example

$$E = 30.3 \times 10^6 \text{ psi} \left(\frac{6895 \text{ Pa}}{1 \text{ psi}} \right)$$

$$= 209{,}000 \times 10^6 \text{ Pa} = 209 \times 10^9 \text{ Pa} = 209 \text{ GPa}$$

Because of the very large value of the modulus of elasticity in SI units for common engineering materials, the unit GPa is normally used.

A geometric interpretation of E may also be made. Figure 3.8 is a stress-strain diagram for a mild steel. This diagram is obtained by plotting the stress against the strain during a tensile test of a material. Modern testing machines are equipped with instrumentation that automatically produce this diagram during a tensile test. A straight line is drawn between the origin and point A. The slope of this line can be calculated as $\Delta\sigma/\Delta\varepsilon$ which is the modulus of elasticity—E. The behavior of the material is described by Hooke's law between 0 and A. The relation between σ and ε is linear over this range. Point A is called the proportional limit. If the material is loaded beyond A, Eq. 3.4 no longer holds. This, however, does not seriously limit the equation, because for many engineering structures loading beyond Point A would be considered a failure.

Many materials do not have a truly linear range on a stress-strain diagram.

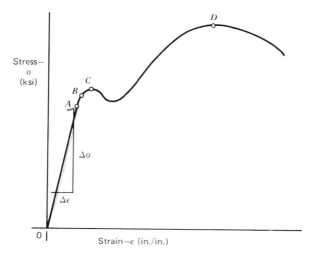

FIG. 3.8 Stress-strain diagram for a ductile material.

Equation 3.4 is widely applied to such materials by approximating the value of E. This amounts to linearizing the stress-strain curve.

While discussing the stress-strain curve, Fig. 3.8, we should note a few other properties of materials that can be obtained from it. We should also note that this particular stress-strain curve does not characterize engineering materials in general. A given material may not exhibit all the properties discussed here. However, defining the properties for Fig. 3.8 will assist us in learning the terms.

ELASTIC LIMIT—POINT B

If the load is increased to any value less than B and then removed, the stress-strain diagram will retrace its path, returning to 0. If the load exceeds B, as shown in Fig. 3.9, the plot will return to the zero stress level parallel to the initial line. It will not return to zero strain, and it results in a permanent set in the material.

YIELD POINT—POINT C (ALSO CALLED YIELD STRENGTH)

If the load reaches this value, then there will be additional deflection without additional load. For ductile steels this point is easily identified. Because of this and because of the close proximity of A, B, and C, its value is normally used as a design value, even though the failure theory is based on A or B. Consequently, material property tables like A3.1 often give this value.

We use the term stress to mean the actual load on a material. Strength refers to the property of the material. In a material's property test the stress at failure becomes the strength.

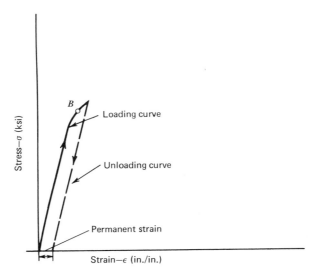

FIG. 3.9 Exceeding the elastic limit.

ULTIMATE STRENGTH—POINT D (ALSO CALLED TENSILE STRENGTH)

This is the maximum stress the material can experience without separation. This stress is calculated using the original area, although for many materials the actual area at this time is considerably reduced. The design stress is often based on the ultimate strength, and it, too, is commonly included in tables like A3.1. The strength based on the true area at failure is called the true tensile strength. It is of interest to metallurgists but is rarely used in design.

Returning again to the experiment described in Example Problem 3.5, we can now develop a direct relation between the load and deflection. From our definitions:

$$\sigma = \frac{P}{A}$$

$$\varepsilon = \frac{\delta}{L}$$

and Hooke's law

$$\sigma = E\varepsilon$$

Then by the appropriate substitution we have:

$$\frac{P}{A} = \frac{E\delta}{L}$$

Solving for δ yields

$$\delta = \frac{PL}{AE} \qquad (3.5)$$

This significant equation need not be memorized, because it is so easily obtained from the definitions of stress, strain, and Hooke's law. It is, however, worthwhile to carefully observe what it says and thereby reinforce or correct our intuition. It says the deflection is directly proportional to two things: load and length. Double the load and you double the deflection. Double the length and you double the deflection. It further says that the deflection is inversely proportional to two things: area and modulus of elasticity. Double the area and you cut the deflection in half. Double the modulus of elasticity—use a "stiffer" material—and you cut the deflection in half.

This relationship emphasizes the fact that every structural member is really a spring—usually very stiff, but nonetheless a spring. Springs are normally described by the relation

$$F = k\delta \qquad (3.6)$$

where k is known as the spring constant whose units are lb/in. Evaluating Eq. 3.6 in terms of Eq. 3.5 yields

$$k = \frac{AE}{L} \qquad (3.7)$$

EXAMPLE PROBLEM 3.6

A member whose cross section is 2×4 (actual size) is loaded as shown in Fig. 3.10. The deformation is 0.120 in. Calculate the stress, strain, and modulus of elasticity.

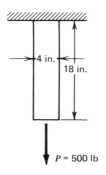

FIG. 3.10

Solution

We calculate the stress:

$$\sigma = \frac{P}{A} = \frac{500 \text{ lb}}{2 \text{ in.} (4 \text{ in.})} = \frac{500}{8} \frac{\text{lb}}{\text{in.}^2}$$

$$\sigma = 62.5 \text{ psi}$$

and then the strain:

$$\varepsilon = \frac{\delta}{L} = \frac{0.120 \text{ in.}}{18 \text{ in.}} = 0.00667 \frac{\text{in.}}{\text{in.}}$$

and finally the modulus of elasticity:

$$E = \frac{\sigma}{\varepsilon} = \frac{62.5 \text{ lb}}{\text{in.}^2 (0.00667)} \frac{\text{in.}}{\text{in.}} = 9380 \text{ psi}$$

We could have calculated E directly from the given data and Eq. 3.5 as follows

$$\delta = \frac{PL}{AE}$$

$$E = \frac{PL}{A\delta} = \frac{(500 \text{ lb})(18 \text{ in.})}{(2 \text{ in.})(4 \text{ in.})(0.120 \text{ in.})} = 9380 \text{ psi}$$

EXAMPLE PROBLEM 3.7

Solve Example Problem 3.6 if the load is 2 kN, the length 0.5 m, the cross section 50 mm × 100 mm, and the deflection 3.00 mm.

Solution

Similar to the above:

$$\sigma = \frac{P}{A} = \frac{2 \text{ kN}}{(50 \text{ mm})(100 \text{ mm})} = 4.00 \times 10^{-4} \frac{\text{kN}}{(\text{mm})^2} \left(\frac{1000 \text{ mm}}{\text{m}} \right)^2$$

$$\sigma = 4.00 \times 10^2 \frac{\text{kN}}{\text{m}^2} = 400 \text{ kPa}$$

$$\varepsilon = \frac{\delta}{L} = \frac{3.00 \text{ mm}}{0.5 \text{ m}} = 6.00 \frac{\text{mm}}{\text{m}} = 6.00 \times 10^{-3} \frac{\text{m}}{\text{m}}$$

$$E = \frac{\sigma}{\varepsilon} = \frac{400 \text{ kPa}}{6.00 \times 10^{-3}} = 6.67 \times 10^4 \text{ kPa}$$

$E = 66.7$ MPa

EXAMPLE PROBLEM 3.8

Determine the spring constant for the 2×4 described in Example Problem 3.6.

Solution

Since

$$F = k\delta$$

$$k = \frac{F}{\delta} = \frac{500 \text{ lb}}{0.120 \text{ in.}} = 4170 \frac{\text{lb}}{\text{in.}}$$

Alternately, by Eq. 3.7

$$k = \frac{AE}{L}$$

$$= \frac{(2 \times 4) \text{ in.}^2}{18 \text{ in.}} \frac{(9380 \text{ lb})}{\text{in.}^2} = 4170 \frac{\text{lb}}{\text{in.}}$$

3.5 CHARACTERISTICS OF OTHER MATERIALS

Figure 3.8 is a typical stress-strain diagram for a ductile material, such as a mild steel. Certain material properties such as yield point are vividly illustrated by this diagram. Of course, not all materials are ductile, and not all ductile materials are described by this diagram. For instance, Fig. 3.11 is a typical stress-strain diagram for a brittle material. This material may have a fairly linear stress-strain curve up to the proportional limit, labeled P. (It may not, too.) However, it does not exhibit a definite yield point, as the ductile material does—point C in Fig. 3.8. In such cases we sometimes estimate a yield strength using the offset method. Typical offsets for metals are strains of 0.001 and 0.002. Much smaller offsets are suggested for materials such as wood or concrete. To determine the yield strength using the offset method, a line is drawn parallel to the linear portion of the stress-strain curve from a point of zero stress and the offset strain, point A in Fig. 3.11, to where it intersects the stress-strain curve, point Y. The stress corresponding to this point is taken as the yield strength.

This method of determining the yield strength can be used for any material that does not exhibit a definite yield point such as bronze or less ductile steels. In

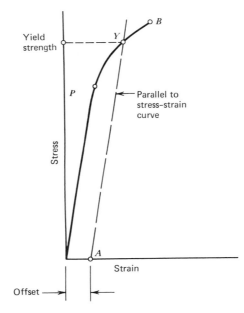

FIG. 3.11 Typical stress-strain curve for a brittle material.

the case of very brittle materials such as steel or concrete it is used in only special circumstances. Accordingly, Table A3.1 does not list yield strengths for these materials. Designs using such materials are more often based on the ultimate strength.

Another characteristic demonstrated in Fig. 3.8 which is not possessed by all materials is a linear stress-strain relationship in the initial stages. Figure 3.12 is a typical stress-strain curve for a material that does not exhibit this linearity. For such materials, special definitions of modulus of elasticity are necessary, and several are common. Recall that the modulus of elasticity is the slope of the stress-strain curve. One way of handling curves like those in Fig. 3.12 is to take the slope of the initial portion of the curve. Accordingly, the line OA is the initial tangent modulus. Second, we might find the slope at some intermediate point B. The line BB' then represents this modulus. It is called the tangent modulus at stress b, the stress associated with this point. The third concept is sort of an average modulus, where the line is not drawn tangent to the curve but from the origin to the point being referenced. The line OC represents this in Fig. 3.12. It is referred to as the secant modulus at stress c.

It is emphasized that the properties given in Table A3.1 are typical of the materials indicated. Some of these vary very little, such as the density of the metals or the modulus of elasticity of steels. Others vary widely, such as the strength of steels or the modulus of elasticity of bronze. We use these properties

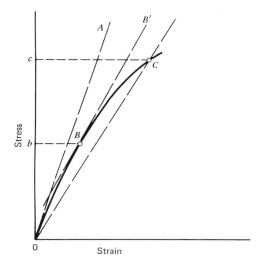

FIG. 3.12

as though they were the property of the material. You should realize that such is generally not the case. More detail specifications which are usually available from material suppliers should be consulted for specific values for particular applications. However, the values from Table A3.1 do represent typical values and are appropriate for many noncritical applications.

3.6 POISSON'S RATIO

Again we return to Example Problem 3.5 concerning the tensile test of a metal specimen. In the previous section we discussed the change in length in the axial direction, δ_A, the direction of the applied load. If we were to measure the diameter as the load is applied, we would find it to be decreasing, as shown in Fig. 3.13. In fact, it would decrease in a very orderly manner. This phenomenon is known as the Poisson effect. It can be described mathematically as

$$\mu = -\frac{\varepsilon_{\text{lateral}}}{\varepsilon_{\text{axial}}} = -\frac{\varepsilon_L}{\varepsilon_A} \tag{3.8}$$

where

$$\varepsilon_A = \frac{\delta_{\text{axial}}}{\text{length}} = \frac{\delta_A}{L}$$

$$\varepsilon_L = \frac{\delta_{\text{lateral}}}{\text{diameter}} = \frac{\delta_L}{D}$$

64 STRESS AND STRAIN

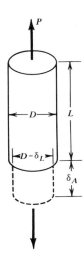

FIG. 3.13 Poisson effect.

and

$$\mu = \text{Poisson's ratio (Greek mu} = \text{English } m)$$

Since strain is dimensionless, the ratio of two strains is likewise dimensionless, and thus Poisson's ratio is also dimensionless.

We previously used μ for $E-6$ referring to microstrain. The Greek letters are used to provide additional symbols so that duplication is minimized—minimized but not eliminated. Usually, the context of the usage will clarify what is intended.

Frequently Eq. 3.8 is given without the negative sign. The negative sign takes into account the fact that the lateral strain will always be of opposite sign of the axial strain. If the length is extended, the diameter will be reduced. If the length is compressed, the diameter will be extended. Later equations using μ always treat it as a positive value, so the negative sign in Eq. 3.8 is really needed.

The Poisson effect describes the strain in the lateral direction if the piece is free to move. If the piece is restrained from moving, a stress will be generated in the lateral direction. This stress can be calculated using Hooke's law.

Poisson's ratio is a property of the material. Since it is a ratio of strains which are dimensionless quantities, it is also dimensionless. Theoretically, it can range in value from 0 to 0.5. Most metals have a value near 0.3, as can be seen in Table A3.1.

EXAMPLE PROBLEM 3.9

A plate is loaded as shown in Fig. 3.14.
(a) Find the stress, axial strain, and lateral strain.
(b) If the plate were fixed in the lateral (4 in.) direction, what stress would be generated in this direction? The material is aluminum.

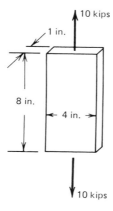

FIG. 3.14

$$E = 11 \times 10^6 \text{ psi} \qquad \mu = 0.30$$

Solution

(a) $\quad \sigma = \dfrac{P}{A} = \dfrac{10{,}000 \text{ lb}}{4(1) \text{ in.}^2} = \mathbf{2500 \text{ psi}}$

Rearranging Eq. 3.4 we have

$$\varepsilon_A = \frac{\sigma}{E} = \frac{2500 \text{ lb-in.}^2}{\text{in.}^2 \; 11 \times 10^6 \text{ lb}} = 0.000227 \; \frac{\text{in.}}{\text{in.}}$$

$$\varepsilon_A = \mathbf{227 \; \mu \text{ strain}}$$

This is the *axial* strain, or the strain in the direction of the load. We find μ from Table A3.1. The lateral strain or strain perpendicular to the load is, by Eq. 3.8,

$$\varepsilon_L = -\mu \varepsilon_A$$

$$= -0.3(0.000227) = -0.0000682 = \varepsilon_L$$

$$\varepsilon_L = \mathbf{-68.2 \; \mu \text{ strain}}$$

The minus sign means there is a reduction in the lateral dimension.
(b) If the side is restrained ε_L will not be allowed to take place. Hence a stress large enough to produce this strain will be generated:

$$\sigma_L = E \varepsilon_L = 11 \times 10^6 \text{ psi}(0.0000682)$$

$$\sigma_L = \mathbf{750 \text{ psi tension}}$$

3.7 THERMAL STRAIN

A common assembly procedure in manufacturing metal components is a shrink fit. In this procedure one part is heated, thereby expanding, and its mating part is cooled, thereby contracting. After sufficient heating and cooling, the two parts can be assembled without interference. As they come back to room temperature, the two parts are locked together. Dealing quantitatively with the above situation involves thermal strains. We define a thermal strain in the same manner as we define other strains:

$$\varepsilon_T = \frac{\delta_T}{L}$$

where δ_T is the change in length due to a change in temperature (ΔT) and L is the original length.

The thermal strain is proportional to the change in temperature

$$\varepsilon_T = \alpha(\Delta T) \tag{3.9}$$

where α (Greek alpha, English a) is the coefficient of thermal expansion. Table A3.1 gives values of α for various materials; α will have the units per °F or per °C in SI.

If thermal strain is not allowed to occur, a stress will be generated which may be calculated from Hooke's law. We should also note that the thermal strain is equal in every direction.

EXAMPLE PROBLEM 3.10

A steel rod is 1.25 in. in diameter and 2.0 ft long. It is heated 200°F. Find its change in diameter and length.

Solution

From Table A3.1 the coefficient of thermal expansion is

$$\alpha = \frac{6.5 \times 10^{-6}}{°F}$$

We calculate the thermal strain as

$$\varepsilon_T = \alpha(\Delta T) = \frac{6.5 \times 10^{-6}}{°F} \times (200°F)$$

$$\varepsilon_T = 1.30 \times 10^{-3}$$

The thermal strain is the same in every direction, so the change in diameter is

$$\delta_{\text{diam}} = \varepsilon_T D = 1.30 \times 10^{-3}(1.25 \text{ in.}) = \mathbf{0.00165 \text{ in.}}$$

And the change in length is

$$\delta_{\text{length}} = \varepsilon_T L = 1.30 \times 10^{-3}(2 \text{ ft}) \frac{12 \text{ in.}}{\text{ft}}$$

$$\delta_{\text{length}} = \mathbf{0.0312 \text{ in.}}$$

EXAMPLE PROBLEM 3.11

A steel plate is $1/4$ in.$\times 2$ in.$\times 12$ in. long, Fig. 3.15. It is placed into a 12-in. opening where it exactly fits. Assuming the opening does not change in size with temperature, find the stress in the plate when the temperature is increased 120°F.

Solution

From Table A3.1

$$\alpha = \frac{6.5 \times 10^{-6}}{°F} \qquad E = 30 \times 10^6 \text{ psi}$$

If the plate were unrestrained it would experience a thermal strain of

$$\varepsilon_T = \alpha(\Delta T) = \frac{6.5 \times 10^{-6}}{°F} \times (120°F) = 7.80 \times 10^{-4}$$

or an extension of

$$\delta = \varepsilon L = 7.80 \times 10^{-4}(12 \text{ in.}) = \mathbf{9.36 \times 10^{-3} \text{ in.}}$$

This extension would occur if the plate were unrestrained. Since the extension does not take place, a load must be applied that would produce an equal compression of the plate.

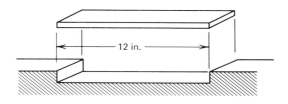

FIG. 3.15

68 STRESS AND STRAIN

$$\delta = \frac{PL}{AE}$$

$$P = \frac{\delta AE}{L} = \frac{9.36 \times 10^{-3} \text{ in. } (2 \text{ in.} \times .25 \text{ in.})}{12 \text{ in.}} \left(30 \times 10^6 \frac{\text{lb}}{\text{in.}^2}\right)$$

$$P = 11,700 \text{ lb}$$

This load would produce a stress:

$$\sigma = \frac{P}{A} = \frac{11,700 \text{ lb}}{(2 \times .25) \text{ in.}^2} = \mathbf{23,400 \text{ psi}}$$

Alternately, the stress could have been calculated directly from the strain.

$$\sigma = E\varepsilon = 30 \times 10^6 \text{ psi } (7.8 \times 10^{-4}) = \mathbf{23,400 \text{ psi}}$$

3.8 SHEAR STRESS AND STRAIN

Consider now a square element of a material in an undisturbed state as shown in Fig. 3.16a. To this element a shear stress τ is applied to each face in the directions shown. By associating the stresses with the respective areas it can be shown that equilibrium conditions are met. This is left to the student. The application of the

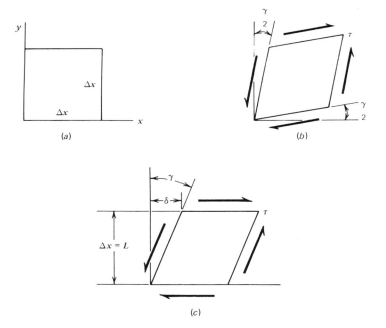

FIG. 3.16a–c Shear strain.

shear stresses will produce the deformation shown in Fig. 3.16b. Figure 3.16c shows the element rotated clockwise through the angle $\gamma/2$ so that the entire deformation can be easily seen as the angle γ (Greek gamma—English g). This angle is defined as shear strain—measured in radians, which is a dimensionless term.

We calculate γ as

$$\tan \gamma = \frac{\delta}{L} = \gamma \quad \text{shear strain} \tag{3.10}$$

remembering that for very small angles

$$\gamma = \tan \gamma$$

This equation for shear strain looks very similar to Eq. 3.3 for normal strain. There is, however, a very important difference. In Eq. 3.3 δ and L are in the same direction. In Eq. 3.10 they are perpendicular. In both cases the deformation, δ, is in the direction of the applied stress.

How is the shear stress, τ, related to the shear strain, γ? By Hooke's law for shear which is

$$\tau = G\gamma \qquad \left(\tau = \frac{P_x}{A} \quad \text{shear stress}\right) \tag{3.11}$$

G is the shear modulus of elasticity, sometimes called the modulus of rigidity. Since γ is dimensionless, G must have the units of τ—psi or Pa. Note that this expression is parallel to Hooke's law for normal stress and strain.

$$\sigma = E\varepsilon$$

It is also very similar in words: The shear stress is proportional to shear strain.

The shear modulus of elasticity is determined experimentally. The method of doing this is discussed in Chapter 5. Like the modulus of elasticity it is a property of the material. In addition to these two elastic properties, E and G, we also have Poisson's ratio, μ. From the theory of elasticity—an advanced subject in engineering mechanics requiring considerable mathematical agility—it can be shown:

$$E = 2G(1+\mu) \tag{3.12}$$

Not aspiring in this text to such agility, we accept the point at face value and go on to evaluate its implications, namely, that the three elastic constants, E, G, and μ, are not independent. If any two are known, the third may be determined.

3.9 ALLOWABLE STRESS

In the first two sections of this chapter equations were developed for finding the normal stress and average shear stress in a structural member. These equations can also be used to select the size of a member if the member's *strength* is known. The strength of a material can be defined in several ways, depending on the material and the environment in which it is to be used. One definition is the ultimate strength or stress. Ultimate strength is the stress at which a material will rupture when subjected to a purely axial load. This property is determined from a tensile test of the material. This is a laboratory test of an accurately prepared specimen which usually is conducted on a universal testing machine. The load is applied slowly and is continuously monitored. The ultimate stress or strength is the maximum load divided by the original cross-sectional area. The ultimate strength for most engineering materials has been accurately determined and is readily available. Appendix A3.1 gives some typical values for common materials.

If a member is loaded beyond its ultimate strength it will fail—rupture. In most engineering structures it is desired that the structure not fail. Thus design is based on some lower value called *allowable stress* or *design stress*. If, for example, a certain steel is known to have an ultimate strength of 110,000 psi, a lower allowable stress would be used for design, say 55,000 psi. This allowable stress would allow only half of the load the ultimate strength would allow. The ratio of the ultimate strength to the allowable stress is known as the *factor of safety*:

$$\text{factor of safety} = \frac{\text{ultimate strength}}{\text{allowable stress}}$$

$$N = \frac{S_u}{S_A} \tag{3.13}$$

We use S for strength or allowable stress and σ for the actual stress in a material. In a design

$$\sigma \leq S_A$$

This so-called factor of safety covers a multitude of sins. It includes such factors as the uncertainty of the load, the uncertainty of the material properties, and the inaccuracy of the stress analysis. It could more accurately be called a factor of ignorance! In general, the more accurate, extensive, and expensive the analysis, the lower the factor of safety necessary.

In many cases failure is predicated on the material's yielding. When this is the case the yield strength replaces the ultimate strength in Eq. 3.13. And, of course, if failure is by shear, then shear strength goes into Eq. 3.13.

In some area, primarily the aircraft industry, the idea of margin of safety (M) is used. It may be defined as

$$M = N - 1 \tag{3.14}$$

In Chapter 13 two concepts that limit the direct application of this material are presented: stress concentrations and fatigue. These are very important restrictions, especially as it is applied to machine design. Stress concentration and

ALLOWABLE STRESS   71

fatigue are sometimes handled by very large factors of safety. Advancements in our understanding of these phenomena and the need to design effectively and efficiently are making the use of these very large factors of safety less and less desirable.

EXAMPLE PROBLEM 3.12

The truss shown (Fig. 3.17) is to be made of circular steel rods whose ultimate strength is 64,000 psi. Using a factor of safety of 4 find the size of member AB.

Solution

Draw a free body diagram (Fig. 3.17b) of the structure and find the reactions.

$$\Sigma M_A = 0 = -4(8) + 7(C_y)$$

$$C_y = \tfrac{32}{7} = 4.57 \text{ kips} \uparrow$$

$$\Sigma F_y = 0 = A_y + C_y$$

$$A_y = -4.57 \text{ kips} \downarrow$$

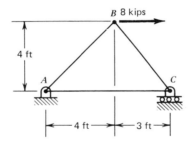

FIG. 3.17a

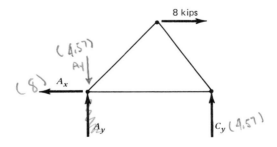

FIG. 3.17b

72 STRESS AND STRAIN

$$\Sigma F_x = 0 = -A_x + 8$$

$$A_x = 8.00 \text{ kips} \leftarrow$$

Draw a free body diagram (Fig. 3.18) of joint B and find the load in AB. Resolve forces into components

$$\Sigma F_y = 0 = -AB\sin 45° + \tfrac{4}{5}BC \doteq 0.707AB + 0.800BC = 0 \qquad (1)$$

$$\Sigma F_x = 0 = -AB\cos 45° - \tfrac{3}{5}BC + 8 \doteq 0.707AB - 0.600BC = -8.00 \qquad (2)$$

Subtracting Eq. (2) from (1)

$$(-0.707 + 0.707)AB + (0.800 + 0.600)BC = 8.00$$

$$1.40 BC = 8.00$$

$$BC = \frac{8.00}{1.40} = 5.71 \text{ kips} \searrow \qquad \text{compression}$$

Substituting back into Eq. (1)

$$-0.707AB + 0.800(5.71) = 0$$

$$AB = \frac{-0.800(5.71)}{-0.707} = 6.46 \text{ kips} \swarrow \qquad \text{tension}$$

This is the load we must design for. Now proceeding to the stress analysis.

$$S_{AL} = \frac{S_{\text{ult}}}{N} = \frac{64{,}000 \text{ psi}}{4} = 16{,}000 \text{ psi}$$

$$\sigma = S_{AL} = \frac{P}{A}$$

Solving for A

$$A = \frac{P}{S_{AL}} = \frac{\pi D^2}{4}$$

And then D

$$D^2 = \frac{4P}{\pi S_{AL}} = \frac{4(6460 \text{ lb}) \text{ in.}^2}{\pi(16{,}000 \text{ lb})} = 0.514 \text{ in.}^2$$

$$D = \sqrt{0.514 \text{ in.}^2} = \mathbf{0.717 \text{ in.}} - \text{minimum diameter}$$

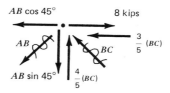

FIG. 3.18

3.10 SUMMARY

A number of new terms and equations are presented in this chapter. The student can minimize the confusion by concentrating on five definitions. Knowing the definition in words allows the equation to be written and vice versa. These terms and their equations are:

Normal stress: $\quad \sigma = \dfrac{P}{A}$

Strain: $\quad \varepsilon = \dfrac{\delta}{L}$

Hooke's law and modulus of elasticity: $\quad \sigma = E\varepsilon$

Coefficient of thermal expansion: $\quad \alpha = \dfrac{\varepsilon_T}{(\Delta T)}$

Poisson's ratio: $\quad \mu = -\dfrac{\varepsilon_L}{\varepsilon_A}$

The first three terms have parallel expressions in shear. Namely,

Average shear stress: $\quad \tau_{AV} = \dfrac{P}{A}$

Shear strain: $\quad \gamma = \dfrac{\delta}{L}$

Hooke's law for shear and the shear modulus of elasticity: $\quad \tau = G\gamma$

Finally we have the concepts of allowable stress and factor of safety given by the equation

$$N = \dfrac{S_u}{S_A}$$

Summaries of essential information are provided at the end of each chapter. You will find it more helpful to make your own summary as you go through the chapter. Give particular emphasis to the topics your instructor emphasizes. Add details and clarifying notes that are meaningful to you. A good exercise to enhance your mastery of the subject is to start with a clean sheet of paper and reproduce your summary as you think your way through the chapter.

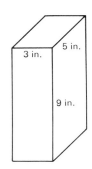

FIG. 3.19 Problem 3.1.

FIG. 3.20 Problems 3.4, 3.5.

PROBLEMS

3.1 The block shown in Fig. 3.19 weighs 150 lb. What bearing stress due to its weight does it exert on the table it is resting on? If it is tipped to the left, where it rests on its side, what bearing stress will it exert?

3.2 A steel block has the dimensions 2 in. × 4 in. × 12 in. It rests on a table. What force does it exert on the table? If it is lying flat on one side, what is the maximum bearing stress it can exert? The minimum?

3.3 A steel block has the dimensions 50 mm × 100 mm × 300 mm. It rests on a table. What force does it exert on the table? If it is lying flat, what is the maximum bearing stress it can exert? The minimum?

3.4 For Fig. 3.20 the load P is 400 lb and the diameter of the shaft is 2 in. Find the normal stress.

3.5 For Fig. 3.20 the load is 2.50 kN and the diameter of the shaft is 5 mm. Find the normal stress.

3.6 What diameter rod is required to carry 1200 lb axial load if the stress is not to exceed 15,000 psi?

3.7 What diameter rod is required to carry 5.00-kN axial load if the stress is not to exceed 100 MPa?

3.8 The two rivets (Fig. 3.21a and 3.21b) carry a total load of 8000 lb. They are 1/2 in. in diameter. Find the shear stress in (a) and (b).

3.9 The two rivets (Fig. 3.21a and 3.21b) carry a total load of 40 kN. They are 12 mm in diameter. Find the shear stress in (a) and (b).

3.10 A brake pad carries a 650-lb shear load (Fig. 3.22). Its dimensions are 2 in. × 3.5 in. × 1/4 in. thick. Find the average shear stress the adhesive must withstand.

3.11 A brake pad carries a 3-kN shear load (Fig. 3.22). Its dimensions are 50

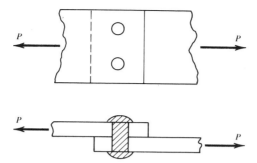

FIG. 3.21a Problems 3.8a, 3.9a, 3.42, 3.43.
Two plates.

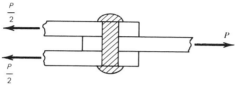

FIG. 3.21b Problems 3.8b, 3.9b, 3.44, 3.45.
Three plates.

mm × 100 mm × 6 mm. thick. Find the average shear stress the adhesive must withstand.

3.12 The truss shown (Fig. 3.23) is constructed of 1/4-in. thick 2×2 angle iron (carbon steel) which has a cross-sectional area of 0.94 in.² Find the stress in members:

(a) AB, AE
(b) AE, BE, CE, EF
(c) BC, CE, EF
(d) EF, CF, CD
(e) CD, DF
(f) All members

3.13 The steel struss shown (Fig. 3.23) is constructed of circular rods with a cross section of 2.00 cm². If the loads given are in kilonewtons rather than pounds and the lengths are in meters, find the stress in members:

(a) AB, AE
(b) AE, BE, CE, EF

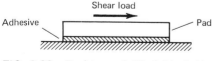

FIG. 3.22 Problems 3.10, 3.11, 3.46.

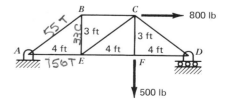

FIG. 3.23 Problems 3.12, 3.13, 3.14, 3.15.

(c) BC, CE, EF (e) CD, DF
(d) EF, CF, CD (f) All members

3.14 For Problem 3.12 find the axial strain and deflection in each member indicated.

3.15 For Problem 3.13 find the axial strain and deflection in each member indicated.

3.16 A punch press is to punch 3/4-in.-diameter slugs out of 3/16-in.-thick aluminum sheet. Assuming an average shear strength of 14,000 psi will be needed, what capacity press is required?

3.17 A punch press is to punch 20-mm-diameter slugs out of 5.0-mm-thick aluminum sheet. Assuming an average shear strength of 96 MPa will be needed, what capacity press is required?

3.18 A 1.0-in.-diameter rod is loaded with 50,000 lb. Its length is 15 in., and its modulus of elasticity is 11,000,000 psi. Find the stress, the strain, and the total elongation.

3.19 A 25-mm-diameter rod is loaded with 200 kN. Its length is 400 mm, and its modulus of elasticity is 71 gPa. Find the stress, the strain, and the total elongation.

3.20 A 0.50-in.-diameter rod is loaded with 10,000 lb. It is 12 in. long and deflects 0.003 in. under this load. Find the stress, the strain, and the modulus of elasticity.

3.21 A 0.50-in.-diameter rod is loaded in tension ($P=15,000$ lb). The total deformation resulting from this load is measured to be 0.008 in. If the rod is 2 ft long, determine the stress, strain, and the modulus of elasticity.

3.22 A 12-mm-diameter rod is loaded in tension (70 kN). The total deformation resulting from this load is measured to be 600 mm. If the rod is 6000 mm long, determine the stress, strain, and the modulus of elasticity.

3.23 A 0.75-in.-diameter rod is loaded in tension ($P=12,000$ lb). The total deformation is measured to be 0.005 in. under this load. If the rod is 1.5 ft. long, determine the stress, the strain, and the modulus of elasticity.

3.24 A tensile test specimen has a circular cross section of 0.505 in. diameter and a gage length of 4.00 in. It is loaded to 18,000 lb and stretches 0.0240 in. Its diameter shrinks 0.0010 in. Under these conditions find the normal stress, axial strain, modulus of elasticity, and Poisson's ratio.

3.25 A tensile test specimen has a circular cross section of 13 mm diameter and a gage length of 100 mm. It is loaded to 80 kN and stretches 0.61 mm. Its diameter shrinks 25.4 μm. Under these conditions find the normal stress, axial strain, modulus of elasticity, and Poisson's ratio.

3.26 A tensile test specimen has a gage length of 8 in. Its circular cross section is 0.505 in. in diameter. It stretches 0.0240 in. when loaded to 32,000 lb. Under these conditions find the axial strain, the normal stress, and the modulus of elasticity. For a Poisson ratio of 0.30 find the change in diameter.

3.27 A 1/8-in.-diameter steel bolt is 6 in. long. Find its spring constant.

3.28 A 3.0-mm-diameter steel bolt is 150 mm long. Find its spring constant.

3.29 A tensile test is made of a steel specimen cut from a pressure cylinder. The gage length is 2.00 in., the width 0.500 in., and the thickness 0.230 in. From the following data plot a stress-strain curve and determine the yield strength, the ultimate strength, and the percentage of elongation at failure.

$A = .115 \text{ in}^2$

σ ε	Load (lb)	Elongation (in.)	Load (lb)	Elongation (in.)
8700	1,000	0.00062	10,000	0.01200
17400	2,000	0.00120	9,900	0.02400
26000	3,000	0.00183	9,900	0.03620
35000	4,000	0.00244	12,000	0.06050
43500	5,000	0.00305	14,000	0.10072
52000	6,000	0.00360	12400 14,200	0.15065
	7,000	0.00421	14,000	0.20016
	8,000	0.00478	10,000	0.25005
	9,000	0.00535	8,250	0.31612
87000	10,000	0.00605		

3.30 A magnesium test bar with a $1/2$ in. $\times 1\frac{1}{2}$ in. cross section is loaded in compression along its 10 in. length by an 8,000-lb load.
(a) Find the change in length of this bar.
(b) Find the change in cross-sectional area.

3.31 A magnesium test bar with a 12 mm $\times$ 38 mm cross section is loaded in compression along its 250 mm length by an 35.0-kN load.
(a) Find the change in length of this bar.
(b) Find the change in cross-sectional area.

3.32 A bar of aluminum has a cross-sectional area of 3.00 in.² It is 12.0 in. long. Find its change in length when it is heated 150°F. What load would be developed in the bar if its ends had been totally restrained during the heating?

3.33 A bar of aluminum has a cross-sectional area of 20 cm.² It is 300 mm long. Find its change in length when it is heated 85°C. What load would be developed in the bar if its ends had been totally restrained during the heating?

78 STRESS AND STRAIN

3.34 A bar of copper has a cross-sectional area of 2.00 in.² It is 20.0 in. long. It is heated 100°F. Find its change in length. What load would be developed in the bar if its ends had been totally restrained during the heating?

3.35 A 3/4-in.-diameter steel rod exactly 6.0 in. long is placed between two rigid plates that are fixed exactly 6.0 in. apart, Fig. 3.24. The rod is heated 400°F in an attempt to force the plates apart. What force would be developed in the rod?

3.36 A 20-mm-diameter steel rod, exactly 150 mm long, is placed between two rigid plates which are fixed exactly 150 mm apart, Fig. 3.24. The rod is heated 250°C in an attempt to force the plates apart. What force would be developed in the rod?

3.37 A 2-in.-diameter steel shaft is to fit into a 2-in.-diameter hole in an aluminum housing. There is an interference of 0.003 in. between the two parts.

(a) If the housing is held at room temperature, how much must the shaft be cooled for assembly?

(b) If the shaft is held at room temperature, how much must the housing be heated for assembly?

(c) If both are heated, how much must they be heated for assembly?

3.38 An aluminum electric power cable is strung between two towers on a hot summer day in Stillwater when it is 100°F. The towers are 300 yd apart, and the 1-in.-diameter cable is pulled to a tension of 1200 lb. Treating the cable as a solid rod, find the tension in it on a January day when the temperature drops to 10°F. Assume the towers to be immovable and the cable length to be exactly 300 yd.

3.39 An aluminum electric power cable is strung between two towers on a hot summer day in Stillwater when it is 38°C. The towers are 300 m apart and the 25-mm-diameter cable is pulled to a tension of 5.4 kN. Treating the cable as a solid rod, find the tension in it on a January day when the temperature drops to $-12°C$. Assume the towers to be immovable and the cable length to be exactly 300 m.

3.40 From a tensile test an aluminum alloy is found to have a modulus of elasticity of 12.0×10^6 psi and a Poisson's ratio of 0.33. Find the shear modulus of elasticity.

3.41 From a tensile test an aluminum alloy is found to have a modulus of

FIG. 3.24 Problems 3.35, 3.36.

elasticity of 83 GPa and a Poisson's ratio of 0.33. Find the shear modulus of elasticity.

3.42 Determine the diameter of the rivets in Fig. 3.21a if the ultimate strength in shear is 60,000 psi and a safety factor of 3 is desired for a load $P = 1500$ lb.

3.43 Determine the diameter of the rivets in Fig. 3.21a if the ultimate strength in shear is 400 MPa and a safety factor of 3 is desired for a load 7.0 kN.

3.44 Determine the diameter of the steel rivets in Fig. 3.21b if the ultimate strength in shear is 80,000 psi and a safety factor of 4 is desired for a load $P = 2500$ lb.   20,000

3.45 Determine the diameter of the steel rivets in Fig. 3.21b if the ultimate strength in shear is 550 MPa and a safety factor of 4 is desired for a load 11.0 kN.

3.46 For Problem 3.10 and Fig. 3.22 assume the adhesive is 1/16 in. thick and has a shear modulus of elasticity of 1.0×10^6 psi. What shear strain would be developed in the adhesive? How far would the brake pad translate in the direction of the load due to the shear strain?

3.47 Define and if appropriate, write an equation for the following terms:
(a) Normal stress
(b) Normal strain
(c) Deformation
(d) Thermal strain
(e) Shear modulus of elasticity
(f) Poisson's ratio
(g) Young's modulus
(h) Hooke's law

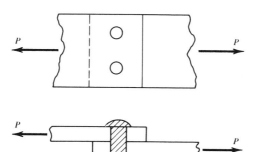

FIG. 3.21a Problems 3.8a, 3.9a, 3.42, 3.43.
Two plates.

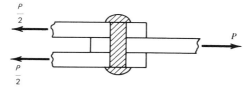

FIG. 3.21b Problems 3.8b, 3.9b, 3.44, 3.45.
Three plates.

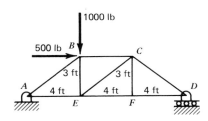

FIG. 3.25   Problems 3.48, 3.49.

3.48 For the truss of Fig. 3.25, based on an ultimate strength of 6000 psi and a factor of safety of 4 find the minimum cross-sectional area of the following members. (Members loaded in compression may buckle, but you may neglect that consideration for the present.)

(a) AB, AE
(b) AE, BE, CE, EF
(c) BC, CE, EF
(d) EF, CF, CD
(e) CD, DF
(f)   All members

3.49 For Fig. 3.25 what size pin is required for joint A based on an allowable shear stress of 20,000 psi?

3.50 For the truss shown in Fig. 3.26, find the axial stress, strain, and deformation for member BC. Its cross section is 2 in.×2 in. square. The members are steel.

3.51 For the truss shown in Fig. 3.27, find the axial strength, strain, and deformation for member BC which has a rectangular cross section of 100 mm×30 mm and is aluminum.

3.52 For Fig. 3.28 find the force Q required to compress the 1.5-in.-diameter block 0.002 in. What is the change in diameter of this block under this load? The block is aluminum.

3.53 The support bracket at B in Fig. 3.28 is shown in detail in Fig. 3.29. Based on a shear strength of 45 ksi and a factor of safety of 3, what size pin is needed if the force Q is 500 lb?

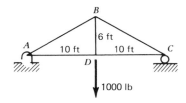

FIG. 3.26   Problem 3.50.

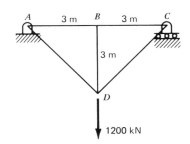

FIG. 3.27   Problem 3.51.

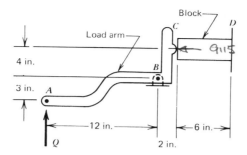

FIG. 3.28 Problems 3.52, 3.53.

3.54 For Fig. 3.30 the bronze link CD is 80 mm long and 20 mm × 10 mm in cross section. Find the strains in link CD in the 20-mm and 80-mm directions under the loading shown.

3.55 The support bracket at point B in Fig. 3.30 is shown in detail in Fig. 3.29. Based on a shear strength of 360 MPa and a factor of safety of 4, what size pin is needed?

3.56 Find the load, stress, strain, and deformation in the member AB, Fig. 3.31. Its cross section is 13 mm × 25 mm and it is aluminum.

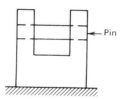

FIG. 3.29 Support bracket B for Figs. 3.28 and 3.30.

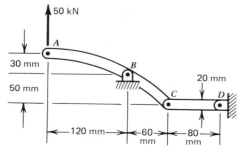

FIG. 3.30 Problems 3.54, 3.55.

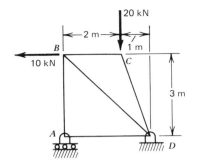

FIG. 3.31   Problem 3.56.

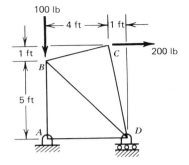

FIG. 3.32   Problem 3.57.

3.57 Find the load, stress, strain, and deformation in member AB, Fig. 3.32. It is a solid rod with a 3/4 in. diameter and is steel.

3.58 Find the diameter of connecting pin needed at B for the linkage shown in Fig. 3.33 based on shear strength of 15 ksi and a safety factor of 3.

3.59 Find the diameter of the pin needed at C for the pump handle shown in Fig. 3.34. Use a shear strength of 80 MPa and a factor of safety of 4.

3.60 Member BD (Fig. 3.35) is a steel pipe, 80 mm O.D. and 70 mm I.D. From the loading shown find its

(a)   Stress

(b)   Change in length

(c)   Change in diameter

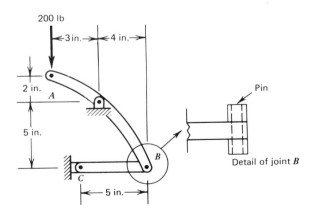

FIG. 3.33   Problem 3.58.

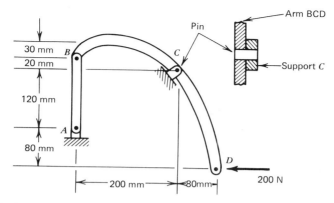

FIG. 3.34 Problem 3.59.

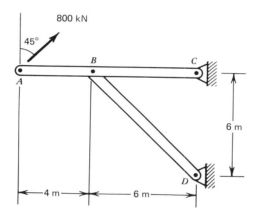

FIG. 3.35 Problem 3.60.

4

Distributed Loads and Properties of Areas

4.1 Distributed load and center of gravity
4.2 Centroid
4.3 Centroid by components
4.4 Second moment of area or moment of inertia
4.5 Parallel axis theorem
4.6 Moment of inertia by components
4.7 Radius of gyration
4.8 Polar moment of inertia
4.9 Summary

There are two approaches to the topic at hand. One is to treat it as a mathematical curiosity. The other is a totally utilitarian one of finding needed values in tables when necessary. Neither approach is adequate for our needs. The first neglects the simplicity with which detailed data may be had and used. The second fails to consider how these phenomena—centroid, moment of area, and moment of inertia—behave, and therefore it fails to understand how they affect stress and deflection. There are also numerous occasions when handbook data are not available. We will strike a middle path—we seek maximum understanding *and* computational ease. We demonstrate the importance of these concepts by the following example. Say we take a 2×12 that is 20 ft long. We wish to support it on each end and walk across it. In the first case we turn it edgewise, that is, 2 in. wide and 12 in. deep. We walk to the center. The deflection is hardly noticeable. Second, we turn the same 2×12 flat, that is, 12 in. wide and 2 in. deep. Now we walk toward the center. The deflection is substantial indeed, and we may even break the board before we reach the center. Same board, same load, and same span, but something is drastically different! What is drastically different is that the stress has been increased by a factor of 6 and the deflection by a factor of 36. This behavior is perfectly predictable, based on an understanding of the moment of inertia of the cross section of the board. In this chapter we will learn to compute this and associated properties and see how these properties vary with the geometry of the cross section.

4.1 DISTRIBUTED LOAD AND CENTER OF GRAVITY

We have classified loads as being distributed or concentrated. We treat concentrated loads as though they act at a point, when in fact all loads are distributed to some degree. The most common distributed load is an object's weight. At this point most of us have the impression that an object's weight may be treated as though it acted at its center of gravity. It turns out that is true for certain situations *but not all*! In this chapter we examine some situations where it is true. In later chapters we will see where it is not.

Consider the rather general distributed load on the simply supported beam in Fig. 4.1. We characterize this general shape mathematically by writing

$$w = f(x)$$

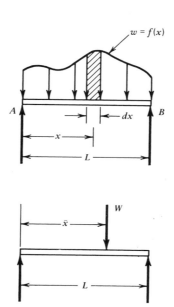

FIG. 4.1 Equivalent loads.

which is to say w is a function of x, or in still plainer terms, the load rate w (i.e., load per unit length of the beam) depends on the position x. We wish to find the reactions at A and B—a statics problem. Normally we would sum the moments of force about A to find B and vice versa. But what is the moment of the distributed load? We pick a small element of the beam—dx. The load on this element is

$$dW = w\,dx$$

The total load on the beam is the sum of all these elements from the left end to the right

$$W = \int dW = \int_0^L w\,dx$$

If $w = f(x)$ is known, this integral can be evaluated mathematically. A more useful form to us is to recognize that this integral is simply the area under the curve w. These areas are normally those we recognize—rectangular, triangular, and the like—and we can easily compute their areas from standard formulas. Thus we write

$$W = \int_0^L w\,dx = \text{area under load rate curve} \qquad (4.1)$$

But we digress; we were after the moment of force, not the force. The moment of the element of dW about A is the magnitude of the force times the distance to it, or

$$dM = x\,dW = x(w\,dx)$$

Then the total moment is

$$M = \int dM = \int x\,dW = \int_0^L x(w\,dx)$$

Again if $w = f(x)$ were known, the above equation could be evaluated. Since we can evaluate the total load by recognizing the area of loading, we ask the question: where would this load W have to act to give the same moment about A? We designate this position as $\bar{x}$, then

$$\bar{x}W = \bar{x}\int dW = \int x\,dW \qquad (4.2)$$

In other words, this says that the moment of the sum of the forces is equal to the sum of the moments of the forces. We solve for $\bar{x}$, the location in question

$$\bar{x} = \frac{\int x\,dW}{\int dW} \qquad (4.3)$$

We call this point the center of gravity and reach the following important conclusion. For the purposes of statics—finding the reactions at A and B (Fig. 4.1)—the distributed load is equivalent to the total load concentrated at the center of gravity. This is because they both give the same moment of force about any point, and they both give the same resultant force. Thus they are said to be *statically equivalent*.

4.2 CENTROIDS

If we think of the distributed load in Fig. 4.1 as sand with a uniform thickness into the paper of t, a uniform density ρ, and depth $y = f(x)$, then the weight per unit length w is

$$w = \rho t y$$

and

$$dW = w\,dx = \rho t y\,dx$$

Equation 4.1 becomes

$$W = \int_0^L w\,dx = \int_0^L \rho t y\,dx$$

$$W = \rho t \int_0^L y\,dx$$

The constants ρ and t have been brought outside the integral, leaving only $y\,dx$. We observe that $y\,dx$ is an element of area, and thus the integral represents only the area under the curve. Hence we have

$$W = \rho t A = \rho \times (\text{volume})$$

which should be familiar to the reader.

Turning to Eq. 4.2 we have

$$\bar{x} W = \bar{x}\,\rho t A$$

and

$$\int x\,dW = \int x \rho t y\,dx = \rho t \int xy\,dx = \rho t \int x\,dA$$

So

$$\bar{x}\,\rho t A = \rho t \int x\,dA$$

Dividing by ρt leaves

$$\bar{x} A = \int x\,dA \qquad (4.4)$$

The expressions $\bar{x}A$ and $x\,dA$ consist of a distance times an area. If we have a distance times a force, we call that a moment of force. Similarly, if we have a distance times an area we call it a moment of area. The idea of moment of area is an interesting mathematical concept. Beyond that, though, it is most useful in the analysis of statics and strength of materials topics—and other topics too—and that is why we spend our time with it. Moment of area has the units of a length times an area—cubic feet, cubic inches, cubic meters, and so on. The SI prefers cubic meters or cubic millimeters. Because these differ by a factor of 10^9, cubic centimeters are also common.

We have used $\bar{x}$ in both Eqs. 4.2 and 4.4. When $\bar{x}$ refers to an area, we call the point it locates the *centroid*. When it refers to a distributed weight, we call it the *center of gravity*. In the case we have examined in which the depth and density are uniform, the two coincide. For the purpose of calculating moment of force (weight) we may treat the distributed load as though it acted at the center of gravity. For the purpose of calculating the moment of area we may treat the entire area as though it acted at the centroid. Similar to Eq. 4.3 we write

$$\bar{x} = \frac{\int x\,dA}{\int dA} \tag{4.5}$$

With respect to the x axis we write

$$\bar{y} = \frac{\int y\,dA}{\int dA} \tag{4.6}$$

The centroids of several common geometric shapes are shown in Table A4.1. The centroids are obtained by evaluating Eqs. 4.5 and 4.6 mathematically. Note that if a body contains an axis of symmetry, the centroid will be on it. Hence the intersection of two axes of symmetry defines the centroid. It should also be noted that the moment of area about the centroid is always zero.

EXAMPLE PROBLEM 4.1

Find the centroid of the triangle shown in Fig. 4.2 with respect to the Y axis.

Solution

The moment of area is

$$\bar{x}A = \int x\,dA$$

$$dA = y\,dx$$

$$\bar{x}A = \int xy\,dx$$

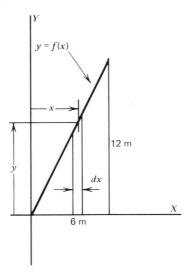

FIG. 4.2

Evaluation of the integral requires that $y=f(x)$ be defined. In this case $y=2x$. We also represent

$$A = \int dA = \int y\, dx$$

$$\bar{x} = \frac{\int x(2x)\, dx}{\int (2x)\, dx} = \frac{2\int_0^6 x^2\, dx}{2\int_0^6 x\, dx}$$

$$\bar{x} = \frac{\left.\dfrac{x^3}{3}\right|_0^6}{\left.\dfrac{x^2}{2}\right|_0^6} = \frac{\dfrac{(6)^3-(0)^3}{3}}{\dfrac{6^2-0^2}{2}} = \frac{72}{18} = \mathbf{4.00\ m}$$

4.3 CENTROIDS OF COMPOSITE AREAS

Tables such as A4.1 can be generated by evaluating areas, as we did in Example Problem 4.1. In practice, this manipulation is rarely made. Most areas for which the centroid must be evaluated may be handled by using tabular information and a finite form of Eqs. 4.5 and 4.6, namely

$$\bar{x} = \frac{\Sigma \bar{x}_i A_i}{\Sigma A_i} \tag{4.7}$$

$$\bar{y} = \frac{\Sigma \bar{y}_i A_i}{\Sigma A_i} \tag{4.8}$$

These equations restate the concept that the moment of the sum of the areas is equal to the sum of the moments of the areas. We apply the above to most problems by recognizing those areas to be the sum of elementary areas whose properties, that is, area and centroid location, are known.

If the areas in Table A4.1 are thought of as solids of uniform thickness, the centroids will coincide with the centers of gravity. Note that a body may be balanced by supporting it at its center of gravity. On this basis we see that the centers of gravity are about where we would expect them to be.

EXAMPLE PROBLEM 4.2

Find the centroid of the area shown in Fig. 4.3 with respect to the y axis.

Solution

We break the area into components whose properties are known and calculate the total area.

$$A_T = A_I + A_{II}$$

$$= 4(3) + 2(6) = 24.0 \text{ ft}^2$$

Next, we compute the moment of area about the Y axis.

$$\bar{x} A_T = \Sigma \bar{x}_i A_i$$

$$= 2[4(3)] + (4+1)[2(6)]$$

$$\bar{x} A_T = 24 + 60 = 84.0 \text{ ft}^3$$

Then the centroid with respect to the Y axis is

$$\bar{x} = \frac{\Sigma x_i A_i}{\Sigma A_i} = \frac{84.0}{24.0} = \mathbf{3.50 \text{ ft}}$$

Note that the centroid at 3.50 ft is between the centroids of the two component areas, as we would expect. This quick check of calculations is always available and should be routinely used.

Tables A4.2 to A4.9 give areas and centroids of common structural shapes as well as other geometric properties. Members in these cross sections are commonly available in structural (mild) steel, high-strength steel, and other materials. We will use Tables A4.2 through A4.6 for aluminum as well. Similar, but not

CENTROIDS OF COMPOSITE AREAS   91

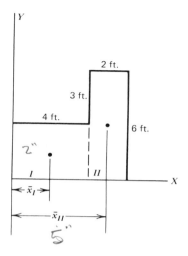

FIG. 4.3

identical, aluminum tables are available. (Of course, the weight per foot—given below—will not be the same.) Table A4.9 is for timber.

These members are identified by indicating the shape, the nominal depth, and the weight per foot of the member. Thus, $W\ 36 \times 230$ refers to the first wide flange (W) beam listed which is nominally 36 in. deep (36) and weighs 230 lb per foot (230). These data are useful in locating the centroids of composite areas that contain these members.

EXAMPLE PROBLEM 4.3

A beam has the cross section shown in Fig. 4.4. Find its centroid with respect to a horizontal axis

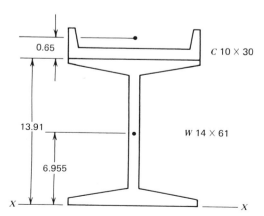

FIG. 4.4

Solution

A horizontal axis X-X is drawn at the bottom of the member.

Geometric properties of the members are found from Tables A4.3 and A4.4. The W 14×61 is a wide flange beam (W), nominally 14 in. deep (14), but actually 13.91 in. (column 3), and weighs 61 lb per foot (61). The second number in its designation is its weight per foot. It has an area of 17.9 in.^{2} (column 2). Its centroid is at its midpoint, so

$$\bar{y}_1 = \frac{13.91}{2} = 6.955 \text{ in.}$$

The C 10×30 is a channel (C), nominally 10 in. deep and actually 10.00 in. deep (column 3). It weights 30 lb per foot (30). Its area is 8.82 in.^{2} (column 2). Its centroid is 0.649 in. above its base (the last column, "x"). Then

$$\bar{y}_2 = 13.91 + 0.649 = 14.56 \text{ in.}$$

The total area is

$$A = \Sigma A_i = A_1 + A_2 = 17.9 + 8.82 = 26.72 \text{ in.}^2$$

The total moment of area is

$$\bar{y}A = \Sigma y_i A_i = 6.955(17.9) + 14.56(8.82)$$

$$\bar{y}A = 252.9 \text{ in.}^3$$

And the centroid is found by

$$\bar{y} = \frac{\Sigma y_i A_i}{\Sigma A_i} = \frac{252.9}{26.72} = \mathbf{9.47 \text{ in.}}$$

Note that this centroid is bounded by the centroids of the component parts.

4.4 SECOND MOMENT OF AREA OR MOMENT OF INERTIA

We have examined the expression, $\int y\,dA$, the moment of area. This expression and the related concept of centroid occur frequently as we analyze the mechanics of materials. A second set of expressions that also occurs commonly is

$$I_x = \int y^2\,dA \tag{4.9}$$

and

$$I_y = \int x^2\,dA \tag{4.10}$$

SECOND MOMENT OF AREA OR MOMENT OF INERTIA 93

Since the distance to the element of area is squared in these expressions, they are properly called the second moment of area but are more commonly called the *moment of inertia*.

We started this chapter by discussing a 2×12 and noting the differences in its deflection and stress, depending on its orientation. The factor that changes with its orientation is its moment of inertia. This property is also pivotal in predicting the buckling of beams and the stress and deflection that occur in torsional loading. So it is a most important concept for the study of strength of materials.

Most often we are interested in the moment of inertia about a centroidal axis. Table A4.1 gives the moment of inertia of various common shapes about a centroidal axis. These were obtained by evaluating Eqs. 4.9 and 4.10. Tables A4.2 through A4.9 give similar properties about centroidal axes for various structural shapes. Only rarely do we actually evaluate Eqs. 4.9 and 4.10. Analytically, however, an understanding of their behavior is very useful.

Note the following regarding these equations:

1. The equations are for the second *moment* of area. A moment requires a reference point or axis, so anytime we talk about a moment of inertia we must talk about it with respect to some axis.
2. The components are an area (dA) times a distance squared (y^2). If we are talking about real areas, the area will be a positive value. (We will use negative values for holes.) The distance y may be positive or negative, but y^2 will be positive regardless. Hence the $\int y^2 \, dA$ will be the product of two positive quantities and will always be positive. On the other hand, the first moment of area may be positive, negative, or zero. We should also note that as material is placed farther and farther from the centroid, its influence on the moment of inertia will be greatly increased, because it varies with the distance squared.
3. The equations are evaluated dimensionally as

$$\int y^2 \, dA \stackrel{D}{=} \text{in.}^2(\text{in.}^2) \stackrel{D}{=} \text{in.}^4$$

Hence moment of inertia has the units of in.^{4} or its equivalent.

EXAMPLE PROBLEM 4.4

Find the moment of inertia of the triangle shown in Fig. 4.2 about the Y axis.

Solution

Since we are taking the moment about the Y axis, we need Eq. 4.10.

$$I_y = \int x^2 \, dA$$

The element of area is

94 DISTRIBUTED LOADS AND PROPERTIES OF AREAS

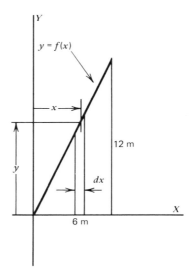

FIG. 4.2

$$dA = y\, dx$$

Note that for moment of inertia we must select an element of area with an infinitesimal dimension (dx) in the direction of the moment arm (x). A horizontal element ($x\, dy$) will not do in this case. Evaluating y in terms of x,

$$y = 2x$$

$$dA = 2x\, dx$$

$$I_y = \int_0^6 x^2(2x\, dx) = \frac{2x^4}{4}\bigg|_0^6 = \frac{2 \cdot 6^4}{4} = \mathbf{648\ m^4}$$

4.5 PARALLEL AXIS THEOREM

We now consider the moment of inertia of the area about the X axis in Fig. 4.5. The axis X_{cg} is a centroidal axis parallel to X, an arbitrary coordinate axis. We write

$$I_x = \int y^2\, dA$$

and note

$$y = y_1 + d$$

where d is the distance from X to the centroidal axis x_{cg}, and y_1 is the distance from the centroidal axis to the element of area.

PARALLEL AXIS THEOREM

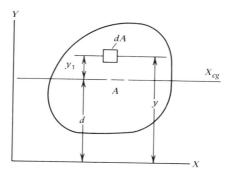

FIG. 4.5 Parallel axes.

By substitution

$$I_x = \int (y_1 + d)^2 \, dA = \int (y_1^2 + 2d \cdot y_1 + d^2) \, dA$$

$$I_x = \int y_1^2 \, dA + \int 2d \cdot y_1 \, dA + \int d^2 \, dA$$

Now we examine each of these three integrals. The first is the moment of inertia of the area about the X_{cg} or centroidal axis

$$\bar{I} = I_{x_{cg}} = \int y_1^2 \, dA$$

(We commonly use $\bar{I}$ to denote the moment of inertia about a centroidal axis.) For the second integral, 2 and d are constants, so we can bring them outside the integral

$$\int 2d \cdot y_1 \, dA = 2d \int y_1 \, dA$$

The remaining integral represents the moment of area about X_{cg} or the centroidal axis. The moment of area about a centroid is always zero, so

$$2d \cdot \int y_1 \, dA = 0$$

For the third integral d^2 is constant, so

$$\int d^2 \, dA = d^2 \int dA = d^2 \cdot A$$

since the remaining integral is simply the area. Returning to the original expression, we have

$$I_x = \int y_1^2 \, dA + \int 2d \cdot (y_1 \, dA) + \int d^2 \cdot dA$$

$$I_x = \bar{I} + Ad^2 \tag{4.11}$$

This is known as the parallel axis theorem. It says that the moment of inertia about any axis is equal to the moment of inertia about a parallel axis through the centroid plus the area times the distance between the two axes squared. This relationship is widely applied in calculating moment of inertia.

EXAMPLE PROBLEM 4.5

Do Example Problem 4.4 using Table A4.1 and the parallel axis theorem.

Solution

From the table we have the centroid 4m to the right of y as shown in Fig. 4.6 and

$$\bar{I} = \frac{bh^3}{36}$$

We have to orient the table to fit our particular problem. In this case h is 6 m and b is 12 m. The base b will always be parallel to the axis about which we are taking the moment. The moment of inertia about the vertical centroidal axis is

$$\bar{I} = \frac{12(6)^3}{36} = 72.0 \text{ m}^4$$

The area is

$$A = \tfrac{1}{2}bh = \tfrac{1}{2}(12)(6) = 36 \text{ m}^2$$

The distance between the Y axis and the parallel centroidal axis (transfer distance) is

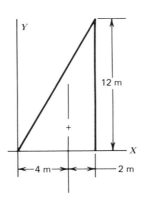

FIG. 4.6

$$d = 4 \text{ m}$$

Then by applying Eq. 4.11 to a vertical axis,

$$I_y = \bar{I} + Ad^2$$

$$= 72 + (36)(4)^2$$

$$I_y = 648 \text{ m}^4$$

This of course agrees with the answer of Example Problem 4.4. We are much more likely to use this simpler, algebra-based method than the integration method.

4.6 MOMENT OF INERTIA OF COMPOSITE AREAS

We have observed that the moment of inertia is an integral function—a summation. In summation we may break the problem into parts and add them in any order. We may do the same thing in integration, and we conclude:

> The moment of inertia of a whole is equal to the sum of the moments of inertia of the parts.

Second, we re-emphasize that we are adding the *moments* of inertia, and for moments to be added, the moment must be about the same axis. We, therefore, have the following general procedure for computing the moment of inertia of composite bodies

1. Break the body into parts whose centroidal moments of inertia may be found.
2. Using tables like A4.1 through A4.9, compute or look up the centroidal moment of inertia for each part.
3. Using the parallel axis theorem, compute the moment of inertia of each part about the desired *common* axis.
4. Sum the moments of inertia of the parts to get the total moment of inertia.

Normally, it is the centroidal moment of inertia that is desired. The following two problems illustrate the method. The second problem presents a tabular form that is very useful in practice for determining all the properties of a cross-sectional area.

EXAMPLE PROBLEM 4.6

Find the moment of inertia about a vertical centroidal axis for Fig. 4.7.

Solution

The area is broken into two areas as in Example Problem 4.2, from which $\bar{x} = 3.50$ ft.

DISTRIBUTED LOADS AND PROPERTIES OF AREAS

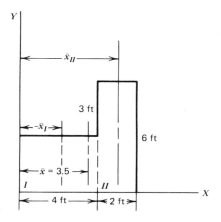

FIG. 4.7

From Table A4.1

$$\bar{I} = \frac{bh^3}{12}$$

Note that in this equation b is the base or side parallel to the axis under consideration.

Then for Area I

$$I_I = \frac{bh^3}{12} + Ad^2 = \frac{3(4)^3}{12} + (3)4(3.5-2)^2$$

The transfer distance, d, is the distance between the centroids of Area I and the composite area; thus $(3.5-2)$.

$$I_I = 16.0 + 27 = 43.0 \text{ ft}^4$$

This is the moment of inertia of Area I about the vertical axis through the composite centroid. For Area II

$$I_{II} = \frac{bh^3}{12} + Ad^2 = \frac{6(2)^3}{12} + (6)2(5-3.5)^2$$

The transfer distance is the distance between the centroids of Area II and the composite area; thus $(5-3.5)$.

$$I_{II} = 4.0 + 27 = 31.0 \text{ ft}^4$$

This is the moment of inertia of Area II about the vertical axis through the

MOMENT OF INERTIA OF COMPOSITE AREAS   99

composite centroid. The total moment of inertia about this axis is the sum of I_I and I_{II}.

$$\bar{I} = I_I + I_{II} = 43.0 + 31.0 = \mathbf{74.0 \text{ ft}^4}$$

EXAMPLE PROBLEM 4.7

Find the centroid and centroidal moment of inertia with respect to the x axis for Fig. 4.8.

Solution

The area is broken into three components. Area *III* is a negative area to be subtracted from Area *II*, a 6 in.×6 in. square. We now introduce a table that is convenient for centroidal and moment of inertia calculations. The components are identified in the initial column.

Components	1 A_i	2 $\bar{y}_i$	3 $y_i A_i$	4 $\bar{I}_i$	5 $d_i = \bar{y}_i - \bar{y}$	6 $A_i d_i^2$
$I_\square$	$6(8)$ $=48$	$+4$	192	$\dfrac{(6)(8)^3}{12}$ $=256$	$4-3.76$ $=0.24$	2.76
$II_\square$	$6(6)$ $=36$	$+3$	108	$\dfrac{(6)(6)^3}{12}$ $=108$	$3-3.76$ $=-.76$	20.8
$III_\triangle$	$\dfrac{-(6)(3)}{2}$ $=-9$	$+2$	-18	$\dfrac{-(3)(6)^3}{36}$ $=-18$	$2-3.76$ $=-1.76$	-27.9
Σ	75.0		282	346		-4.32

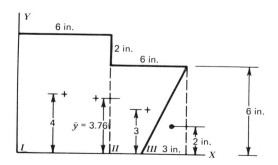

FIG. 4.8

Columns **1, 2,** and **3** are filled in first. It is good to write the factors down and to perform the calculations, even when they are done mentally. This will make it easier to locate the errors which are inevitable. Column **1** is the area of each component. Notice that Area *III* is negative. Column **2** is the distance from the reference axis X to the centroid of the component areas. The distance $\bar{y}_i$ may be positive or negative. Column **3**, $\bar{y}_i A_i$, is simply the product of the elements in the first two columns with the appropriate algebraic sign. We are now ready to compute the centroid location using the sums of columns **1** and **3**.

$$\bar{y} = \frac{\Sigma \bar{y}_i A_i}{\Sigma A_i} = \frac{282}{75.0} = 3.76 \text{ in.}$$

Column **4** is the moment of inertia of each component area about its *own* centroidal axis. We proceed to fill in column **4** using Table A4.1. For structural shapes these values may be found directly from tables. Notice the negative Area *III* will have a negative value of $\bar{I}$. Note that in every case b, the base, is the side parallel to the axis about which we are taking moments. Column **5** is the distance between the centroid of the component area and the centroid of the composite or whole area. To apply the parallel axis theorem the transfer distances d_i are computed. They may be positive or negative. These can be computed by examining the geometry or by the equation $d_i = \bar{y}_i - \bar{y}$. Calculating by one method and checking by the second is recommended. Column **6** is the second term in the parallel axis theorem, Ad^2. Since d is squared, its sign is not significant. The sign of the area will determine the sign of Ad^2. Since Area *III* is negative, the corresponding value of Ad^2 will also be negative.

To apply the parallel axis theorem we could add the elements of columns **4** and **6**, that is,

$$I_i = \bar{I}_i + A_i d_i^2$$

The total $\bar{I}$ would then be the sum of the I_i. Evaluating this we have

$$\bar{I} = \Sigma I_i = \Sigma (\bar{I}_i + A_i d_i^2)$$

$$\bar{I} = \Sigma \bar{I}_i + \Sigma A_i d_i^2$$

The two summations on the right side of the equation represent the sums of columns **4** and **6**. Hence it is not necessary to add the elements of columns **4** to **6**.

$$\bar{I} = \Sigma \bar{I}_i + \Sigma A_i d_i^2$$

$$= 346 - 4.32 = \mathbf{342 \text{ in.}^4}$$

MOMENT OF INERTIA OF COMPOSITE AREAS

In this problem the Ad^2 term plays a minor role. That is not characteristic, as it often provides the major portion of the resultant moment of inertia.

EXAMPLE PROBLEM 4.8

Find the moment of inertia of the cross section in Fig. 4.4 about its horizontal centroidal axis.

Solution

The centroid was found in Example Problem 4.3. It is recomputed here using the table. The values for columns **1**, **2**, and **3** are found from Tables A4.3 and A4.4 as before.

Components	1 A_i	2 $\bar{y}_i$	3 $y_i A_i$	4 $\bar{I}_i$	5 $d_i = \bar{y}_i - \bar{y}$	6 $A_i d_i^2$
I	17.9	$\dfrac{13.91}{2} = 6.955$	124.8	641	$6.955 - 9.47$ $= 2.52$	113.2
⊔	8.82	$13.91 + 0.65$ $= 14.56$	128.1	3.94	$14.56 - 9.47$ $= 5.09$	228.5
Σ	26.72		252.9	644.9		341.7

We compute $\bar{y}$ by dividing the sum of column 3 by the sum of column 1.

$$\bar{y} = \frac{\Sigma y_i A_i}{\Sigma A_i} = \frac{252.9}{26.72} = \mathbf{9.47 \text{ in.}}$$

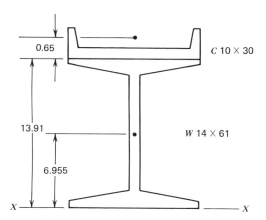

FIG. 4.4

Column **4** lists the moment of inertia of each component about its *own* centroidal axis. Be sure to orient the part properly in reading these tables. For the $W\ 14 \times 61$ we need 641 from column **7** of Table A4.3; not 107 from column **10**. From A4.4 we need for the $C\ 10 \times 30$, 3.94 from column **10** and not 103 from column **7**.

Our column **5** is the difference between the component centroids and the composite centroid and these are calculated as shown. Column **6** is column **5** squared times column **1**.

The resultant moment of inertia of the composite area is the sum of column **4** plus the sum of column **5**.

$$\bar{I} = \Sigma(\bar{I}_i + A_i d_i^2)$$

$$= \Sigma \bar{I}_i + \Sigma A_i d_i^2$$

$$\bar{I} = 644.9 + 341.7 = \mathbf{987\ in.^4}$$

4.7 RADIUS OF GYRATION

For the purpose of calculating the moment of area, the entire area may be considered as concentrated at the centroid:

$$\text{moment of area} = \bar{y} A$$

An analogous idea for the second moment of area, or moment of inertia, is called the radius of gyration. We define a k such that

$$I = k^2 A$$

which is to say the moment of inertia may be calculated by considering the entire area A as being concentrated at a distance k from the axis. Solving for k:

$$k = \sqrt{\frac{I}{A}} \qquad (4.12)$$

where k is the radius of gyration.

The concept of radius of gyration has a number of applications in mechanics. When we study columns we find that the load that can be carried by a column depends on it. Tables A4.2 through A4.7 give the values of the radius of gyration, which is labeled r in these tables. Of course, there is enough information in the table to calculate k (or r), but it is used frequently enough to justify the separate listing.

4.8 POLAR MOMENT OF INERTIA

Consider the area shown in Fig. 4.9. We have examined the moments of inertia about the X and Y axes. These are called the rectangular moments of inertia defined by Eqs. 4.9 and 4.10 as

$$I_x = \int y^2 \, dA \tag{4.9}$$

$$I_y = \int x^2 \, dA \tag{4.10}$$

The moment of inertia about a third axis may also be considered, namely, the Z axis or the axis perpendicular to the paper. This moment is also the distance to an element of area squared times the area and is called the *polar moment of inertia*. We designate it as J. In differential form it is

$$dJ = \rho^2 \, dA$$

and in integral form

$$J = \int \rho^2 \, dA \tag{4.13}$$

The polar moment of inertia is conceptually similar to the rectangular moment of inertia. The concepts of radius of gyration and the parallel axis theorem are applicable. Thus we write

$$k = \sqrt{\frac{J}{A}} \tag{4.14}$$

and

$$J_0 = \bar{J} + Ad^2 \tag{4.15}$$

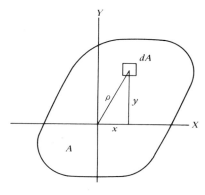

FIG. 4.9 Polar moment of inertia.

There are a couple of special relationships worth observing. Referring to Fig. 4.9 we note

$$\rho^2 = x^2 + y^2$$

then

$$J = \int \rho^2 \, dA = \int (x^2 + y^2) \, dA$$

$$= \int x^2 \, dA + \int y^2 \, dA$$

$$J = I_x + I_y \tag{4.16}$$

Thus we see that the polar moment of inertia is the sum of the rectangular moments of inertia where the three axes in question intersect at a point.

We now evaluate Eq. 4.13 for the important case of a solid circular cross section. This is an important property for determining the stress and deflection in shafts, which we consider in the next chapter. Figure 4.10 is of a circle of a radius R. At a distance ρ from the origin a circular element of area of width $d\rho$ has been drawn. Its area is its circumference times its width—$d\rho$, which is small compared to the radius:

$$dA = 2\pi\rho \, d\rho$$

then

$$J = \int \rho^2 \, dA = \int \rho^2 2\pi\rho \, d\rho$$

We factor out the constant 2π and integrate from 0 to R.

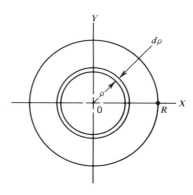

FIG. 4.10 Polar moment of inertia of a circular cross section.

$$J = 2\pi \int_0^R \rho^3 \, d\rho = 2\pi \frac{\rho^4}{4} \bigg|_0^R$$

$$J = \frac{\pi}{2}[R^4 - 0^4] = \frac{\pi R^4}{2} \tag{4.17a}$$

Or in terms of the diameter D

$$J = \frac{\pi}{2}\left(\frac{D}{2}\right)^4 = \frac{\pi D^4}{32} \tag{4.17b}$$

Polar moment of inertia has the same units as rectangular moment of inertia, that is, in.^{4}

EXAMPLE PROBLEM 4.9

Find the centroidal polar moment of inertia of a 3-in.-diameter circular cross section. Also find the polar radius of gyration.

Solution

From Eq. 4.17b

$$J = \frac{\pi D^4}{32} = \frac{\pi (3)^4}{32} = \mathbf{7.95 \text{ in.}^4}$$

We can find k directly from Eq. 4.14. Here we solve it algebraically first:

$$k = \sqrt{\frac{J}{A}} = \sqrt{\frac{\pi D^4}{32} \cdot \frac{4}{\pi D^2}} = \sqrt{\frac{D^2}{8}}$$

$$k = \frac{\sqrt{2}}{4} \cdot D$$

Then substituting in the value for D:

$$k = \frac{\sqrt{2}}{4} \cdot (3) = \mathbf{1.06 \text{ in.}}$$

4.9 SUMMARY

A number of equations are associated with this chapter, but each is generally best understood as a concept or definition rather than merely an equation or formula.

106 DISTRIBUTED LOADS AND PROPERTIES OF AREAS

Moment of area:
$$M_x = \int y \, dA$$

Centroid:
$$\bar{y} = \frac{\int y \, dA}{\int dA} = \frac{\Sigma \bar{y}_i A_i}{\Sigma A_i}$$

Moment of inertia:
$$I_x = \int y^2 \, dA$$

Parallel axis theorem:
$$I_x = \bar{I} + A d^2$$

Radius of gyration:
$$k = \sqrt{\frac{I}{A}}$$

Polar moment of inertia:
$$J = \int \rho^2 \, dA$$

$$k = \sqrt{\frac{J}{A}}$$

$$J_0 = \bar{J} + A d^2$$

$$J = I_x + I_y$$

Many students find the table in Example Problems 4.7 and 4.8 easier to work with. Whichever approach is used, it should be exercised until the student is comfortable and confident with it. *Every* topic which follows depends on understanding the moment of inertia.

PROBLEMS

4.1 For the triangle shown in Fig. 4.11, locate by integration the centroid with respect to the Y axis.

4.2 For the triangle shown in Fig. 4.11 locate by integration the centroid with respect to the X axis.

4.3 Find the centroid of Fig. 4.12 with respect to the X axis.

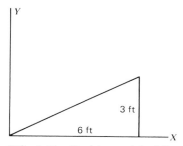

FIG. 4.11 Problems 4.1, 4.2, 4.5, 4.6, 4.7, 4.8.

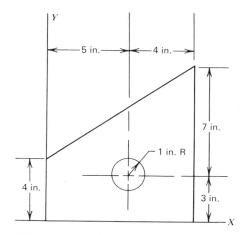

FIG. 4.12 Problems 4.3, 4.4, 4.9, 4.10.

4.4 Find the centroid of Fig. 4.12 with respect to the Y axis.

4.5 By integration, find the moment of inertia of the triangle in Fig. 4.11 about the Y axis.

4.6 By integration, find the moment of inertia of the triangle in Fig. 4.11 about the X axis.

4.7 Find the moment of inertia of the triangle in Fig. 4.11 about the y axis using Table A4.1 and the parallel axis theorem.

4.8 Find the moment of inertia of the triangle in Fig. 4.11 about the X axis using Table A4.1 and the parallel axis theorem.

4.9 Find the moment of inertia about a centroidal axis parallel to the X axis for Fig. 4.12.

4.10 Find the moment of inertia about a centroidal axis parallel to the Y axis for Fig. 4.12.

4.11 Find the centroid and centroidal moment of inertia for the cross section shown in Fig. 4.13.

(a) About a horizontal axis.

(b) About a vertical axis.

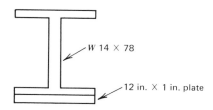

FIG. 4.13 Problem 4.11.

108 DISTRIBUTED LOADS AND PROPERTIES OF AREAS

4.12 Find the polar moment of inertia and radius of gyration for a 5-cm-diameter circular cross section.

4.13 A 4-in.-diameter shaft is hollowed out with a 3-in. diameter. Find the polar moment of inertia and radius of gyration. Treat the hole as a negative area.

4.14 Solve for J in Problem 4.13 by evaluating Eq. 4.13.

4.15 Find $\bar{I}_x$, $\bar{I}_y$, and $\bar{J}_0$ for the Fig. 4.14.

4.16 Find J_A and the polar radius of gyration about A for Fig. 4.14.

4.17 Find the moment of inertia and radius of gyration about the X axis for Fig. 4.15.

4.18 Find the moment of inertia and radius of gyration about the Y axis for Fig. 4.15.

4.19 Find the polar moment of inertia and radius of gyration about the centroid for Fig. 4.15.

4.20 For Fig. 4.16 find the:
   (a) Horizontal centroidal axis.
   (b) The moment of inertia about it.
   (c) The radius of gyration about it.

4.21 For Fig. 4.16 find the:
   (a) Vertical centroidal axis.
   (b) The moment of inertia about it.
   (c) The radius of gyration about it.

4.22 Find the reactions at A and B for Fig. 4.17.

4.23 Find the reactions at A and B for Fig. 4.18.

4.24 Find the reactions at A and B for Fig. 4.19.

4.25 Find the reactions at A and B for Fig. 4.20.

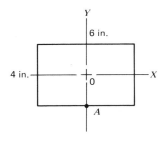

FIG. 4.14   Problems 4.15, 4.16.

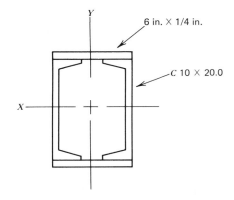

FIG. 4.15   Problems 4.17, 4.18, 4.19.

PROBLEMS 109

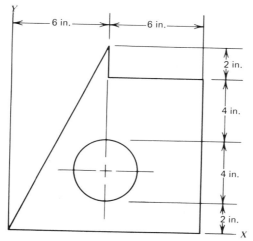

FIG. 4.16 Problems 4.20, 4.21.

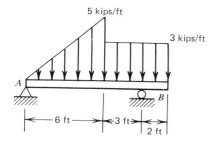

FIG. 4.17 Problem 4.22.

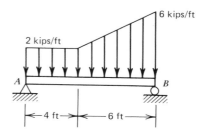

FIG. 4.18 Problem 4.23.

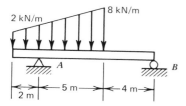

FIG. 4.19 Problem 4.24.

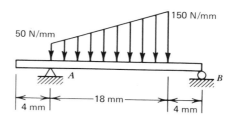

FIG. 4.20 Problem 4.25.

110 DISTRIBUTED LOADS AND PROPERTIES OF AREAS

4.26 Locate the vertical centroidal axis and determine the moment of inertia about it for the cross section shown in Fig. 4.21.

4.27 Locate the horizontal centroidal axis and determine the moment of inertia about it for the cross section shown in Fig. 4.22.

4.28 For Fig. 4.23 locate the horizontal centroidal axis. Find the moment of inertia and radius of gyration about it.

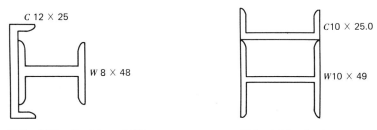

FIG. 4.21 Problem 4.26.   FIG. 4.22 Problem 4.27.

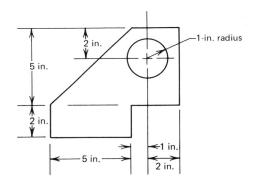

FIG. 4.23 Problem 4.28.

5

Torsional Loads

5.1 Power transmission
5.2 Shear stress
5.3 Longitudinal shear stress
5.4 Hollow shafts
5.5 Angular twist
5.6 Shaft couplings
5.7 Summary

Many modern machines involve a transmission of mechanical energy. The trail of man's technical advances is well marked by the development of mechanical power transmission devices. Many of these date to antiquity, but for others the ink on the patent is hardly dry. The family automobile contains many, such as gears, V-belts and pulleys, geared belts, clutches, rigid shafts, flexible shafts, and U-joints. All these devices are connected to a shaft. Thus a shaft is extremely common in transmitting power.

When power is transmitted by a shaft, a torque is applied to the shaft. A torque is merely a moment of force. If this moment tends to twist the shaft about its longitudinal axis, we call it a *torque* (see Fig. 5.1). If it tends to bend the shaft about an axis perpendicular to the longitudinal axis, we all it a *bending moment*. Both, however, are simply moments of force.

5.1 POWER TRANSMISSION

If 500-hp is to be transmitted at 3600 rpm, what size shaft is to be used? Can this same shaft transmit 1000 hp? Why are so many transmission shafts hollow? In this chapter we answer these and similar questions. Before doing this, however, it is necessary to understand thoroughly the terms and concepts we are going to use.

TORSIONAL LOADS

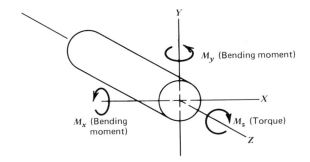

FIG. 5.1 Torque and bending moment.

These are all basic physics concepts, but it is usually worthwhile to review them.

Let's start with the idea of *work*. In mechanics, work is defined as a force (F) acting through a distance(s) where both have the same direction. So

$$W = F \cdot s \tag{5.1}$$

The units are foot-pounds. What if the force is moving in a circular path, as in Fig. 5.2, say, driving a wrench around a bolt? If the force is everywhere tangent to the path, we have

$$dW = F(ds)$$

for the work done as the force F moves the incremental distance ds. Then

$$ds = r(d\theta)$$

and

$$dW = F(r\,d\theta)$$

or, regrouping,

$$dW = (F \cdot r)(d\theta)$$

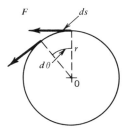

FIG. 5.2 A force acting along a circular path.

The $F \cdot r$ is the moment of force (or torque) of F about 0; $d\theta$ is the incremental angular displacement. We see, then, that the work done by a torque ($T = F \cdot r$) is the product of the torque times its angular displacement.

$$dW = T\, d\theta$$

Integrated through a finite angle, this yields

$$W = T\theta \tag{5.2}$$

The units are foot-pounds for torque and none for the angle—if radian measure is used—resulting in foot-pounds for work.

In the SI the units for torque are $N \cdot m$ (newton-meter). The angular measurement in radians is the same, so for work we have simply the newton-meter or its recognized equivalent, the joule (J).

Power is defined as the time rate of doing work. This may be found by taking the derivative of Eq. 5.2 with respect to time:

$$P = \frac{dW}{dt} = \frac{d}{dt}(T\theta)$$

$$P = T\frac{d\theta}{dt} \quad \text{if } T \text{ is constant}$$

$d\theta/dt$ is the rate of change of angular position with time or angular velocity. Normally, the Greek letter ω (omega) is used for angular velocity. We write

$$P = T\omega \tag{5.3}$$

and say power equals torque times angular velocity. This equation, like 5.1 and 5.2, is a basic physical equation—in a sense a definition—and is valid for any set of units. Typical units would be

$$P = T\omega \underline{\underline{D}} \text{ft-lb}\left(\frac{\text{rad}}{\text{s}}\right) = \frac{\text{ft-lb}}{\text{s}}$$

Remember that radian is a dimensionless quantity—a ratio of lengths—and all basic physical equations involving angles require radian measure.

Now we have power in foot-pounds per second—or any other consistent set of units you wish to use, though these units are often used. In the English-speaking world, the term horsepower (hp) is also commonly used especially if mechanical power is involved. It is an artificial unit defined simply as

$$1 \text{ hp} = 550 \text{ ft-lb/s} \tag{5.4a}$$

If we multiply the right-hand side by 60 s/min we have

$$1 \text{ hp} = 33{,}000 \text{ ft-lbs/min} \tag{5.4b}$$

114 TORSIONAL LOADS

There are many special equations, useful in special situations, for horsepower in various textbooks. However, all the student *needs* to deal with power in English units is Eq. 5.3—a definition of power—and either Eq. 5.4a or 5.4b—a definition of the unit horsepower. Other forms can be had by dimensional algebra. An important thing to note when using the special formulas is that not only must the formula be known, but also the units that go with each term. Of course, if one routinely solves a particular type problem, then an appropriate special form is certainly in order.

Evaluating Eq. 5.3 dimensionally for the SI gives

$$P = T\omega \underline{\underline{D}} (N \cdot m) \left(\frac{rad}{s} \right)$$

$$P \underline{\underline{D}} \frac{N \cdot m}{s} = \frac{J}{s} = W$$

The above says the units for power are a newton-meter per second, or using the equivalent for work, a joule per second. This, of course, yields the watt (W), the unit for power so common in the electrical industries, even where the English system of units is dominant. In fact, this unit of power is so dominant that we frequently invert these relations and speak of work or energy as watts times time. For example, in paying my electric bill I buy a quantity of energy described as kilowatt hours (kW·h). Although it is easily derived, it is convenient to note

$$745.7 \text{ watts} = 1 \text{ hp}$$

At this juncture, the advantages of a scientifically developed units system like the SI strongly emerge. In fact, those of us who cut our "dimensional teeth" by finding how many stone-furlongs per fortnight were required to raise 400 barrels of olive oil a day to a height of 12 fathoms find the simplicity of the SI—well, dull.

A power shaft from a lawn tractor is shown in Fig. 5.3. Consider 8 hp coming into the shaft from the engine, 3 hp taken off at the second pulley for the mower, and 5 hp taken off at the third pulley to drive the tractor. To size the shafts for this system, the power transmitted by each section must be known. This can be determined by applying the principle of conservation of power—a corollary to the principle of conservation of energy. (This, of course, neglects energy losses in the system.) If a section is passed at *A* and we apply the principle—power in equals power out—we find the shaft carries 8 hp between pulleys 1 and 2.

Applying the principle to section *B*, we get 5 hp transmitted by the shaft between pulleys 2 and 3. By the methods covered earlier in this section, the power transmitted can be related to torque if the shaft speed is known. Presently we will relate torque in the shaft to stress and angular deflection.

Figure 5.3*b* and 5.3*c* are thinly veiled free body diagrams. If first of all, the input torque is found, then Figures 5.3*b* and 5.3*c* are free body diagrams and the

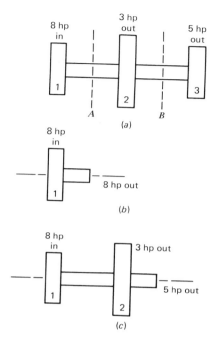

FIG. 5.3a-c Power shaft and pulleys.

torque in each shaft can be found by summing the moment of force about the axis of the shaft.

EXAMPLE PROBLEM 5.1

The following equation is often used

$$T = \frac{63,000\,P}{N}$$

where T is in inch-pounds, P is in horsepower, and N is in revolutions per minute. Prove that it is valid from basic equations.

Solution

From Eq. 5.3

$$P = T\omega$$

$$T = \frac{P}{\omega}$$

116 TORSIONAL LOADS

Substitution in the desired units and doing the dimensional algebra:

$$T = \frac{P(\text{hp})(\text{min})}{N(\text{rev})} \frac{\text{rev}}{2\pi \text{ rad}} \frac{550 \text{ ft-lb}}{\text{s hp}} \frac{60 \text{ s}}{\text{min}} \frac{12 \text{ in.}}{\text{ft}}$$

$$= \frac{550(60)}{2\pi}(12)\frac{P}{N}\text{in-lb}$$

$$T = 63{,}000 \frac{P}{N} \text{ (in.-lb)}$$

EXAMPLE PROBLEM 5.2

The power shaft illustrated in Fig. 5.3 is rotating at 2400 rpm. Find the torque in each portion of the shaft.

Solution

First find the input torque.

$$T_1 = \frac{P_1}{\omega} = \frac{8 \text{ hp}}{} \frac{550 \text{ ft-lb}}{\text{s hp}} \cdot \frac{\text{min}}{2400 \text{ rev}} \cdot \frac{\text{rev}}{2\pi \text{ rad}} \cdot \frac{60 \text{ s}}{\text{min}}$$

$$= \frac{8(550)(60)}{2400(2\pi)} \frac{\text{ft-lb}}{\text{rad}} = 17.5 \text{ ft-lb}$$

We assume clockwise as viewed from the right end.

Ignoring the power analysis of Fig. 5.3 (which would give identical results), we draw a free body diagram (Fig. 5.4a) with the shaft cut between pulleys 1 and 2.

$$\Sigma M_{\text{axis}} = -17.5 + T_A$$

$$T_A = 17.5 \text{ ft-lb}$$

To find the torque taken out at pulley 2 we calculate

$$T_2 = \frac{P_2}{\omega} = \frac{(3)(550)(60)}{2400(2\pi)} = 6.57 \text{ ft-lb}$$

Notice only the 3 hp replaces the 8 hp from above. Alternately, with 17.5 stored in the calculator, we write

$$T_2 = T_1(\tfrac{3}{8}) = 6.57 \text{ ft-lb}$$

Now drawing a second free body diagram (Fig. 5.4b) with the shaft cut between pulleys 2 and 3.

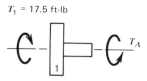

FIG. 5.4a

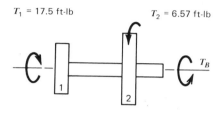

FIG. 5.4b

$$\Sigma M_{\text{axis}} = 0 = -17.5 + 6.57 + T_B$$

$$T_B = 10.9 \text{ ft-lb}$$

5.2 SHEAR STRESS

We now come to the important development of the relationship between shear stress and torsional load. It is developed using the calculus—because the development is exact (a good application of basic mathematics) and because it helps the student understand shear stress, especially the all-important stress distribution. There are a number of assumptions the reader should note carefully, even if unimpressed by the mathematical development. These assumptions constitute *limitations* on the resulting equation.

We consider a shaft loaded in pure torsion as shown in Fig. 5.5. The assumptions are:

1. The cross section is circular—so the equation will not apply to rectangular or other noncircular cross sections.
2. The member is straight
3. On twisting, plane sections do not warp. This rules out "large" deflections.
4. Hooke's law is obeyed, that is, stress is proportional to strain. This rules out stresses beyond the proportional limit.

The torque T applied to the body results in the twisting of the cylinder rotating point B to B'. This motion results in the shear strain being proportional to the distance from the center 0. Following Hooke's law, the stress must also be proportional to the distance from the center. The resulting distribution is illustrated in Fig. 5.5b. Note the stress is zero at the center and maximum at the surface. We may express this in equation form as

$$\tau = \frac{\rho}{c} \tau_{\max} \qquad (5.5)$$

118 TORSIONAL LOADS

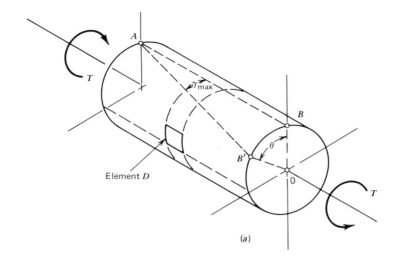

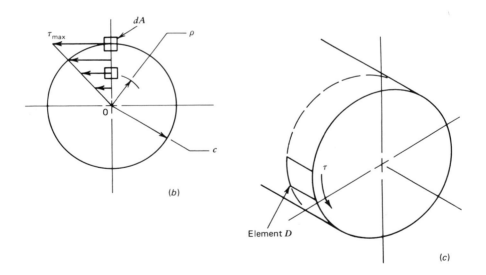

FIG. 5.5a–c Pure torsion.

where ρ is the distance from the center to an element and c is the radius of the shaft. We wish to relate this stress τ (the Greek tau, English t) to the applied torque T. Considering each element of area (dA) to have a stress τ acting on it, we have an element of force

SHEAR STRESS

$$dF = \tau \, dA$$

which produces an element of moment of force (or torque) about 0 as

$$dT = \rho \, dF = \rho \tau \, dA$$

Since this is the torque of τ acting over dA about 0, the total torque may be found by summing these elements of torque over the entire area. Hence:

$$T = \int dT = \int_{\text{area}} \rho \tau \, dA$$

We previously indicated the stress distribution of Eq. 5.5. Substituting this for τ in the above yields

$$T = \int_{\text{area}} \rho \left(\frac{\rho}{c} \tau_{\text{max}} \right) dA$$

$$T = \frac{\tau_{\text{max}}}{c} \int_{\text{area}} \rho^2 \, dA$$

Since τ_{max} and c are constants, they may be factored outside the integral. The integral contains the distance to an element (ρ) squared times the area of the element (dA). This is the second moment of area or the moment of inertia—in this case, the polar moment of inertia. It was shown in Chapter 4 that for a circular section this is

$$J = \int \rho^2 \, dA = \frac{\pi D^4}{32} \tag{5.6}$$

where D is the diameter of the shaft.
Hence we have

$$T = \frac{\tau_{\text{max}}}{c} J$$

or solving for τ_{max}:

$$\tau_{\text{max}} = \frac{Tc}{J} \tag{5.7}$$

This is the fundamental relation for shear stress due to torsion in circular members. Combining this with Eq. 5.5 yields a similar equation for the shear stress at any distance from the center.

$$\tau = \frac{T\rho}{J} \tag{5.8}$$

120 TORSIONAL LOADS

Check the units of the equation.

$$\tau = \frac{Tc}{J} \overset{D}{=} \frac{\text{in.-lb} \times \text{in.}}{\text{in.}^4} \overset{D}{=} \frac{\text{lb}}{\text{in.}^2}$$

So psi, Pa, or any other previous unit of stress is acceptable. Note the relationship being described. If torque is doubled—shear stress will be doubled. As the diameter varies we consider

$$\tau_{max} = \frac{T(D/2)(32)}{\pi D^4}$$

$$\tau_{max} = \frac{16T}{\pi D^3} \qquad (5.9)$$

which shows that the shear stress varies inversely as the diameter cubed. Hence doubling the shaft diameter reduces the shear stress by a factor of 8. Also, review what load is causing this stress condition—a torque about the axis of the shaft. And note the stress pattern produced—that shown in Fig. 5.5b with zero shear stress at the axis and a maximum on the surface.

EXAMPLE PROBLEM 5.3

A 50-mm-diameter shaft carries a torque of 700 N·m. Find the maximum shear stress in the shaft.

Solution

$$\tau = \frac{Tc}{J} \qquad (5.7)$$

First we evaluate J using Eq. 4.17

$$J = \frac{\pi D^4}{32} = \frac{\pi (50 \text{ mm})^4}{32} = 614{,}000 \text{ mm}^4$$

Or in cm^4

$$J = 61.4 \text{ cm}^4$$

$$\tau = \frac{Tc}{J} = \frac{700 \text{ N} \cdot \text{m } (25 \text{ mm})}{61.4 \text{ cm}^4} \left(\frac{100 \text{ cm}}{\text{m}}\right)^4 \left(\frac{\text{m}}{1000 \text{ mm}}\right)$$

$$\tau = 28.5E6 \frac{\text{N}}{\text{m}^2} = \mathbf{28.5 \text{ MPa}}$$

5.3 LONGITUDINAL SHEAR STRESS

Refer once again to Fig. 5.5a and examine element D on the surface. This area of the shaft is shown enlarged in Fig. 5.5c with stress element τ acting on it. If another section is made to the left of the element D, we isolate an element shown in Fig. 5.6a. This left-hand face would have the same shear stress acting on it as acts on the right-hand face, namely τ.

We have extracted a square element Δx by Δx. Assuming a depth of unity, we examine the equilibrium of this element. Summing the forces vertically yields zero, as indicated in Fig. 5.6a. We then sum the moments of force about the lower left-hand corner, point P. The force due to the shear stress is

$$F = \tau \Delta x$$

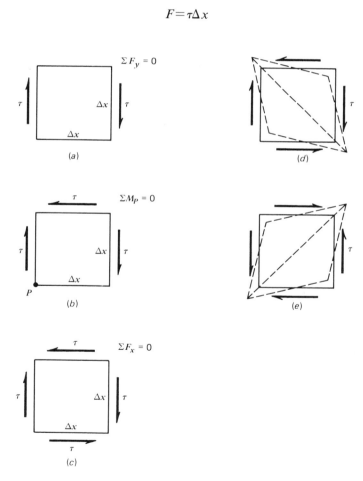

FIG. 5.6a–e Shear stress element.

and the moment of force

$$M = (\Delta x)F = \Delta x (\tau \Delta x)$$

At this point this moment appears unbalanced and would cause the element to rotate clockwise. What counteracts this rotation? A shear stress on the top horizontal surface. Since the moment arms and areas are equal, Δx, the magnitude of this stress must also be τ. However, it must provide a counterclockwise moment and thus must act to the left, as shown in Fig. 5.6b.

Finally, we sum forces horizontally. This requires that there also be a shear stress to the right on the bottom surface. Its magnitude is also τ.

Figure 5.6c indicates that when a shear stress acts on one face of an element of area, it must act on all four faces.

Figure 5.6d shows the distortion generated in the element by these stresses. Note that the stresses must meet at the corners for equilibrium. The only other possible condition for shear stress that satisfies equilibrium is shown in Fig. 5.6e. Note it also has the stresses meeting at the corners.

The significance of this is that although shear stresses are generated on the face of a transverse plane by an applied torque, they are simultaneously generated longitudinally as well. This is illustrated in Fig. 5.7. This longitudinal stress was very evident when wooden shafts were commonly used to transmit power. These shafts were constructed with the grain parallel to the longitudinal axis. Of course, the wood was very weak in shear parallel to the grain, and so failure of these shafts was often due to longitudinal shear.

EXAMPLE PROBLEM 5.4

A wooden shaft can carry 1000 psi in shear stress perpendicular to the grain and 150 psi in shear stress parallel to the grain. What torque can a 3-in.-diameter shaft carry?

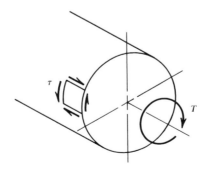

FIG. 5.7 Transverse and longitudinal shear stresses.

Solution

$$\tau = \frac{Tc}{J}$$

Solving for T and doing the algebra:

$$T = \frac{\tau J}{c} = \frac{\tau \pi D^4 (2)}{D(32)} = \frac{\pi \tau D^3}{16}$$

Then, substituting for τ and D,

$$T = \pi \left(1000 \frac{\text{lb}}{\text{in.}^2}\right) \frac{(3 \text{ in.})^3}{16} = \frac{\pi(1000)(3)^3}{16} \text{ lb-in.}$$

$$T_{\text{trans}} = 5300 \text{ in.-lb}$$

This is the load that can be carried without shearing perpendicular to the grain. To find the load that can be carried without shearing parallel to the grain, we use 150 psi instead of 1000 psi for τ.

$$T_{\text{long}} = \frac{\pi(150)}{16}(3)^3 \text{ lb-in.} = 795 \text{ in.-lb}$$

The smaller of the two will limit the load, so the allowable torque is

$$T = 795 \text{ in.-lb}$$

5.4 HOLLOW SHAFTS

In the developing Eq. 5.7 it was assumed that the cross section of the shaft was circular. It was *not* assumed that it was solid. Hence Eq. 5.7 applies equally well to hollow shafts. In fact, a hollow shaft is a preferred shape for a shaft for economical use of the material.

Figure 5.8a shows the stress distribution accompanying Eq. 5.7. We have already referred to the fact that this stress is zero at the center and varies linearly to a maximum at the surface. Careful observation reveals that material near the center of the shaft is not highly stressed, but that near the surface of the shaft it is. Efficient use of the material requires the removal of lower stressed areas. This is illustrated in Fig. 5.8b where only the more highly stressed material remains.

Applying Eq. 5.7 to a hollow shaft requires the evaluation of J for a hollow shaft (the value c is unchanged). Recall that J, the polar moment of inertia, is

$$J = \int \rho^2 \, dA$$

and that an integral is simply a summation.

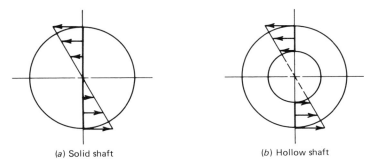

(a) Solid shaft (b) Hollow shaft

FIG. 5.8 a–b Stress distribution.

Hence:

$$J_{\text{hollow}} = J_{\text{solid}} - J_{\text{hole}}$$

This treats the hole as negative area.

Carrying out the above, we have

$$J = \frac{\pi D_s^4}{32} - \frac{\pi D_H^4}{32}$$

$$J_{\text{hollow}} = \frac{\pi}{32}\left(D_s^4 - D_H^4\right) \tag{5.8}$$

Note the following:

$$\left(D_s^4 - D_H^4\right) \neq (D_s - D_H)^4$$

Because the hollow shaft uses the material very efficiently, it is usually found where the weight of the shaft is critical, the material is expensive, or the production volume is high. In the last case, material savings may be offset by higher production costs. The optimum shape, according to this theory, is one with a maximum J, which is obtained by a very thin film at an infinite radius. Other modes of failure—as well as certain impracticalities—preclude this optimized shape. This theory would suggest that a tin can with the ends removed would make an excellent shaft. You've probably never observed such a shaft in use. The reason is that failure will occur in a mode not predicted by this theory—namely, surface buckling. The study of surface buckling is beyond the scope of this text.

EXAMPLE PROBLEM 5.5

A 3-in.-diameter shaft is made lighter by taking 1.5 in. out of the center. What material savings is achieved by removing one-half of the diameter? What reduction in strength follows?

Solution

First we calculate the volume of a solid shaft of diameter D_2

$$V_s = \frac{\pi D_2^2}{4} \cdot L$$

Then the volume of a hollow shaft, where the hole diameter is D_1, is

$$V_H = \frac{\pi D_2^2}{4} \cdot L - \frac{\pi D_1^2}{4} \cdot L = \frac{\pi L}{4}(D_2^2 - D_1^2)$$

For the shaft in this problem

$$D_1 = \frac{D_2}{2}$$

and

$$V_H = \frac{\pi L}{4}\left[D_2^2 - \left(\frac{D_2}{2}\right)^2\right] = \frac{\pi L}{4}\left(D_2^2 - \frac{D_2^2}{4}\right)$$

$$V_H = \frac{3}{4}\left(\frac{\pi D_2^2}{4} \cdot L\right) = \frac{3}{4} V_s$$

This is 3/4 of the volume of the solid shaft; hence 25% of the weight is saved. We describe strength in terms of the torque that can be carried, assuming constant material properties. From Eq. 5.7

$$\tau = \frac{Tc}{J}$$

(We frequently write τ for τ_{max} without misunderstanding.) For a solid shaft

$$T_s = \frac{\tau J}{c} = \frac{\tau \pi D_2^4}{32\left(\frac{D_2}{2}\right)} = \frac{\tau \pi D_2^3}{16}$$

And for a hollow one where c is still $D_2/2$

$$T_H = \frac{\tau J_H}{c} = \frac{\tau(2)}{D_2}\left[\frac{\pi}{32} \cdot (D_2^4 - D_1^4)\right]$$

Again

$$D_1 = \frac{D_2}{2}$$

126 TORSIONAL LOADS

$$T_H = \frac{\pi\tau}{16D_2}\left[D_2^4 - \left(\frac{D_2}{2}\right)^4\right]$$

$$= \frac{\pi\tau}{16D_2}\left(D_2^4 - \frac{D_2^4}{16}\right) = \frac{\pi\tau}{16D_2}\left(\frac{15}{16}D_2^4\right)$$

$$T_H = \frac{15}{16}\left(\frac{\tau\pi D_2^3}{16}\right)$$

The term inside the bracket is T_s, the torque that can be carried by a solid shaft, so

$$T_H = \tfrac{15}{16}T_s$$

Hence the strength is reduced by $1/16$ or 6%, and the weight is reduced 25%. Note that we have solved the problem in general for any shaft with the center half of its diameter removed.

5.5 ANGULAR TWIST

Return once more to Fig. 5.5a repeated here. Because of the applied torque T, B will rotate to point B' representing the twist of the member. Line $0B$ rotates to $0B'$, forming the angle θ, and line AB to AB', forming the angle γ_{max}. For clarity, the section $0BB'A$ is redrawn as a rotated segment from the shaft in Fig. 5.9.

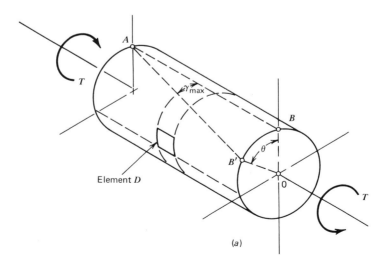

(a)

FIG. 5.5a

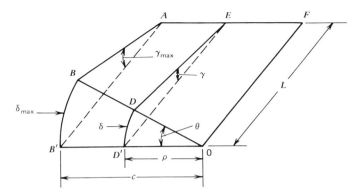

FIG. 5.9 Distortion of shaft segment due to torque.

With this figure we can also examine some of our assumptions. Additional points D, D', E, and F are added to aid our discussion. The D' represents the rotation of D due to the applied load. There is no rotation of AEF. We label the maximum rotation δ_{max}, which is the arc length BB' at a radius c from the center 0. The displacement of any arbitrary point D to D' at a distance of ρ from 0 may be described as

$$\delta = \delta_{max} \frac{\rho}{c}$$

This displacement is proportional to the distance from the center. We now calculate the angle of rotation from the definition of an angle, namely arc length divided by radius:

$$\theta = \frac{\delta_{max}}{c}$$

and

$$\theta = \frac{\delta}{\rho}$$

The angle θ, the angle of twist, does not vary with the distance ρ. We get the same value for θ regardless of the distance from the center used.

Notice, again, that the line segment AEF does not rotate. Hence the angle θ will vary from zero at the far end of the segment to some maximum value at the section we are considering—and the variation is linear. In summary, θ, the angle of twist, depends on the position along the length of the shaft but not on the position along the radius.

Next, we turn our attention to the angle γ. This angle is the shear strain in the shaft. This angle can be calculated by dividing the arc length, δ, by the length L, since we are only considering very small angles of γ. Thus

$$\gamma = \frac{\delta}{L}$$

Recalling from above

$$\delta = \delta_{max} \frac{\rho}{c}$$

We get

$$\gamma = \frac{\delta_{max}}{L} \frac{\rho}{c}$$

Defining

$$\gamma_{max} = \frac{\delta_{max}}{L}$$

then

$$\gamma = \gamma_{max} \frac{\rho}{c}$$

This says the shear strain is proportional to the distance from the center, which was one of our original assumptions. This can be seen from Fig. 5.9 by observing that while δ changes with ρ, the length of the element L does not. We now have

$$\gamma_{max} = \frac{\delta_{max}}{L}$$

or

$$\delta_{max} = \gamma_{max} L$$

and

$$\theta = \frac{\delta_{max}}{c}$$

or

$$\delta_{max} = \theta c$$

Thus, by equating the two values of δ_{max},

$$\gamma_{max} L = \theta c$$

and solving for θ gives

$$\theta = \frac{\gamma_{max} L}{c}$$

Hooke's law for shear is

$$\gamma = \frac{\tau}{G}$$

where G is the shear modulus of elasticity—a material property. Substituting into the above gives

$$\theta = \frac{\tau_{max} L}{cG}$$

From Eq. 5.7

$$\tau_{max} = \frac{Tc}{J}$$

so

$$\theta = \frac{Tc}{J} \frac{L}{cG}$$

$$\theta = \frac{TL}{JG} \qquad (5.9)$$

This is the equation for the twist in a circular member. We examine the units

$$\theta \overset{D}{=} \frac{(\text{in.-lb}) (\text{in.}) (\text{in.}^2)}{\text{in.}^4 (\text{lb})} \overset{D}{=} \text{dimensionless} = \text{radians}$$

As always, naturally occurring angular measurements are in radians—a dimensionless ratio of lengths.

Now let's see what the equation says. The twist varies directly with the applied torque; double the load—double the twist. It varies directly with the length; double the length—double the twist. It varies inversely with the shear modulus of elasticity; doubling G cuts θ in half. It varies inversely with the polar moment of inertia, J, which varies with the diameter to the fourth power. Double the diameter and reduce the twist by a factor of 16!

It is beneficial to our understanding and recollection of Eq. 5.9 to compare it with Eq. 3.5.

$$\delta = \frac{PL}{AE} \qquad (3.5)$$

$$\theta = \frac{TL}{JG} \qquad (5.9)$$

The following analogies are noted.

Deflection

$$\delta - \text{in inches}$$

$$\theta - \text{in radians}$$

Load

$$P - \text{axial in pounds}$$

$$T - \text{torque in inch-pounds}$$

Length

$$L - \text{in inches}$$

$$L - \text{in inches}$$

Geometric property

$$A - \text{area in in.}^2$$

$$J - \text{polar moment of inertia in in.}^4$$

This analogy illustrates how moment of inertia for rotational problems is similar to area for linear ones.

Material property

$$E - \text{modulus of elasticity in psi}$$

$$G - \text{shear modulus of elasticity in psi}$$

Equation 5.9 is for a uniform member. If we go through the same argument with a length of dL instead of L and $d\theta$ instead of θ we obtain

$$d\theta = \frac{TdL}{JG} \tag{5.10}$$

which is equation 5.9 in differential form.

This equation must be integrated along L to obtain the twist. This integration can take into account variations in T, J, and G with L.

$$\theta = \int \frac{TdL}{JG} \tag{5.11}$$

These equations allow us to handle problems such as a tapered shaft.
In finite form Eq. 5.11 becomes

$$\theta = \sum \frac{T_i L_i}{J_i G_i} \qquad (5.12)$$

This can be used to calculate the total angle of twist in a shaft composed of multiple uniform sections and/or several discretely applied torques.

EXAMPLE PROBLEM 5.6

A 3-in.-diameter shaft is 18 in. long. It transmits a torque of 200 ft-lb. Find its deflection.

$$G = 12 \times 10^6 \text{ psi}$$

Solution

From Eq. 5.9

$$\theta = \frac{TL}{JG} = \frac{(200 \text{ ft-lb})(18 \text{ in.})}{\left(\frac{\pi (3 \text{ in.})^4}{32}\right)\left(12E6 \frac{\text{lb}}{\text{in.}^2}\right)} \times \frac{12 \text{ in.}}{\text{ft}}$$

$$\theta = 4.53\,E-4 \text{ radians}$$

or in degrees

$$\theta = 4.53E-4 \text{ rad}\left(\frac{180°}{\pi \text{ rad}}\right) = 2.59\,E-2°$$

EXAMPLE PROBLEM 5.7

The shaft shown in Fig. 5.10 is aluminum. Find the rotation of the left end.

$$G = 4.0E6 \text{ psi}$$

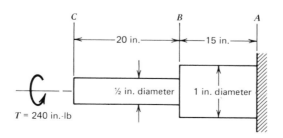

FIG. 5.10

Solution

From Eq. 5.11

$$\theta = \int \frac{T\,dL}{JG} = \int_B^A \frac{T\,dL}{J_{AB}G} + \int_C^B \frac{T\,dL}{J_{CB}G}$$

breaking the integral into parts. Carrying out the integration would give

$$\theta = \frac{TL_{AB}}{GJ_{AB}} + \frac{TL_{BC}}{GJ_{BC}}$$

Since T and G are common, they are factored out:

$$\theta = \frac{240 \text{ in.-lb}}{4 \times 10^6 \frac{\text{lb}}{\text{in.}^2}} \left\{ \frac{15 \text{ in.}}{\left(\frac{\pi (1 \text{ in.})^4}{32}\right)} + \frac{20 \text{ in.}}{\left(\frac{\pi (0.5 \text{ in.})^4}{32}\right)} \right\}$$

$\theta = 0.205$ rad

$$\theta = 0.205 \text{ rad} \times \frac{180°}{\pi \text{ rad}} = \mathbf{11.7° = \theta}$$

5.6 SHAFT COUPLINGS

Shaft couplings occur in two forms—rigid and nonrigid. The nonrigid allow for nonalignment of the shafts. A common coupling of this type is the universal joint found in the drive shaft of many automobiles. The analysis of flexible couplings is beyond the scope of this text. They are properly covered in texts on machine design. Rigid couplings are not only within our scope but also offer the opportunity to extend the present concepts.

A rigid coupling can consist of a flange with bolt holes, as shown in Fig. 5.11. In analyzing this, two assumptions previously used are made. One, we assume the stress on the bolt is proportional to its distance from the center, as we did in developing the shear stress equation for torsional loading. Also, we assume the shear stress on each bolt is uniform, as we did in Chapter 3 when discussing average shear stress. Because the bolt diameter is typically small compared to the bolt circle diameter, this assumption produces little error. These assumptions would allow us to say

$$\tau_{max} = \frac{Tc'}{J}$$

where c' is the distance to the center of the bolts in the largest bolt circle in the coupling and J is the polar moment of inertia of all the bolt areas about their

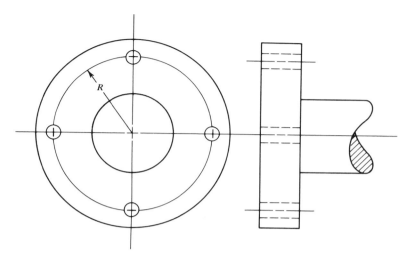

FIG. 5.11a

collective centroid—which coincides with the center of the shaft. This, however, is not as functional as the following

We define F_i

$$F_i = \tau_i A_i$$

where A_i is the area of the bolt, τ_i is the average stress on the bolt, and F_i is the force on it (FIG. 5.11b).

Then

$$T_i = r_i F_i = r_i \tau_i A_i$$

where T_i is the torque resisted by an individual bolt at a distance r_i from the centroid of the coupling. For a coupling with N bolts of the same size A on a single bolt circle of radius R, we have

$$T = \Sigma r_i \tau_i A_i = NRA\tau \tag{5.13}$$

where T is the total applied torque.

For two bolt circles we have

$$T = N_1 R_1 A_1 \tau_1 + N_2 R_2 A_2 \tau_2 \tag{5.14}$$

and

$$\frac{\tau_1}{R_1} = \frac{\tau_2}{R_2} \tag{5.15}$$

Combining Eqs. 5.14 and 5.15 allows this type of problem to be solved.

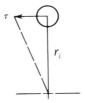

FIG. 5.11b

This analysis assumes the entire torque is carried in shear. Since the bolts are tightened, there will be a friction force between the flange plates, reducing the torque carried by the bolts. In fact, using high-tensile-strength bolts, the bolts may be tightened until the entire load is carried by friction and the bolts are not in shear at all.

EXAMPLE PROBLEM 5.8

A bolt has an allowable shear stress of 5000 psi. There are four of these 1.5-in.-diameter bolts on an 8-in.-diameter bolt circle, as shown in Fig. 5.11. What torque can this coupling carry?

Solution

Using Eq. 5.13:

$$T = NRA\tau$$

$$T = 4 \ (4 \text{ in.}) \left(\frac{\pi (1.5)^2}{4} \text{ in.}^2 \right) 5000 \frac{\text{lb}}{\text{in.}^2}$$

T = 141,000 in.-lb

Note that the shank area of the bolt is used, since it is in shear—not the threads. The area could have been obtained from Table A4.8 where it is identified as the gross area.

5.7 SUMMARY

Preliminary concepts and equivalencies are:

$$P = T\omega$$

$$1 \text{ hp} = \frac{550 \text{ ft-lb}}{\text{s}} = \frac{33,000 \text{ ft-lb}}{\text{min}}$$

$$1\text{ W} = 1\frac{\text{J}}{\text{s}} = 1\frac{\text{N}\cdot\text{m}}{\text{s}}$$

Shear stress distribution is linear, as seen in Fig. 5.5b. Its maximum value is given by

$$\tau_{max} = \frac{TC}{J}$$

This acts transversely and longitudinally.
For a circular cross section

$$J = \frac{\pi D^4}{32}$$

The twist is given by

$$\theta = \frac{TL}{JG}$$

For bolted couplings

$$T = N_1 R_1 A_1 \tau_1 + N_2 R_2 A_2 \tau_2$$

$$\frac{\tau_1}{R_1} = \frac{\tau_2}{R_2}$$

PROBLEMS

5.1 An automobile engine delivers 120 hp at 4000 rpm. What is the torque?

5.2 A motor delivers 90 kW at 4000 rpm. What is the torque?

5.3 A crank arm 10 in. long requires a 35-lb force perpendicular to it to turn it. It must go through five full turns every second. What is the minimum horsepower of a motor driving it?

5.4 A crank arm 25 mm long requires a 150-N force perpendicular to it to turn it. It must go through five full turns every second. What is the minimum power of a motor driving it?

5.5 A $\frac{1}{8}$-horsepower drill motor drives a 4-in.-diameter disk sander at 1800 rpm. What is the maximum tangential force it can exert at the edge of the disk?

5.6 A 100-W drill motor drives a 100-mm-diameter disk sander at 1800 rpm. What is the maximum tangential force it can exert at the edge of the disk?

5.7 The allowable shear stress in a 1-in.-diameter shaft is 12,000 psi. What torque can this shaft transmit?

136   TORSIONAL LOADS

5.8 The allowable shear stress in a 25-mm-diameter shaft is 80 MPa. What torque can this shaft transmit?

5.9 Find the maximum shear stress in a $\frac{1}{2}$-in.-diameter bolt when a 50-lb. force is applied to the end of an 8-in. wrench.

5.10 Find the maximum shear stress in a 12-mm-diameter bolt when a 200-N force is applied to the end of a 200-mm wrench.

5.11 A 2-in.-diameter shaft rotates at 300 rpm and has an allowable stress of 10,000 psi. What horsepower can be transmitted?

5.12 A 50-mm-diameter shaft rotates at 300 rpm and has an allowable stress of 70 MPa. What power can be transmitted?

5.13 Design a solid shaft to transmit 80 hp at 200 rpm using an allowable stress of 8000 psi.

5.14 Design a solid shaft to transmit 60 kW at 200 rpm using an allowable stress of 50 MPa.

5.15 Solve Problem 5.7 if the shaft has a $\frac{3}{4}$-in.-diameter longitudinal hole in it.

5.16 Solve Problem 5.8 if the shaft has a 20-mm-diameter longitudinal hole in it.

5.17 Solve Problem 5.9 if the bolt is hollow with a $\frac{1}{8}$-in. wall thickness.

5.18 Solve Problem 5.10 if the bolt is hollow with a 3-mm wall thickness.

5.19 Solve Problem 5.11 for a hollow shaft with an inside diameter of a 1.5 in.

5.20 Solve Problem 5.12 for a hollow shaft with an inside diameter of 38 mm.

5.21 Design a hollow shaft to transmit 80 hp at 200 rpm using an allowable stress of 8000 psi and $D_i = 0.90 D_o$.

5.22 Design a hollow shaft to transmit 60 kW at 200 rpm using an allowable stress of 55 MPa and $D_i = 0.90 D_o$.

5.23 An 8-ft shaft with a diameter of 1.5 in. is twisted 5°. What torque is required? What stress would this torque produce ($G = 10,000,000$ psi)?

5.24 A 2.5-m long aluminum shaft with a diameter of 40 mm is twisted 5°. What torque is required? What stress would this torque produce?

5.25 Solve Problem 5.23 if the shaft is hollow with $D_i = 1.0$ in.

5.26 Solve Problem 5.24 if the shaft is hollow with $D_i = 25$ mm.

5.27 A 10-ft shaft has a diameter of 2 in. The allowable stress is 8000 psi, and the shear modulus of elasticity is 12,000,000 psi. Find the allowable torque. Find the angle of twist if this load is applied.

5.28 A 3.0-m shaft has a diameter of 60 mm. The allowable stress is 55 MPa, and the shear modulus of elasticity is 85 GPa. Find the allowable torque. Find the angle of twist if this load is applied.

5.29 Solve Problem 5.27 for a hollow shaft with $D_i = 1.5$ in.

5.30 Solve Problem 5.28 for a hollow shaft with $D_i = 40$ mm.

5.31 A 2-in.-diameter solid steel shaft which is 5 ft long must withstand a torque of 20,000 in.-lb at an angular velocity of 1570 rad/min. Determine the horsepower transmitted, the stress produced in the shaft, and the angle of twist of the shaft.

5.32 A 50-mm-diameter solid steel shaft which is 1.5 m long must withstand a torque of 2.3 kN·m at an angular velocity of 1570 rad/min. Determine the horsepower transmitted, the stress produced in the shaft, and the angle of twist of the shaft.

5.33 A solid aluminum shaft 3 in. in diameter and 6 ft long must transmit 60 hp at an angular velocity of 1520 rad/min. Determine the required torque, stress produced in the shaft, and the angle of twist of the shaft.

5.34 A solid bronze shaft 75 mm in diameter and 2.0 m long must transmit 45 kW at an angular velocity of 1520 rad/min. Determine the required torque, stress produced in the shaft, and the angle of twist of the shaft.

5.35 Solve problem 5.31 if the shaft is hollow with a 1.75 in. I.D.

5.36 Solve problem 5.32 if the shaft is hollow with a 40 mm I.D.

5.37 Solve problem 5.33 if the shaft is hollow with a 2.5 in. I.D.

5.38 Solve problem 5.34 if the shaft is hollow with a 60 mm I.D.

5.39 A flange coupling similar to Fig. 5.11 has four 1/2-in.-diameter bolts on a 6-in.-diameter bolt circle. For an allowable shear stress of 6000 psi what torque can be transmitted by the connection?

5.40 If the above flange also has four $\frac{1}{2}$-in. bolts at a 4-in.-diameter bolt circle, what torque can be transmitted?

5.41 The wooden shaft shown in Fig. 5.12 can carry 1200 psi perpendicular to the grain and 200 psi parallel to the grain in shear stress. Considering only torsional loading, can the 50.0 lb load shown be carried?

5.42 The coupling shown in Fig. 5.13 has 6 bolts of $\frac{3}{8}$-in. diameter. The bolt circle has a 3-in. diameter. The hollow shaft is 3 ft long and 2 in. in diameter with a wall thickness of 1/4 in. 200 hp are transmitted at 3000 rpm. Find the maximum stress in the bolts, the maximum stress in the shaft, and the twist of the steel shaft.

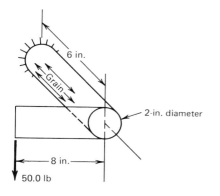

FIG. 5.12 Problem 5.41.

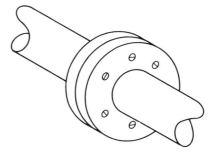

FIG. 5.13 Problem 5.42.

138 TORSIONAL LOADS

5.43 A 3.0-in. shaft has a torque of 20,000 in.-lb applied. Calculate the shear stress at the center of the shaft and every 0.25 in. along the radius. Make a plot of shear stress versus radius.

5.44 A 60-mm shaft has a torque of 2.0 kN·m applied. Calculate the shear stress at the center of the shaft and every 5 mm along the radius. Make a plot of shear stress versus radius.

5.45 A hollow shaft, 4.0 in. O.D. and 3.5 in. I.D., has a shear stress of 9000 psi on the inside surface. Find the maximum shear stress on this shaft.

5.46 A hollow shaft, 100 mm O.D. and 80 mm I.D. has a shear stress of 60 MPa on the inside surface. Find the maximum shear stress on this shaft.

5.47 For the structure shown in Fig. 5.14 find the torque in section BC. Find the twist of the arm at B with respect to the fixed end at C. The aluminum shaft has a 3-in. diameter.

5.48 Find the twist of the arm at A in Problem 5.47.

5.49 A 3-in. diameter hollow magnesium shaft, Fig. 5.15, is 8 ft long and is to be rotated 2° clockwise at the end connected to the linkage. Find the force P, applied as shown, necessary to cause this rotation and find the maximum shear stress produced in the shaft. The shaft has a wall thickness of $\frac{1}{4}$ in.

5.50 A 4-in. diameter hollow shaft (Fig. 5.16) with a wall thickness of $\frac{1}{4}$ in. and a length of 6 ft is connected to the lever shown. Assuming the far end of the shaft to be fixed, find the rotation of the aluminum shaft and the maximum shear stress produced in it.

5.51 An oil well drilling string is a 4-in. standard weight steel pipe, 500 ft long. What will be the angle of twist of the string when an applied torque of 3000 ft-lb is used for drilling? What will the stress in the pipe be?

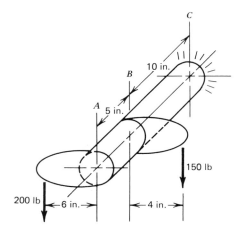

FIG. 5.14 Problems 5.47, 5.48.

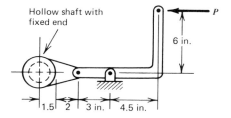

FIG. 5.15 Problem 5.49.

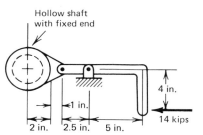

FIG. 5.16 Problem 5.50.

6

Internal Forces

6.1 Axial loads
6.2 Beams
6.3 Internal shear and bending moment
6.4 Beam graphs
6.5 Relation between load, shear, and bending moment
6.6 General procedure for beam diagrams
6.7 Summary

The methods of statics have been primarily used to determine external reactions. However, in the analysis of trusses and other structures the loads carried by members and the loads at joints have been determined. These loads are *internal* loads. In this chapter internal loads are examined more generally. In selecting members for a machine or structure, a stress analysis must be made. Finding the internal load is the necessary first step in any stress analysis. In strength of materials the relation of the stress to the internal load is determined, but first the internal load must be determined.

Three principles, examined previously, are relied on in determining the internal loads. The first of these is an extension of Newton's first law.

1. If a body is at rest and therefore in equilibrium, all its parts—real or imagined—are at rest and in equilibrium.
2. An imaginary cutting plane may be passed through any body in equilibrium without disturbing its equilibrium. Passing the cutting plane will result in two sections, each of which is independently in equilibrium.
3. When a cutting plane is passed through a member, additional forces and moments of forces may be necessary for equilibrium. These forces, which are

AXIAL LOADS   141

internal forces, occur in equal but opposite pairs according to Newton's third law. One set will act upon each of the pair of free bodies created by the section.

These principles are illustrated in the sections that follow.

6.1 AXIAL LOADS

The problem shown in Fig. 6.1 was solved in Chapter 2. In that chapter it was found that the loads on member CE were as shown in Fig. 6.1b. A common design problem is to select a member (i.e., its size and shape) that will function as member CE. Ultimately, stresses must be calculated. However, as indicated, the first step in that analysis is to find the internal loads. Notice that the 267-lb loads are external to the member CE but they are internal to the entire structure.

The steps for finding the internal load are as follows:

1. Draw a free body diagram for the entire structure.
2. Solve for all external reactions, if possible.
3. Draw free body diagrams of the members.
4. Solve for all reactions external to the members.

The first four steps were carried out in Example Problems 2.3 and 2.5.

5. Pass a cutting plane through the member at the section of interest and draw a free body diagram of the simpler portion (Fig. 6.1b and c).

In this case, where the axial load is uniform, any location is satisfactory.

6. Apply Newton's first law to determine the internal loads.

$$\Sigma F_x = 0 = P - 267$$

$$P = 267 \text{ lb}$$

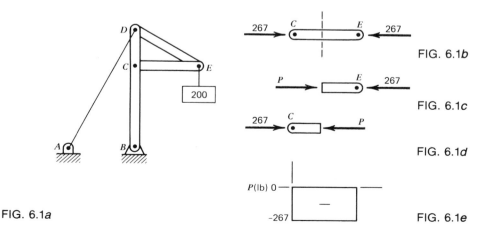

FIG. 6.1a

FIG. 6.1b

FIG. 6.1c

FIG. 6.1d

FIG. 6.1e

142   INTERNAL FORCES

Thus it is found that the internal load is 267-lb compression. In strength of materials this load is related to stress, which provides a basis for selecting a member adequate to carry the load. Fig. 6.1d shows the alternate segment of the member. Note that this force P is in the opposite direction on this body. This is in accordance with Newton's third law. Note further that although the force P acts in opposite directions on the two parts, in both cases it is a compressive load.

Sometimes it is desirable to make a plot of the internal load along the member. This is shown in Fig. 6.1e. When the loading is very simple, as in this case, it is unnecessary, but in more complex load situations it can be desirable, as shown in Example Problem 6.1.

EXAMPLE PROBLEM 6.1

Construct an axial load diagram for member ABC with the loading as shown in Fig. 6.2a.

Solution

A cutting plane is passed between A and B resulting in the free body diagram Fig. 6.2b. Alternately the right end may be used as shown in Fig. 6.2c. Notice, the internal reactions (P_1, V_1, M_1) differ only by sense in the two diagrams.

The internal reactions V_1 and M_1 are ignored for the present, since we are only interested in the axial load. Summing the forces on the left-hand free body diagram yields:

$$\Sigma F_x = 0 = -300 + P_1$$

$$P_1 = 300 \text{ lb tension}$$

Since this section is typical of any location between A and B, this will be the axial load for all points between A and B.

Moving to a section between B and C and drawing the free body diagram, Fig. 6.2d

$$\Sigma F_x = 0 = +P_2 - 100$$

$$P_2 = 100 \text{ lb compression}$$

This value is typical of the section from B to C.

Plotting these values yields the result shown in Fig. 6.2e—a picture of the axial loads in the member—from which the maximum load and its location are easily identified.

BEAMS   143

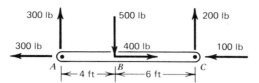

FIG. 6.2a

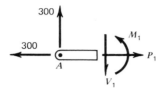

FIG. 6.2b

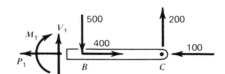

FIG. 6.2c

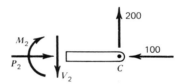

FIG. 6.2d

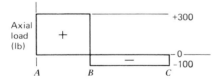

FIG. 6.2e

6.2 BEAMS

When referring to beams one normally pictures the cross members in a building structure, and that is a correct picture. However, the analysis of beams has a much broader application. Beam analysis is used when the dominant loads are perpendicular to the longitudinal axis of a member. For example, the leaf spring-axle assembly for a trailer, Fig. 6.3a, is pin supported at each end, where it is connected to the trailer frame. The load from the axle comes into the middle of the spring. These loads are perpendicular to the length of the spring, Fig. 6.3b,

144 INTERNAL FORCES

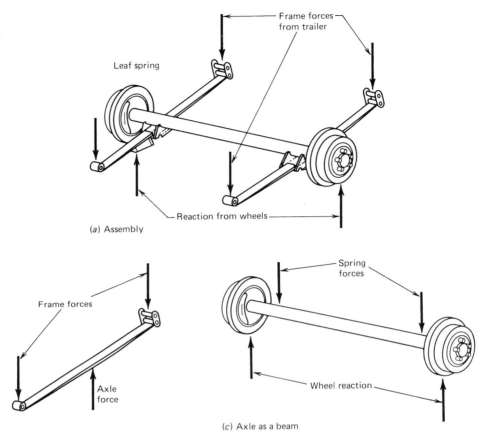

FIG. 6.3a–c Trailer axle and spring assembly.

and the spring is analyzed as a beam. The axle is also analyzed as a beam. Fig. 6.3c shows that the loads from the wheels and the springs are perpendicular to the axle.

Loading in beams is treated as concentrated or distributed. Figure 6.4 shows examples of both. The distributed loads of Fig. 6.4b and 6.4c are uniform. Nonuniform distributed loads are also common in beams. Figure 6.4e shows such a load.

Beams are often classified according to the method of support. Types of supports and their classifications are shown in Fig. 6.4. Only Fig. 6.4a, b, and d are statically determinate; that is, their reactions may be found from the methods of statics. The others are called statically indeterminate. They require an understanding of the beam deflection characteristics to determine the reactions. Methods of handling statically indeterminate beams are taken up in Chapter 14.

INTERNAL SHEAR AND BENDING MOMENT **145**

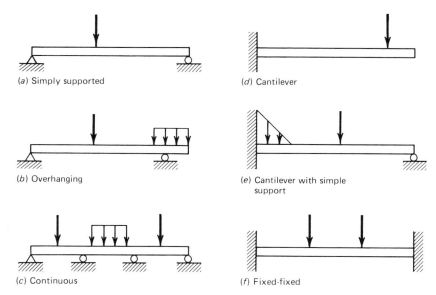

FIG. 6.4a–f Beams classified by supports.

Figure 6.4a shows a simply supported beam with a concentrated load. It is statically determinate. Figure 6.4b shows a simply supported, overhanging beam with a concentrated load and a uniformly distributed load. It, too, is statically determinate. Figure 6.4c shows a beam with redundant simple supports. This beam is called *continuous*. It is not statically determinate. It has two concentrated loads and a uniformly distributed load. The beam in Fig. 6.4d is rigidly supported on one end and free on the other. It is called a *cantilever beam*, and it is statically determinate. The load is concentrated. The next beam, Fig. 6.4e, is fixed on one end and simply supported on the other. It is not statically determinate. It has a concentrated load and a uniformly varying distributed load. Figure 6.4f shows a beam with both ends fixed. It is not statically determinate, and the two loads are concentrated.

6.3 INTERNAL SHEAR AND BENDING MOMENT

A simply supported beam whose reactions have been found is shown in Fig. 6.5a. Thus the first four steps outlined in Section 6.1 have been completed. It is desired to find the internal load at D. Step 5 is carried out by passing a cutting plane perpendicular to the member at point D and drawing a free body diagram of the right-hand portion, as in Fig. 6.5b. Step 6 requires the application of Newton's first law, yielding:

$$\Sigma F_y = 0 = -V + 200$$

$V = 200$ N ↓

146   INTERNAL FORCES

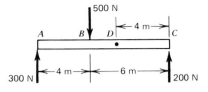

FIG. 6.5a

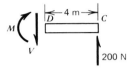

FIG. 6.5b

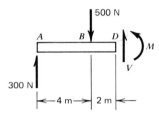

FIG. 6.5c

and

$$\Sigma M_D = 0 = -M + 4(200)$$

$$M = 800 \text{ N} \cdot \text{m}$$

Thus the *internal* shear load and the *internal* bending moment (moment of force) have been found.

The left-hand portion of the beam could also have been drawn. The right-hand was used because it was simpler. However, it can be instructive to examine the left-hand side, which is shown in Fig. 6.5c. Applying Newton's first law to this free body diagram yields

$$\Sigma F_y = 0 = 300 - 500 + V$$

$$V = 200 \text{ N} \uparrow$$

$$\Sigma M_D = 0 = -6(300) + 2(500) + M$$

$$M = 800 \text{ N} \cdot \text{m}$$

As expected, analysis of the left-hand portion produces the same values for shear and bending moment. However, there is an important difference—the signs are opposite. This is in spite of the fact that there is a single state of shear and a single state of bending in the beam. This leads to the conclusion that the vector sign system used to describe forces and bending moments up to this point is inadequate to describe internal forces. The following sign system is arbitrarily but widely used.

Figure 6.6a shows an axial force pulling on the surface of a member. This is called *tension* or a *positive axial force*. Note that it may act either to the left or right—positive or negative vectors—but if it acts away from the surface it is a tensile, or positive, internal force. Alternately, if it acts into the surface, it is a compressive, or negative, internal force.

Figures 6.7 and 6.8 give two ways of thinking about internal shear load. The first shows the force acting on a face. A downward shear on a right-hand face or an upward shear on a left-hand face is considered positive. We can also think of positive shear as that combination of forces which tends to move the left-hand portion of the beam up with respect to the right-hand portion, as shown in Fig. 6.8. The opposite of these conditions is negative shear. Notice that once again the vector sign of a force is inadequate for characterizing the sign of an internal force.

Two ways of thinking of the bending moment sign are given in Figures 6.9 and 6.10. We may think of a positive bending moment as counterclockwise on a right-hand face or clockwise on a left-hand face. Or we may think of the sign in

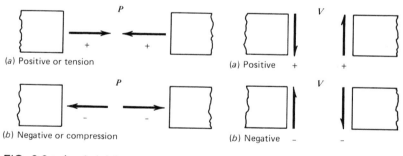

FIG. 6.6a-b Axial forces.   FIG. 6.7a-b Shear forces.

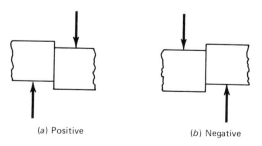

FIG. 6.8a-b Tendency to shear.

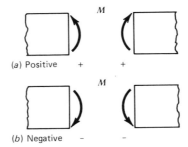

FIG. 6.9a-b Bending moment.

terms of what it tends to do to the geometry of the beam. If the beam were bent concave upward such that it would hold water, it would be positive bending. You may remember this more easily if you think of a smiling, happy face—therefore, "positive." Contrary to this we have a frowning, sad face—therefore, "negative." Still another way of thinking about the sign of bending is based on the stresses produced in the beam. Positive bending compresses the uppermost fibers and stretches the bottom ones. negative bending produces just the reverse.

Most students remember the sign convention for axial loads and bending moments quite easily. The shear force sign system usually requires rote memorization.

Based on this sign convention the shear in the previous problem (Fig. 6.5) is negative regardless of which portion of the beam is used. The bending moment on

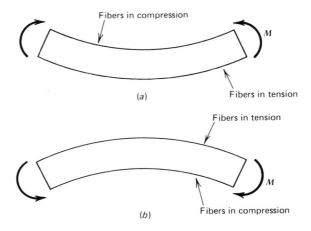

FIG. 6.10 a-b Tendency to bend.

the other hand, is positive. In this problem the internal shear and bending moment at a single point, D, have been found. In a similar manner other points on the beam may be examined, as is illustrated in Example Problem 6.2.

EXAMPLE PROBLEM 6.2

Find the shear and bending moment every 2 feet across the beam, Fig. 6.11a.

Solution

Pass a cutting plane 2 ft from the left end and draw the free body diagram (Fig. 6.11b). The shear force is assumed to be downward, which is a negative direction for *vectors*. Summing the forces is a vector equation thus

$$\Sigma F_y = 0 = 300 - V_2$$

$$V_2 = +300 \text{ lb}$$

The positive sign resulting from the equation means that the assumed direction for the vector was correct (i.e., downward) and hence a negative *vector*. However, based on the sign system for shear, Fig. 6.7, a downward force on a right-hand face is positive *shear*.

$$V_2 = 300 \text{ lb positive shear}$$

Summing the moments about the right-hand end

$$\Sigma M_2 = 0 = -2(300) + M_2$$

$$M_2 = +600 \text{ ft-lb}$$

Again the positive sign means the assumed direction was correct. The sign system in Fig. 6.9 indicates this is also a positive bending moment.

$$M = 600 \text{ ft-lb positive bending}$$

At 4 ft the downward load of 500 lb occurs. The shear cannot be examined at this point; therefore, it will be examined slightly to the left of the load and then slightly to the right. Just to the left of the load excludes the 500-lb load from the free body diagram, Fig. 6.11c

$$\Sigma F_y = 0 = 300 - V_{4L}$$

$$V_{4L} = 300 \text{ lb positive shear}$$

$$\Sigma M_4 = 0 = -4(300) + M_{4L}$$

150 INTERNAL FORCES

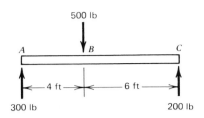

FIG. 6.11a

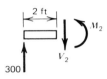

FIG. 6.11b

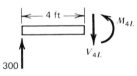

FIG. 6.11c

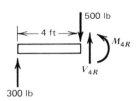

FIG. 6.11d

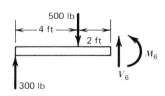

FIG. 6.11e

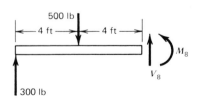

FIG. 6.11f

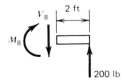

FIG. 6.11g

$$M_{4L} = 1200 \text{ ft-lb positive bending}$$

Just to the right of the load includes the 500-lb load, Fig. 6.11d

$$\Sigma F_y = 0 = +300 - 500 + V_{4R}$$

$$V_{4R} = 200 \text{ lb negative shear}$$

$$\Sigma M_4 = 0 = -4(300) + M_{4R}$$

$$M_{4R} = 1200 \text{ ft-lb positive bending}$$

The shear change is from 300-lb positive to 200-lb negative shear—a change of 500 lb, which exactly equals the load applied at the point. The bending moment did not change from one side of the 500-lb load to the other.

At 6 ft, Fig. 6.11e

$$\Sigma F = 0 = +300 - 500 + V_6$$

$$V_6 = 200 \text{ lb negative shear}$$

$$\Sigma M_6 = 0 = -6(300) + 2(500) + M_6$$

$$M_6 = 800 \text{ ft-lb positive bending}$$

The left-hand free body diagram is shown (Fig. 6.11f) next for a cutting plane at 8 ft. Alternately, we may draw and analyze the simpler, right-hand free body diagram, Fig. 6.11g. Notice that force and moment are in equal but opposite pairs.

$$\Sigma F_y = -V_8 + 200 = 0$$

$$V_8 = 200 \text{ lb negative shear}$$

$$\Sigma M_8 = 0 = -M_8 + 2(200)$$

$$M_8 = 400 \text{ ft-lb positive bending}$$

6.4 BEAM GRAPHS

In Example 6.2 the shear and bending moment were found at four different locations. Additional locations could, of course, be found. One way to represent the pattern of shear and bending moment being developed is to graph them. Figure 6.12 is such a graph. If additional points were sampled the points would be found on the graph matching the broken lines. These graphs are very valuable

152 INTERNAL FORCES

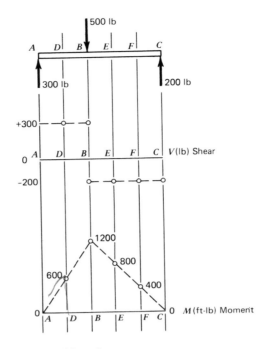

FIG. 6.12 Graph of shear and bending moment.

aids in the design and analysis of beams, and they are examined further in this chapter.

In Example Problem 6.2 the shear and bending moment were examined every 2 ft. By applying algebraic principles, this process can be simplified. A general section is examined at some position x for each portion of the beam bounded by discrete loads. Example Problem 6.3 illustrates this method. This analysis yields *equations* for the shear and bending moment for each section of the beam. The equations can be evaluated for any location for a number; they can be substituted into more detailed analysis; they can be programmed into a computer; they can be examined functionally to determine how the shear and bending moment behave in the beam (such as where do maximums or minimums occur?); and finally, they can be plotted as in Fig. 6.12.

EXAMPLE PROBLEM 6.3

Find equations for the shear and bending moment in segments A-B and B-C for the beam shown in Fig. 6.11a.

Solution

Note that the sections are bounded by discrete loads. A cutting plane is

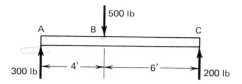

FIG. 6.11a

passed at a distance x to the right of A. The free body diagram is drawn, Fig. 6.13a, and

$$\Sigma F_y = 0 = 300 - V_1$$

$$V_1 = 300 \text{ lb positive shear}$$

This says the shear is constant, independent of position, between A and B. To find the moment M_1 we may sum the moments of force about any point. However, it is usually desirable to sum moments about a point at the section in question. This eliminates V_1, the shear at the section, from the expression for the moment. Thus

$$\Sigma M_x = 0 = -300x + M_1$$

$$M_1 = 300x \text{ positive bending moment}$$

This equation is limited to values of x between 0 and 4, that is, $0 < x < 4$. The above equations allow the shear and bending moment to be evaluated for any position between A and B substituting in the appropriate value of x. Referring to Fig. 6.12, these two equations are those for the curves plotted as shear and bending moment between A and B.

Now pass a cutting plane between B and C and proceed (Fig. 6.13b)

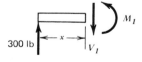

FIG. 6.13a

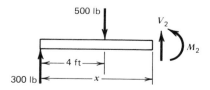

FIG. 6.13b

$$\Sigma F_y = 0 = +300 - 500 + V_2$$

$V_2 = 200$ lb negative shear

$$\Sigma M_x = 0 = -300x + 500(x-4) + M_2$$

$$M_2 = 300x - 500x + 2000$$

$M_2 = 2000 - 200x$ positive bending

These two equations are valid for

$$4 < x < 10$$

and are those for the curves plotted in Fig. 6.12 between B and C.

Shear is constant for the section and the bending moment varies as follows:

At $x = 4$ ft:

$$M_4 = 2000 - 200(4) = \mathbf{1200 \text{ ft-lb}}$$

At $x = 10$ ft:

$$M_{10} = 2000 - 200(10) = \mathbf{0}$$

In this example we guessed the sense of V_1, V_2, M_1 and M_2 based on our intuition and we were right. In general, when finding equations, we will be better off not to guess. Instead, choose vectors representing positive internal shear and bending moment, as in Figs. 6.7a and 6.9a. Use the left-hand free body diagram. Then the sign from the equation for shear or bending will correspond to the physical sense given in Figs. 6.7 to 6.10.

6.5 RELATION BETWEEN LOAD, SHEAR, AND BENDING MOMENT

The graphs plotted in Fig. 6.12 are more formally known as shear and bending moment diagrams. These diagrams greatly aid in the analysis and design of beams—which includes a very large portion of the members in buildings, machines, and the like. Students often have great difficulty with these diagrams. It is true that considerable detail is involved, but the fundamental principle is quite simple. In our study we will eventually apply this principle in four successive steps. It will, therefore, be worthwhile to master it early.

The plotting of the graph in Fig. 6.12 was tedious and required many calculations. We will come to obtain graphs of this type very simply and quickly. As a basis for doing so we will first develop fundamental relationships that exist between the load rate, shear, and bending moment. By load rate, we mean the load per unit length of the beam.

RELATION BETWEEN LOAD, SHEAR, AND BENDING MOMENT

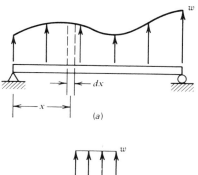

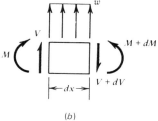

FIG. 6.14a–b

Consider a beam under some general loading as in Fig. 6.14a. (The loading is shown upward, positive, to establish a sign system. Furthermore, the supports shown are schematic, so don't worry about the beam's lifting up.) A very small element of length dx is cut from the beam, and a free body diagram is drawn, Fig. 6.14b. The load on the left end is a shear V and a bending moment M. Over the distance dx the shear changes an infinitesimal amount dV and the bending moment dM. The distance dx is sufficiently short that w may be treated as constant over the interval. Applying equilibrium conditions to the element yields

$$\Sigma F_y = 0 = +V + w\,dx - (V + dV)$$

$$w\,dx = dV$$

Then

$$w = \frac{dV}{dx} \tag{6.1}$$

which is the slope of the shear (V) curve. This says the slope of the shear curve is equal to the instantaneous value of the load rate. Alternately, this may be stated as

$$dV = w\,dx$$

and then integrated to give

$$\Delta V = \int w\,dx = \text{area under } w \text{ curve} \tag{6.2}$$

This says that the change in shear is equal to the area under the load rate diagram. The above principles are illustrated by the graphs in Fig. 6.12, hereafter called load rate, shear, and bending moment diagrams. There is no loading between A and B in Fig. 6.12, that is, $w=0$. According to Eq. 6.1 the slope of the shear diagram for the segment A-B should also be zero, which, of course, is the case. Equation 6.2 indicates that the change in the shear is equal to the area under the load curve. Since the load between A and B is zero, this area, and thus the change in shear, is also zero.

Referring again to the free body diagram of the element dx, Fig. 6.14b, we sum the moment of the forces about the right end, giving

$$\Sigma M_R = 0 = -M - V\,dx - (w\,dx)\frac{dx}{2} + (M + dM)$$

$$-V\,dx - \frac{w(dx)^2}{2} + dM = 0$$

The term $w(dx/2)^2$ is an order of magnitude smaller than $V\,dx$, so it is disregarded, giving

$$V = \frac{dM}{dx} = \text{slope of the } M \text{ curve} \qquad (6.3)$$

This says that the slope of the bending moment curve is equal to the instantaneous value of the shear. Alternately, it may be stated as

$$dM = V\,dx$$

Integrating gives

$$\Delta M = \int V\,dx = \text{area under } V \text{ curve} \qquad (6.4)$$

This says the change in bending moment equals the area under the shear curve. If we refer again to Fig. 6.12, it is observed that the area under the shear curve between A and D is 300 lb $\times$ 2 ft = 600 ft-lb. According to Eq. 6.4 the change in the bending moment should then be 600 ft-lb, as it is. A similar area and a corresponding similar change in M is found from D to B, and so across the beam. Equation 6.3 indicates that the slope of the bending moment diagram at any point equals the corresponding value of the shear. The shear is constant and positive from A to B, and accordingly, the slope of the bending moment is also constant and positive.

A very important relationship is established here between load rate (w), shear (V), and bending moment (M), that is, each curve is the integral of the previous one. Alternately, each curve is the derivative of the following one. The integral of $x^n dx$ is as follows:

$$\int x^n\,dx = \frac{x^{n+1}}{n+1} + C$$

The degree (n) of a curve increases by one with each integration. Thus if the load rate is constant—zero degree—the shear will be first degree, linear, and the bending moment will be second degree, parabolic. The complexity here is considerably reduced when it is realized that the bending moment has the same relationship to the shear as the shear has to the load rate.

These principles allow the shear and moment diagrams to be plotted directly from the solved free body diagram of the entire beam. This method quickly and efficiently gives a picture of the shear and bending moment in the entire beam which is essential in comprehensive design problems. The method is detailed in the following section.

The derivative and integral define the geometric relationships. In practice, we do not normally exercise the calculus at this point. However, if the concepts of the derivative and integral are well understood, they will be of great assistance in obtaining shear and moment diagrams. On the other hand, as the concepts of these diagrams are mastered, they can enhance our understanding of the derivative and integral.

6.6 GENERAL PROCEDURE FOR BEAM DIAGRAMS

To construct shear and bending moment diagrams the following rules should be followed:

1. Draw a free body diagram of the beam.
2. Solve for the support reactions using Newton's first law.

SHEAR DIAGRAM

3. Starting at the left end of the beam, reproduce vertical loads as they occur. Upward forces make a positive change in shear; downward forces make a negative change.
4. Apply Eq. 6.2 to determine the change in shear between discrete forces. Note that $\int w\,dx$ equals the area under the load rate curve. (We use the word "curve" in its general sense. The "curve" may be a straight line, zero valued, or any other function.) Downward load rates are negative; upward ones are positive.
5. The shear curve will be one degree higher than the load rate curve. Its specific shape is determined by applying Eq. 6.1. Note that dV/dx is the slope of the shear curve. The curve will close at the right end.

(The next three steps are merely repetitions of steps 3 through 5.)

BENDING MOMENT DIAGRAM

6. Starting at the left end, reproduce discrete bending moments as they occur. The sign of the bending moment is determined by the effect it would have on the beam to its right; that is, if it will produce positive bending in the beam to its right, it will cause a positive change in the bending moment diagram. (Students hardly ever get this right the first time—better read it again.)

158 INTERNAL FORCES

7. Apply Eq. 6.4 to determine the change in bending moments between discrete forces or bending moments. The integral $\int V\,dx$ is the area under the shear curve.

8. The bending moment curve will be one degree higher than the shear curve. Its specific shape is determined by applying Eq. 6.3. The derivative dM/dx is the slope of the moment curve. The curve will close at the right end.

EXAMPLE PROBLEM 6.4

Construct the shear and bending moment diagrams for the cantilever beam shown in Fig. 6.15.

Solution

Steps 1 and 2 are left to the student.

Executing step 3, we start at the left end. The 200-N downward force is reproduced on the shear diagram, as shown in Fig. 6.15b.

Applying step 4, we note that between A and B the load rate is zero ($w=0$), thus there is no area under the load rate curve and therefore no change in shear. Alternately, the value of the load rate is the slope of the shear curve, both being zero in this case. This is step 5.

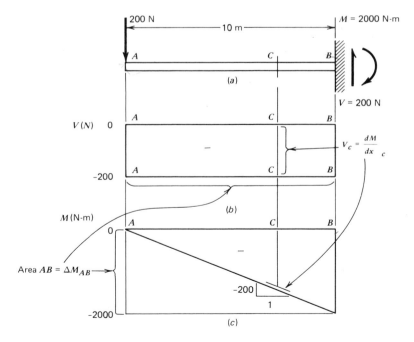

FIG. 6.15a–c

GENERAL PROCEDURE FOR BEAM DIAGRAMS   159

The 200-N concentrated reaction at B is reproduced, closing the diagram.

The shear diagram must always close. If it does not close, you have either made an error in its construction, or the beam was not in equilibrium to start with.

The shear is negative 200 N throughout the beam.

In step 6 (Fig. 6.15c), there is no bending moment at A, so this starts at zero. The change between A and B equals the area under the shear curve:

$$\Delta M_{A-B} = -200(10) = -2000 \text{ N} \cdot \text{m}$$

Notice the area and change are negative.

Since the shear is constant, according to Eq. 6.3, the slope of M is constant—hence the straight line. The slope at C, dM/dx, is equal to the value of shear at C.

The concentrated moment at B, 2000 N·m, would bend the beam to the right of it (if it existed) positively; thus it is treated as a positive 2000-N·m load, which closes the diagram.

The bending moment diagram must always close. If it does not close, you have either made an error in its construction, or the beam was not in equilibrium to start with.

EXAMPLE PROBLEM 6.5

Construct the shear and bending moment diagram for the uniformly loaded, simply supported overhanging beam shown in Fig. 6.16a.

Solution

1. We draw the free body diagram (Fig. 6.16 b) of a statically equivalent beam by replacing the distributed 20 kip/ft load with a concentrated load of 200 kips at the center of the beam. This free body diagram is statically equivalent, but not equivalent!

2. $\Sigma M_A = 0 = -5(200) + 8(B)$

$$B = \frac{5(200)}{8} = \textbf{125 kips}$$

$$\Sigma F_y = 0 = A + B - 200$$

$$A = 200 - B = 200 - 125 = \textbf{75.0 kips}$$

We proceed to step 3, reverting back to the original load diagram. We do not use the statically equivalent beam with the concentrated 200-kip load.

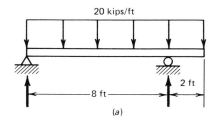

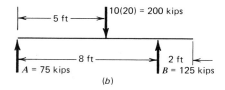

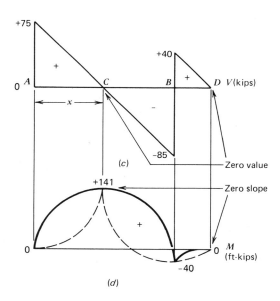

FIG. 6.16a–d

The shear diagram goes up 75 units at A, Fig. 6.16c. Between A and B the area under the load rate curve is:

$$\Delta V = 8(-20) = -160 \text{ kips}$$

Thus the shear just left of B is

$$\Delta V_{BL} = +75 - 160 = -85 \text{ kips}$$

GENERAL PROCEDURE FOR BEAM DIAGRAMS   161

Since the load rate is constant—a zero degree curve—the shear will be a first-degree or linear curve. Thus we draw a straight line from $+75$ to -85. Another way of arriving at the same conclusion is to note

$$w = \frac{dV}{dx} = \text{slope of shear}$$

This is to say that the instantaneous value of the load rate is the slope of the shear diagram. Since the load rate is constant and negative, the slope of the shear must also be constant and negative.

Applying step 3 again, at B the shear changes $+125$ kips. So

$$V_{BR} = -85 + 125 = +40.0 \text{ kips}$$

This is a step change.

From B to the right end of the beam we find

$$\Delta V = (2)(-20) = -40 \text{ kips}$$

This brings us back to 0 at the right end, and the diagram closes as it always should. Applying Eq. 6.1 gives us a straight line for this portion of the curve. The diagram should be labeled as shown, indicating peak values, units, and positive and negative areas.

We proceed to the moment diagram and steps 6, 7, and 8. There is no concentrated moment at the left end, so we start at zero. We notice the shear curve changes from positive to negative between A and B. We must find this crossover point. We call the point C and say it is x feet to the right of A. The shear starts at $+75$ and must drop to zero over the distance x, so we write

$$\Delta V = wx$$
$$-75 = (-20)x$$

The constant load rate of 20 kips/ft is downward, which we call negative; therefore, we write "-20" in the above equation. Then

$$x = \frac{-75}{-20} = \mathbf{3.75 \text{ ft}}$$

Locating this point, we proceed with the moment diagram from A to C (Fig. 6.16d)

$$\Delta M = \text{area of triangle} = 1/2 \, bh$$

$$\Delta M = \tfrac{1}{2}(3.75)(+75) = 140.6 = +141 \text{ ft-kips}$$

The shear curve is first degree, so the moment curve will be second degree.

Two possibilities of second-degree curves are shown, one solid and one broken. The solid is chosen by applying Eq. 6.3, which says

$$V = \frac{dM}{dx}$$

So the slope must go from a highly positive value ($+75$) to zero. When a zero is included, this selection is much easier.

From C to B we have

$$\Delta M = \tfrac{1}{2} bh = \tfrac{1}{2}(8 - 3.75)(-85)$$

$$\Delta M = -180.6 \text{ ft-kips}$$

and

$$M_B = +141 - 181 = -40.0 \text{ ft-kips}$$

The curve required to connect $+141$ and -40 is again second degree. We select the solid one, zero slope at C, because of the zero value of shear at C. Finally, from B to the right end, D:

$$\Delta M = \tfrac{1}{2} bh = \tfrac{1}{2}(2)(+40)$$

$$= +40.0 \text{ ft-kips}$$

and

$$M_D = -40 + 40 = 0$$

The final value is zero, and once again the curve closes. The solid is chosen because of the zero shear at the right end—zero shear gives zero slope on the moment curve.

With these diagrams at a glance we see the maximum shear load is -85 kips and the maximum bending moment is $+141$ ft-kips. We also know where these values occur. These and other values in the diagram are necessary to design (select) the beam.

EXAMPLE PROBLEM 6.6

For the beam shown in Fig. 6.17 construct the shear and moment diagrams.

Solution

An equivalent free body diagram is drawn (Fig. 6.17b) and the reactions are found.

GENERAL PROCEDURE FOR BEAM DIAGRAMS   163

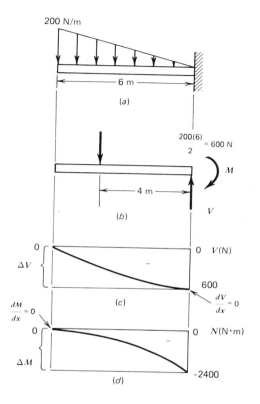

FIG. 6.17a-d

$$V = 600 \text{ N}$$

$$M = 2400 \text{ N} \cdot \text{m} \qquad \text{clockwise}$$

Returning to the original loading, note the initial shear is zero. Thus the shear diagram, Fig. 6.17c, starts at zero. The change from the left end to the right equals the area under the load rate curve:

$$\Delta V = \frac{1}{2}\left(-200 \frac{\text{N}}{\text{m}}\right) 6\text{m}$$

$$\Delta V = -600 \text{ N}$$

The load rate curve is first degree, so the shear curve will be second degree—parabolic. The load rate goes from -200 to 0 so the slope of the shear curve must do the same, since

$$w = \frac{dV}{dx}$$

Of the two options available, we pick the one with zero slope at the right end. Also note the shear curve has a negative slope everywhere. The 600-N reaction at the right end closes the shear diagram.

For the moment diagram we start at zero on the left end, Fig. 6.17d. The change in the moment from the left to the right end is equal to the area under the shear curve. The shear curve is second degree with a zero slope at the right end, and the area under it is negative. We can find this area from the information in Table A4.1 for a parabola. Note that to use this table we must include a vertex (point with zero slope for our purposes). The table gives

$$A = \tfrac{1}{3}bh$$

for the smaller area within the bh rectangle. We have the larger or complimentary part, thus

$$A = \tfrac{2}{3}bh$$

and

$$\Delta M = \tfrac{2}{3}(6 \text{ m})(-600 \text{ N})$$

$$\Delta M = -2400 \text{ N} \cdot \text{m}$$

This means the change from the left end to the right end is -2400 N·m. Since the shear is a second-degree curve, the moment, which is its integral, is a third-degree curve—which looks like a second-degree curve. We pick from our two options by noting that the shear is zero on the left end, hence the slope of the moment curve must be zero on the left end:

$$V = 0 = \frac{dM}{dx}$$

Finally, the concentrated moment, the reaction at the right end, produces positive bending on a beam to its right (if it existed) and consequently produces a positive change of 2400 N·m, closing the moment diagram. Notice that the bending moment is everywhere negative, as we would intuitively anticipate from the beam loading.

6.7 SUMMARY

Internal loads may be found from the methods of statics. The internal load sign system is given in Figs. 6.5 to 6.9. Relations between load rate, shear, and bending moment are:

$$w = \frac{dV}{dx} = \text{slope of } V$$

$$V = \frac{dM}{dx} = \text{slope of } M$$

In integral form:

$$V = \int w\, dx = \text{area under } w$$

$$M = \int V\, dx = \text{area under } V$$

Beam diagrams are graphical interpretations of the above.

The step-by-step procedure of Section 6.6 gives a method for constructing beam diagrams. The efficient and rapid construction of shear and moment diagrams is necessary for the analysis and design of beams.

PROBLEMS

6.1 to 6.8 For Figs. 6.18 through 6.25, respectively, calculate the shear and bending moment at 2-ft or 2-m intervals as appropriate. Indicate whether they are positive or negative.

6.9 to 6.16 Plot the value of shear and bending moment at 2 ft or 2-m intervals as appropriate for Figs. 6.18 through 6.25, respectively.

6.17 to 6.24 Write an algebraic expression for shear and bending moment for each section as necessary to completely describe the beam. Use Figs. 6.18 through 6.25, respectively.

6.25 to 6.32 Directly plot the load, shear, and bending moment diagrams for Figs. 6.18 through 6.25, respectively, by the method of Section 6.6.

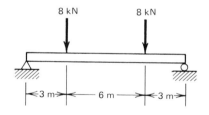

FIG. 6.18 Problems 6.1, 6.9, 6.17, 6.25.

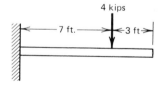

FIG. 6.19 Problems 6.2, 6.10, 6.18, 6.26.

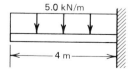

FIG. 6.20 Problems 6.3, 6.11, 6.19, 6.27.

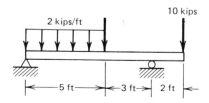

FIG. 6.21 Problems 6.4, 6.12, 6.20, 6.28.

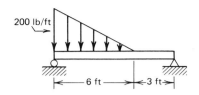

FIG. 6.22 Problems 6.5, 6.13, 6.21, 6.29.

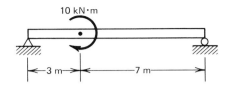

FIG. 6.23 Problems 6.6, 6.14, 6.22, 6.30.

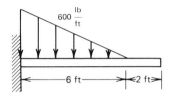

FIG. 6.24 Problems 6.7, 6.15, 6.23, 6.31.

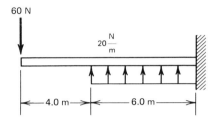

FIG. 6.25 Problems 6.8, 6.16, 6.24, 6.32.

6.33 For the beam shown in Fig. 6.26 construct the shear and moment diagrams.

6.34 Construct the shear and moment diagrams for the beam shown in Fig. 6.27.

6.35 Construct the shear and moment diagrams for the beam shown in Fig. 6.28.

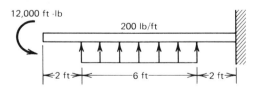

FIG. 6.26 Problem 6.33.

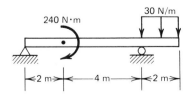

FIG. 6.27 Problem 6.34.

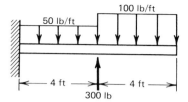

FIG. 6.28 Problem 6.35.

7
Stress and Strain from Bending Loads

7.1 Stress and strain distribution
7.2 Flexure formula
7.3 Beam design using the flexure formula
7.4 Summary

In Chapter 3 we examined the relationship between an axial load and the stresses it produces; in Chapter 5 it was torque and its resultant stresses. In this chapter we examine the relationship between the bending moment and the stresses it produces. This relationship, called the flexure formula, may well be the most important of the three, as it so often governs the design in structures. At the risk of boring the reader we will once again emphasize that stress equations are of little value until the load causing the stress has been found—a statics problem. In the case of bending moment, it can be a rather involved statics problem, as illustrated in Chapter 6.

In this chapter we spend considerable effort in developing the governing equation. The arguments are about as exciting as tomato soup for lunch. However—like tomato soup—they are nourishing. The understanding of the bending phenomenon, the generation of the flexure formula, and most importantly, the stress distribution described by this equation is essential for the intelligent application of this principle. It is tempting to forego the development, going directly to Eq. 7.5 to "plug it in" to problems. And indeed that will work for many of the problems of this chapter. However, it will be totally inadequate when we get to Chapter 10.

Your soup's getting cold!

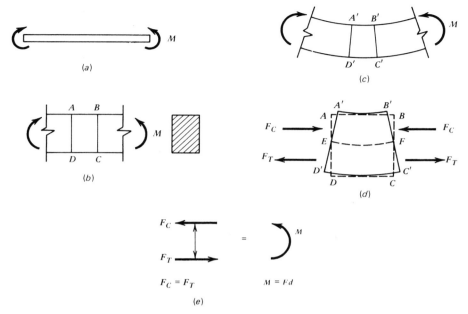

FIG. 7.1a–e Beam in pure bending.

7.1 STRESS AND STRAIN DISTRIBUTION

To develop the flexure formula we consider an initially straight beam under pure bending, Fig. 7.1a. This requirement of pure bending does not prove restrictive, as we will be able to apply the results to most loading situations by the method of superposition. We will show the cross section of the beam as rectangular, although the result is not restricted to that. A smaller section of the beam and its cross section are shown in its initially straight condition in Fig. 7.1b. In Fig. 7.1c the deformation that results from the applied bending moment M is shown. If this deformation is not intuitively obvious, load a ruler in a similar manner until it is clear. Notice that for this deformation to take place, the initial element BC rotates to a new position $B'C'$. We assume it remains straight. This rotation is emphasized in Fig. 7.1d by superimposing the element $A'B'C'D'$ on its original shape $ABCD$ and further enlarging the figure.

Notice the fiber AB has shortened to $A'B'$ while the fiber DC has been lengthened to $D'C'$. Thus the top portion of the element is shortened or compressed, but the bottom portion is stretched or placed in tension. Also note that some transition element, labeled EF in Fig. 7.1d, does not change length at all. This element is known as the neutral axis. All the fibers above the neutral axis are shortened. All the fibers below it are stretched.

The shortening of the fibers above EF requires a compressive force shown as F_C in Fig. 7.1d. The stretching of fibers below EF requires a tensile force, labeled F_T in Fig. 7.1d. The force F_C is equal to the force F_T, and they combine to form a couple equal to the applied bending moment M as illustrated in Fig. 7.1e. Recall

170 STRESS AND STRAIN FROM BENDING LOADS

that a couple is a system of forces having a resultant moment but no resultant force.

We have assumed that the line element BC remains straight and rotates to $B'C'$ as shown in Fig. 7.1b–d. In Fig. 7.2 the element is rotated so that AD is in line with $A'D'$. This will conveniently allow us to examine the total elongation of each fiber. Thus BB' is the change in length of fiber AB, and CC' is the change in length of fiber DC. We denote by δ the elongation of any fiber at any distance y from the neutral axis. The maximum elongation, δ_{max}, of any fiber is equal to CC' and occurs on the extreme fiber at a distance c from the neutral axis. We observe that the elongation of any fiber is proportional to its distance from the neutral axis, and thus we can write

$$\delta = \frac{y}{c} \delta_{max} \tag{7.1}$$

This simply says that the deformation of fibers is a linear function of position, being zero at the neutral axis and a maximum at the extreme fiber.

The strain of any fiber can be found from the definition of strain, namely

$$\varepsilon = \frac{\delta}{L}$$

Since all the fibers have the same original length L, we divide Eq. 7.1 by L to get

$$\frac{\delta}{L} = \frac{y}{c} \frac{\delta_{max}}{L}$$

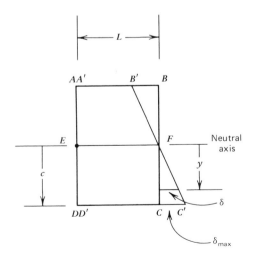

FIG. 7.2 Deformation is linearly distributed.

which yields

$$\varepsilon = \frac{y}{c} \varepsilon_{max} \qquad (7.2)$$

Thus we see that the strain is also linearly distributed, being zero at the neutral axis and a maximum at the extreme fiber.

We decide now to limit our discussion to linearly elastic materials stressed below their proportional limits, so we may apply Hooke's law

$$\sigma = E\varepsilon$$

or

$$\varepsilon = \frac{\sigma}{E}$$

And for Eq. 7.2

$$\frac{\sigma}{E} = \frac{y}{c} \frac{\sigma_{max}}{E}$$

giving

$$\sigma = \frac{y}{c} \sigma_{max} \qquad (7.3)$$

This important relationship says that stress, like deflection and strain, is also linearly distributed. It is zero at the neutral axis, that is, the neutral axis is unstressed and unstrained. It is a maximum at the extreme fiber. This stress distribution is shown in Fig. 7.3. Notice that it is compressive above the neutral axis and tensile below. The resultant force from the compressive stress is equal to

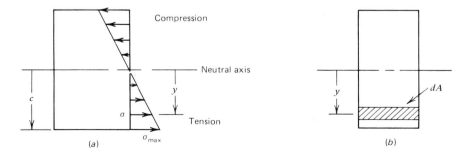

FIG. 7.3a–b Stress distribution.

F_C, Fig. 7.1d, and from the tensile stress F_T. The resultant of the two together is zero force and a couple equal to the applied moment M.

We have seen that pure bending results in linear distributions for total elongation, strain, and stress. All are zero at the neutral axis and maximum at the extreme fiber.

EXAMPLE PROBLEM 7.1

A beam with the cross section shown in Fig. 7.4 has a maximum stress of 120 MPa compression due to pure bending at fiber A. Its modulus of elasticity is 76 GPa. Calculate the stress and strain 30 mm below and 20 mm above the neutral axis.

Solution

The stress distribution is as shown in Fig. 7.5. At 30 mm below the neutral axis

$$\sigma_{-30} = \frac{y}{c} \sigma_{max}$$

$$\sigma_{-30} = \tfrac{30}{40}(120) = \mathbf{90.0 \text{ MPa } C}$$

From Hooke's law the strain is

$$\varepsilon_{-30} = \frac{\sigma}{E} = \frac{90.0 \text{ MPa}}{76 \text{ GPa}} = \mathbf{1.18}E-\mathbf{3}$$

Similarly, for 20 mm

$$\sigma_{+20} = \tfrac{20}{40}(120) = \mathbf{60.0 \text{ MPa } T}$$

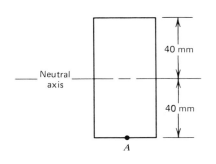

FIG. 7.4

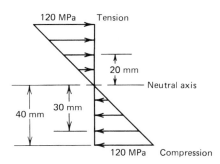

FIG. 7.5

Note that on opposite sides of the neutral axis the signs of the stresses are also opposite

$$\varepsilon_{+20} = \frac{\sigma}{E} = \frac{60.0 \text{ MPa}}{76 \text{ GPa}} = \mathbf{0.789E-3}$$

7.2 FLEXURE FORMULA

Our goal in this section is to develop a relationship between the applied bending moment and the stress it produces. Prior to that, however, the location of the neutral axis must be determined. We accomplish this by observing that the resultant force on the cross section of the beam is zero. Referring to Fig. 7.3, we note that the stress on an element of area dA at a distance y from the neutral axis is σ. The element of force dF acting on the area is

$$dF = \sigma \, dA$$

This is illustrated in Fig. 7.6.

The total force can be found by summing these elements of force over the entire cross section. Thus

$$F = \int dF = \int \sigma \, dA$$

As we said, the resultant force is zero. Then substituting Eq. 7.3 we have

$$\int \sigma \, dA = \int \left(\frac{y}{c}\right) \sigma_{\max} \, dA = 0$$

For a given cross section $\sigma_{\max}$ and c are constants and may be brought outside the integral:

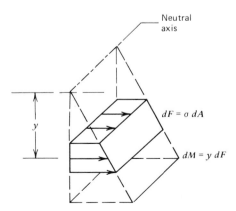

FIG. 7.6 Element of force.

174   STRESS AND STRAIN FROM BENDING LOADS

$$\frac{\sigma_{max}}{c} \int y \, dA = 0 \tag{7.4}$$

As σ_{max} and c are constant, the only condition in which Eq. 7.4 can be satisfied is if

$$\int y \, dA = 0$$

Referring back to Fig. 7.3b, we see that this integral is y, the distance from the neutral axis to the element dA times dA. Thus this integral is simply the first moment of area—one of those exciting but abstract concepts from Chapter 4. (I promised it would be useful!) Now, when is the moment of area equal to zero? "When it's about a centroidal axis!" said the class in one accord. This leads to the very important (and convenient) conclusion that the location of the neutral axis and our old friend the centroidal axis are one and the same.

Returning to Fig. 7.6 we note that the moment of the element of force dF about the neutral axis is

$$dM = y \, dF$$

The total moment is the summation or integral of these moments over the cross section. This total moment equals the applied moment M.

$$M = \int dM = \int y \, dF$$

Using the following equalities from before

$$dF = \sigma \, dA$$

$$\sigma = \frac{y}{c} \sigma_{max}$$

we have

$$M = \int y \left(\frac{y}{c} \sigma_{max} \right) dA$$

Again factoring out the constants, σ_{max} and c

$$M = \frac{\sigma_{max}}{c} \int y^2 \, dA$$

Examining the integral, we see it to be the distance squared from the neutral (or centroidal) axis to the element dA. This expression is the second moment of area or moment of inertia of the cross section about its own centroid. This was designated I in Chapter 4. Hence

$$M = \frac{\sigma_{max}}{c} I$$

Solving for σ_{max} yields

$$\sigma_{max} = \frac{Mc}{I} \quad (7.5)$$

This is known as the flexure formula. Combining it with Eq. 7.3 gives

$$\sigma = \frac{My}{I} \quad (7.6)$$

which gives the stress on any fiber at a distance y from the neutral axis and once again shows this stress distribution to be linear.

Let's sum up what we have learned. Pure bending produces a linear stress distribution that is tension on one surface and compression on the other. There is an unstressed fiber called the neutral axis that coincides with the centroid of the cross section (for beams that are initially straight). The maximum stress occurs on the fiber furthest from the neutral axis and can be determined by Eq. 7.5. Note that it varies directly with the applied load M and with the distance to the extreme fiber c. It varies inversely with the moment of inertia I.

Also note that for Eq. 7.5, M is the moment about the centroidal axis, c is the distance to the extreme fiber from the centroidal axis, and I is the moment of inertia about the centroidal axis. That is, all the terms on the right-hand side of Eq. 7.5 refer to the same axis, namely, *the centroidal axis* of the cross section about which the bending moment acts. The equation is only valid for materials that are homogeneous, isotropic, and loaded below their proportional limit.

EXAMPLE PROBLEM 7.2

A 2×6 (actual size) turned on edge must carry a bending moment of 1200 ft-lb. (a) What is the maximum stress produced in the beam (Fig. 7.7)? (b) Solve the above problem for a finished 2×6 (S4S).

Solution

(a) We use our new friend, the flexure formula

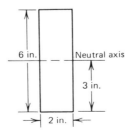

FIG. 7.7

$$\sigma = \frac{Mc}{I} \qquad (7.5)$$

In this equation M, c, and I all are referenced to the same axis. If the bending moment M is about a horizontal axis, c will be the distance from that same axis to the extreme fiber, and I will be the moment of inertia about that same axis. Thus

$$I = \frac{bh^3}{12} = \frac{2(6)^3}{12} = 36.0 \text{ in.}^4$$

Note that in this formula b, the base, is the side parallel to the axis about which we are finding the moment of inertia. Therefore, b is parallel to the axis about which the bending moment is acting. The centroid and therefore the neutral axis is at the middle of the cross section. The distance from the neutral axis to the extreme fiber, c, is 3 in. Then applying the flexure formula

$$\sigma = \frac{1200 \text{ ft-lb } (3 \text{ in.})}{36.0 \text{ in.}^4} \left(\frac{12 \text{ in.}}{\text{ft}} \right)$$

$\sigma = 1200$ psi

The stress distribution produced by this moment is that shown in Fig. 7.8, or just the opposite, depending on the direction of the applied moment. When the cross section is symmetric the magnitude of the stresses will be equal at the top and bottom surfaces.

(b) From Table A4.9 we find

$$h = 5.5 \text{ in.}$$

$$I = 20.8 \text{ in.}^4$$

Note that I for the finished size is only a little more than half of the full size:

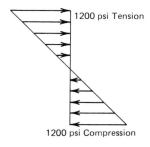

FIG. 7.8

$$\sigma = \frac{Mc}{I} = \frac{1200 \text{ ft-lb } (5.5 \text{ in.})}{20.8 \text{ in}^4 \ 2} \cdot \frac{12 \text{ in.}}{\text{ft}}$$

$\sigma = \mathbf{1900 \text{ psi}}$

Note further that the stress in the finished size is more than 50% more than in the full cut.

EXAMPLE PROBLEM 7.3

The beam shown in Fig. 7.9 is a $S\ 10 \times 35$. Find the maximum bending stress.

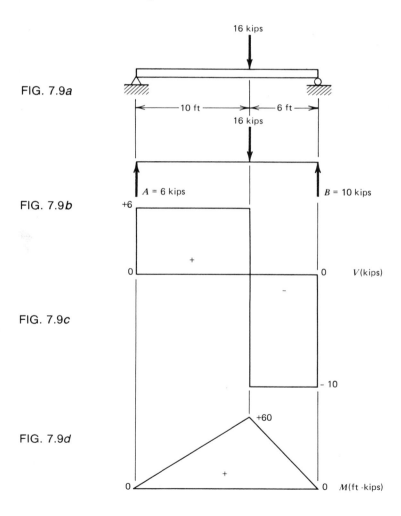

FIG. 7.9a

FIG. 7.9b

FIG. 7.9c

FIG. 7.9d

Solution

This problem well illustrates the general stress analysis problem.

1. *Solve the statics problem*

Draw a free body diagram and find the reactions (Fig. 7.9b)

$$\Sigma M_A = 0 = -10(16) + 16(B)$$

$$B = \frac{16(10)}{16} = 10.0 \text{ kips}$$

$$\Sigma F_y = 0 = A - 16 + B$$

$$A = 16 - (10) = 6.00 \text{ kips}$$

2. *Find the internal loads*

This can best be done in this case by drawing shear and bending moment diagrams, Figs. 7.9c and 7.9d.

We observe from the moment diagram that the maximum moment is 60.0 ft-kips.

3. *Find the geometric properties of the cross section*

This beam is symmetric, and the centroid can be found by inspection. In other cases it must be calculated.

The moment of inertia can also be easily determined in this case from Table A4.2. Note that two values of I are available, I_x and I_y. We must use the one that corresponds to the axis of the bending moment. Our problem does not state how this beam is oriented. We assume it is turned so that it will carry the greatest load, that is, so that it presents the greatest I. The result will be as good as the assumption.

$$I = 145.8 \text{ in.}^4 \qquad c = 5.00 \text{ in.}$$

4. *Find the stress*

Equation 7.5

$$\sigma = \frac{Mc}{I}$$

$$\sigma = \frac{60.0 \text{ ft-kips } (5.00 \text{ in.})}{145.8 \text{ in.}^4} \cdot \frac{12 \text{ in.}}{\text{ft}}$$

$$\sigma = 24.7 \text{ ksi}$$

The bending moment was positive 60 ft-kips. This will produce compression in the top fiber and tension in the bottom. Thus 24.7 ksi compression is produced in the top of the beam and 24.7 ksi tension in the bottom at the section 10 ft from the left support.

EXAMPLE PROBLEM 7.4

The cross section shown in Fig. 7.10 carries a positive bending moment of 200 ft·lb. Find the stress on the top and bottom fibers.

Solution

By Eq. 7.6 the stress on the top fiber is

$$\sigma_T = \frac{My}{I} = \frac{200 \text{ ft·lb } (4 \text{ in.})}{533 \text{ in.}^4} \cdot \frac{12 \text{ in.}}{\text{ft}}$$

$$\sigma_T = 18.0 \text{ psi} \quad \text{Compression}$$

Since the bending is positive, the top fiber is in compression. The stress on the bottom fiber is

$$\sigma_B = \frac{My}{I} = \frac{200 \text{ ft·lb } (8 \text{ in.})}{533} \cdot \frac{12 \text{ in.}}{\text{ft}}$$

$$\sigma_B = 36.0 \text{ psi} \quad \text{Tension}$$

The resulting stress distribution is shown in Fig. 7.11. Note that the stress at the neutral axis is zero and that the stress is greatest on the fiber most distant from the neutral axis.

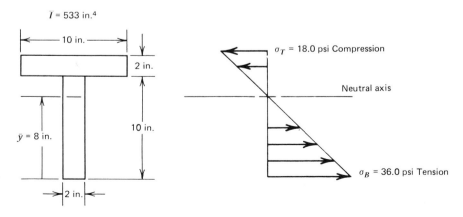

FIG. 7.10   FIG. 7.11

7.3 BEAM DESIGN USING THE FLEXURE FORMULA

When we speak of designing a beam we usually mean selecting the "most appropriate" beam that will satisfactorily carry the required load. This selection is normally from a finite number of choices, since we will normally use commonly available materials such as those given in Table A4.2 through A4.9. Most appropriate usually means the most economical, which means the beam that uses the least material. For uniform beams of a given material, the amount of material used is directly proportional to the cross-sectional area. Thus our problem boils down to finding the beam of least cross-sectional area that can carry the required load. This is indicated by the area itself, or by the weight per unit length of the beam.

To select a beam for a given load situation, a full static analysis must be made and critical internal loads determined, as in steps 1 and 2 in Example Problem 7.3. Given these loads, we must now select a member that will have the appropriate geometric properties to keep the stress below some predetermined value—the design stress. Structural steel often carries a design stress of 24,000 psi (165 MPa) for static loads.

As we seek to apply Eq. 7.5 in design we notice it contains two geometric factors, c and I. Lumping these two together and solving for them we have

$$\frac{I}{c} = \frac{M}{\sigma}$$

If, for instance, the moment is 720 in.-kips and the allowable stress is 24 ksi, we would have

$$\frac{I}{c} = \frac{720 \text{ in.-kips in.}^2}{24 \text{ kips}} = 30.0 \text{ in.}^3$$

Actually, I/c should be equal to or greater than 30.0 in.^{3} to keep the stress below 24 ksi. The I is available from Tables A4.2 to A4.9, and c may be easily calculated from the other geometric data—usually one-half the depth. Hence we may seek a combination of I divided by c that would be greater than 30 in.^{3} Obviously, it would require a number of trials to arrive at a satisfactory number, and then we would not be sure we had the best beam unless we had exhausted the tables. This rather tedious and frustrating exercise is avoided by including in the tables an additional column which gives the quotient I divided by c.

We give this quotient the name section modulus and designate it as

$$Z = \frac{I}{c} \tag{7.7}$$

BEAM DESIGN USING THE FLEXURE FORMULA

Although this is designated S in Tables A4.2 to A4.9, we will stick with Z to avoid confusing it with the beam shape or the material's strength. Equation 7.5 may then be written as

$$\sigma = \frac{M}{Z} \qquad (7.8)$$

This equation is essential for design and convenient for analysis.

In our design problem we now look in the tables for a section modulus Z greater than 30.0 in.^{3} for beams that can satisfactorily carry the load. After finding several candidates, we make a final choice based on the one that has the least area or weight per foot.

EXAMPLE PROBLEM 7.5

Find the best (lightest) S or W beam to carry a 720-in.-kip bending load if the allowable stress is 24 ksi.

Solution

For Eq. 7.8

$$\sigma = \frac{M}{Z}$$

Solving for the section modulus Z

$$Z = \frac{M}{\sigma} = \frac{720 \text{ in.-kip in.}^2}{24 \text{ kip}} = 30.0 \text{ in}^3$$

From Table A4.2 reading in the S ($S = Z$) column

$S\ 12 \times 31.8$ has $Z = 36.4$ so it is satisfactory

From A4.3

$W\ 14 \times 26$ has $Z = 35.1$ which is more than satisfactory

Either of the above would carry the load, but the $W\ 14 \times 26$ is preferred because it uses less material.

Note that in this problem optimizing the solution involves a different parameter (weight per foot) from the one used to control the load-carrying capability (section modulus).

EXAMPLE PROBLEM 7.6

Select a standard or wide flange beam to carry 2000 lb/ft over a 12-ft length when supported as a cantilever, Fig. 7.12a. The yield strength is 36 ksi, and a safety factor of 4 is intended.

Solution

To select this beam we must know the load (maximum bending moment) it is to carry. Once again, a statics problem must be solved first. We make a sketch of the problem and then draw a free body diagram, Fig. 7.12b, of a statically equivalent beam for the purpose of finding the reactions V and M.

Applying Newton's first law yields the results indicated.

Having these reactions, we sketch shear and moment diagrams for the beam, remembering to return to the original loading. These are as shown in Figs. 7.12c and 7.12d. Their development is left to the student.

Having solved a fairly substantial statics problem, we are now ready to apply the solution to a fairly simple strength of materials problem.

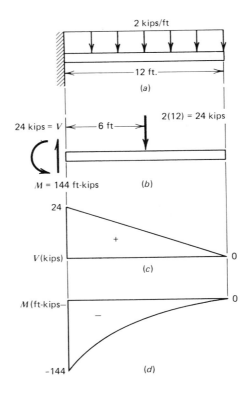

FIG. 7.12a–d

We examine the moment diagram to find the maximum moment on the beam. It is 144 ft-kips and occurs at the support. It turns out that the maximum moment will always be at the support on a cantilever beam when all the loading is perpendicular to the beam and in the same sense. However, in general, the moment diagram is needed, and this shortcut should not be taken until the student is very sure that he has the proper load.

We have a yield strength of 36 ksi and a factor of safety of 4, so

$$S_{AL} = \frac{S_y}{N} = \frac{36}{4} = 9.0 \text{ ksi}$$

Using Eq. 7.8,

$$\sigma = \frac{M}{Z}$$

$$Z = \frac{M}{\sigma} = \frac{144 \text{ ft-kips}}{9.0 \frac{\text{kips}}{\text{in.}^2}} \frac{12 \text{ in.}}{\text{ft}} = 192 \text{ in.}^3$$

This is the required section modulus. We go to Tables A4.2 and A4.3, seeking members with values of Z greater than this. For a standard we select S 24×100 having a section modulus of 199 in.^{3} For a wide flange there are several choices. The best is W 27×84 having a section modulus of 212 in.^{3} Since it is 6% lighter than the standard beam, it is the one we specify.

7.4 SUMMARY

The neutral axis is an unstressed fiber at the centroid. The flexure formula states

$$\sigma = \frac{MC}{I}$$

giving a linear stress distribution which is zero at the centroid and maximum at the extreme fiber. Note that this bending stress is a normal stress.

The moment M, the distance c, and the moment of inertia I are all with respect to the same centroidal axis.

For design purposes it is often more convenient to use the section modulus

$$Z = \frac{I}{c}$$

Remember that Tables A4.2 to A4.9 use S for Z. The design equation becomes

$$\sigma = \frac{M}{Z}$$

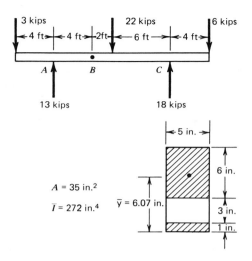

FIG. 7.13 Problems 7.1, 7.7, 7.15, 7.21.

PROBLEMS

7.1 The beam shown in Fig. 7.13 has the cross section indicated. Find the bending moment at B and the maximum stress due to bending at B.

7.2 The beam shown in Fig. 7.14 has the cross section indicated. Find the bending moment at B and the maximum stress due to bending at B.

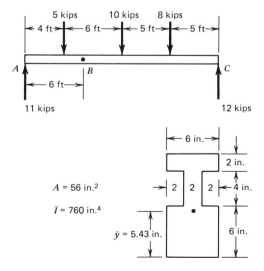

FIG. 7.14 Problems 7.2, 7.8, 7.16, 7.22.

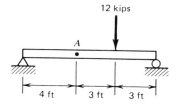

FIG. 7.15 Problems 7.3, 7.17, 7.23.

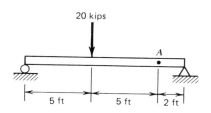

FIG. 7.16 Problems 7.4, 7.18, 7.24.

7.3 Find the maximum stress due to bending at A if the beam is a $W\ 8\times 40$, Fig. 7.15.

7.4 Find the maximum stress at A due to bending if the beam is a $W\ 12\times 36$, Fig. 7.16.

7.5 The beam shown in Fig. 7.17 is a $W\ 14\times 38$. Find the maximum bending stress at A.

7.6 The beam shown in Fig. 7.18 is a $W\ 4\times 13$. Find the maximum bending stress at A. $R_1 = 219$ lb and $R_2 = 31.0$ lb.

7.7 For a bending moment of 80 in.-kips and the cross section of Fig. 7.13 find the stress on the bottom fiber, 1 in. above the bottom fiber, 4 in. above the bottom fiber, at the centroid, 2 in. below the top fiber, and on the top fiber. Plot this stress distribution.

7.8 For a bending moment of 96 in.-kips and the cross section of Fig. 7.14 find the stress on the top fiber, 2 in. and 4 in. below the top fiber, at the centroid, 2 and 4 in. above the bottom fiber, and at the bottom fiber. Plot this stress distribution.

7.9 A 2×4 (actual size) turned on edge has a bending load of 60 ft-lb on it. (a) Find the maximum stress due to bending. (b) Find the stress if the beam is finish cut.

7.10 A 30mm$\times$90mm beam turned on edge has a bending load of 90 N·m on it. Find the maximum stress due to bending.

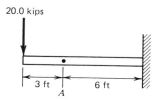

FIG. 7.17 Problems 7.5, 7.19, 7.25.

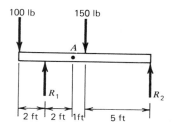

FIG. 7.18 Problems 7.6, 7.20, 7.26.

7.11 For the indicated loading and cross section in Fig. 7.19 find the maximum stress due to bending at point A. (B-3 refers to the loading in B and the cross section of 3.)

(a) $A-1$ (e) $B-1$ (i) $C-1$ (m) $D-1$
(b) $A-2$ (f) $B-2$ (j) $C-2$ (n) $D-2$
(c) $A-3$ (g) $B-3$ (k) $C-3$ (o) $D-3$
(d) $A-4$ (h) $B-4$ (l) $C-4$ (p) $D-4$

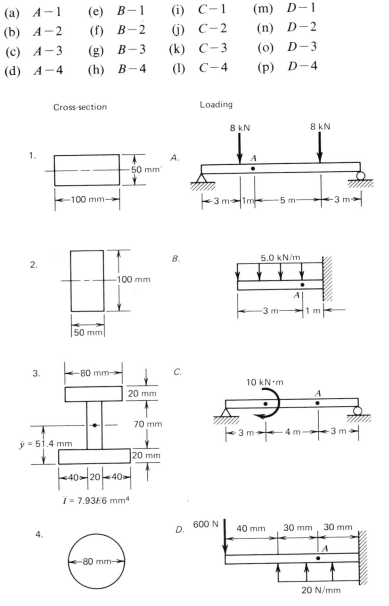

FIG. 7.19 Cross sections and loadings for Problems 7.11 to 7.14.

7.12 For the indicated loading and cross section in Fig. 7.19 find the maximum stress due to bending that occurs anywhere in the beam. (*B*-3 refers to the loading in *B* and the cross section of 3.)

(a) $A-1$ (e) $B-1$ (i) $C-1$ (m) $D-1$
(b) $A-2$ (f) $B-2$ (j) $C-2$ (n) $D-2$
(c) $A-3$ (g) $B-3$ (k) $C-3$ (o) $D-3$
(d) $A-4$ (h) $B-4$ (l) $C-4$ (p) $D-4$

7.13 For the indicated loading in Fig. 7.19 select a beam with a square cross section based on an allowable stress of 140 MPa.

(a) *A*   (c) *C*
(b) *B*   (d) *D*

7.14 For the indicated loading in Fig. 7.19 select a beam that is twice as deep (*h*) as it is wide (*b*) based upon an allowable stress of 140 MPa.

(a) *A*   (c) *C*
(b) *B*   (d) *D*

For each of the following problems calculate the maximum stress due to bending that occurs anywhere in the beam. The loading and cross section are as indicated in the following problems.

7.15 Problem 7.1.
7.16 Problem 7.2.
7.17 Problem 7.3.
7.18 Problem 7.4.
7.19 Problem 7.5.
7.20 Problem 7.6.

For each of the following problems select the best *S* or *W* beam using an allowable stress of 24,000 psi. The loading is as indicated in the figures listed below:

7.21 Figure 7.13
7.22 Figure 7.14
7.23 Figure 7.15
7.24 Figure 7.16
7.25 Figure 7.17
7.26 Figure 7.18
7.27 The cross section of a 12-ft cantilever beam is 2 in. wide and 8 in. deep.

The cantilever has a 200-lb load at the free end. (a) Find the maximum bending stress. (b) Find the maximum stress if the beam is finish cut (S4S).

7.28 A cross section of a 4-m cantilever beam is 60 mm wide and 180 mm deep. The cantilever has a 1.5-kN load at the free end. Find the maximum bending stress.

7.29 Compare the stress due to bending in a 2×4 (actual size) laid flat with that when it is laid on edge.

8

Beam Deflection

8.1 Deflection in design
8.2 Relation between bending moment and curvature
8.3 Load rate to shear to bending moment to slope to deflection
8.4 Deflection by multiple integration
8.5 Deflection by graphical integration
8.6 Deflection formulas
8.7 Deflection by superposition
8.8 The moment-area method
8.9 Summary

8.1 DEFLECTION IN DESIGN

In the previous chapter we investigated the stresses that occur in a beam because of bending. A great number of design problems are governed by the stresses that are present. However, there are many design problems that are governed by deflection rather than stress. For example, if a ceiling beam or joist deflects too much, the ceiling plaster will crack. In this case deflection would govern the design. In intricate, high-speed machinery excessive deflection of parts can interfere with and even prevent the desired motion. In such cases deflection would govern the design.

Since deflection of a beam will sometimes govern a design, we must understand how deflection is related to load. There are, however, other reasons—which are equally important—for studying this relationship. Our attention in this book has been on statically determinate structures, that is, structures whose loads can be determined by the methods of statics. Many problems of concern to us can be adequately handled by such an analysis—but not all. Those that cannot be—called statically indeterminate structures—require load-deflection characteristics for solution. In Chapter 3 the relation between axial load and axial deflection was developed. In Chapter 5 the relation between torque and rotation was developed. These, along with the relations we will develop here, allow us to handle statically indeterminate problems.

A third reason for studying load-deflection relations is that they are necessary for the analysis of dynamic loading and vibration problems. Analysis of such problems requires mathematics beyond that normally required in technology programs, so most technologists are not likely to solve such problems independently. However, they are likely to be involved in these solutions.

8.2 RELATION BETWEEN BENDING MOMENT AND CURVATURE

We will examine several methods of finding the deflection of a beam. Fundamental to all of them is the relationship between the applied bending moment and the curvature of the beam, which we will now develop. This relationship is extremely important, for it allows us to bridge the gulf between the loads on a beam and its resulting geometry. An initially straight beam with a constant bending moment is shown in Fig. 8.1a. Consider an element of this beam whose length is Δx, Fig. 8.1b. The uniform bending moment M results in the deformation shown in Fig.

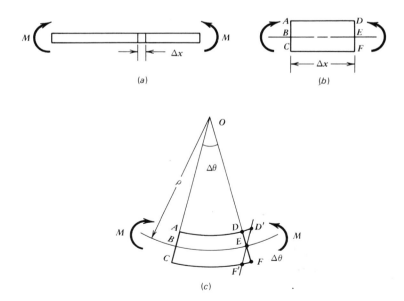

FIG. 8.1 Elements of a deflected beam.

RELATION BETWEEN BENDING MOMENT AND CURVATURE 191

8.1c. As we discussed in Chapter 7, we assume this bending takes place by the rotation of plane sections that remain plane. Thus *ABC* and *DEF* rotate as shown, shortening the upper fibers *AD* and stretching the lower fibers *CF*. The fiber *BE*, the neutral axis, does not change length. We extend the rotated planes *ABC* and *DEF* until they intersect. This point of intersection is the center of curvature, and we designate it as *O*. The distance *OB* is the radius of curvature which we designate as ρ (Greek rho—say row—equivalent to r). This means that if we rotate a compass about *O* with a radius of ρ we could trace out the path *BE*. That is, the arc *BE* is part of the circle whose radius is ρ and whose center is at 0.

We now construct a line *D'EF'* parallel to *ABC*. The segments *AD'*, *BE*, and *CF'* will all represent the original length of the fibers. Note that *BE* is unchanged, being the neutral axis. *DD'* represents the shortening of *AD*, and *FF'* represents the stretching of *CF*. Notice that the segment *BOE* is similar to the segment *F'EF*. (They have a common side *OEF* and parallel sides *OB* and *EF'*). Hence the included angles are equal, which we label as $\Delta\theta$. Recalling that an angle is defined as its arc length divided by its radius, we observe for the segment *BOE*

$$\Delta\theta = \frac{BE}{OB} = \frac{BE}{\rho}$$

and for the segment *F'EF*

$$\Delta\theta = \frac{FF'}{EF}$$

and setting the two equal we get

$$\frac{BE}{\rho} = \frac{FF'}{EF}$$

or

$$\frac{1}{\rho} = \frac{FF'}{BE} \cdot \frac{1}{EF}$$

Observe that *FF'* is the extension of the original element *CF*, and *BE* is equal to *CF'*. Hence this ratio (FF'/BE) is the deformation over the original length or the strain in the element *CF*. Also, observe that *EF* is the distance from the neutral axis to the extreme fiber. This was labeled c in the previous chapter. This yields

$$\frac{1}{\rho} = \frac{\varepsilon}{c}$$

Applying Hooke's law:

$$\varepsilon = \frac{\sigma}{E}$$

BEAM DEFLECTION

we get

$$\frac{1}{\rho} = \frac{\sigma}{E} \cdot \frac{1}{c}$$

and the flexure formula

$$\sigma = \frac{Mc}{I}$$

gives

$$\frac{1}{\rho} = \frac{Mc}{EI} \cdot \frac{1}{c}$$

Giving finally,

$$\frac{1}{\rho} = \frac{M}{EI} \tag{8.1}$$

Recall that ρ is the radius of curvature. Its inverse, $1/\rho$, is called the curvature. We have, then, this most important relationship between the curvature and the bending moment. It has some applications of its own which we will see in the following problems. More importantly, it serves as the basis for analyzing beam deflections, which we will explore in the following sections. Notice what the equation says—The curvature is proportional to the bending moment. The greater the bending moment, the greater the curvature. This means a tighter curve, that is, one with a smaller radius. The curvature is inversely proportional to the modulus of elasticity and moment of inertia. Using a stiffer material yields less curvature—greater radius. Using a greater moment of inertia yields less curvature.

EXAMPLE PROBLEM 8.1

A band saw turns over a 500-mm-diameter wheel. The steel saw is 0.800 mm thick and 10 mm wide. What stress is developed in the blade as it is bent over the wheel?

Solution

$$\frac{1}{\rho} = \frac{M}{EI} \tag{8.1}$$

Solving for M

$$M = \frac{EI}{\rho}$$

then

$$\sigma = \frac{Mc}{I} = \frac{EI}{\rho} \cdot \frac{c}{I}$$

$$\sigma = \frac{Ec}{\rho}$$

The modulus of elasticity, E, is 207 GPa for steel. The distance from the neutral axis to the extreme fiber, c, is half the thickness or 0.400 mm. The radius of curvature, ρ (rho), is half the diameter of the pulley or 250 mm. Hence

$$\sigma = \frac{207E9 \text{ Pa } (0.4E-3 \text{ m})}{(0.250 \text{ m})}$$

$$= 3.31E8 \text{ Pa}$$

$$\sigma = 331 \text{ MPa}$$

8.3 LOAD RATE TO SHEAR TO BENDING MOMENT TO SLOPE TO DEFLECTION

In Chapter 6 the relationship between load rate and shear was developed

$$w = \frac{dV}{dx} \tag{6.1}$$

Also developed was the relation between shear and bending moment

$$V = \frac{dM}{dx} \tag{6.3}$$

and in the previous section the moment was related to curvature

$$\frac{1}{\rho} = \frac{M}{EI} \tag{8.1}$$

We extend this chain of relations a second step by noting the mathematical expression for curvature

$$\frac{1}{\rho} = \frac{\dfrac{d^2y}{dx^2}}{\left[1 + \left(\dfrac{dy}{dx}\right)^2\right]^{3/2}} \tag{8.2}$$

194 BEAM DEFLECTION

This rather awful looking creature is the exact mathematical expression for curvature. For our purposes we simplify this expression by noting that for most engineering structures, large deflections are not acceptable. Accordingly,

$$\frac{dy}{dx} \simeq 0$$

Squaring the term will make it even smaller. Thus the denominator in Eq. 8.2 approaches 1.0, yielding

$$\frac{1}{\rho} = \frac{d^2y}{dx^2} \tag{8.3}$$

We then observe by substitution

$$\frac{d^2y}{dx^2} = \frac{M}{EI} \tag{8.4}$$

A mathematician would say we linearized the equation. (As a general rule, one should never pass up the opportunity to linearize an equation.)

Equation 8.4 may also be written as

$$\frac{d}{dx}\left(\frac{dy}{dx}\right) = \frac{M}{EI} \tag{8.5}$$

By definition,

$$\frac{dy}{dx} = \theta \tag{8.6}$$

which is the slope of the beam. Equation 8.5 may be written, therefore, as

$$\frac{d\theta}{dx} = \frac{M}{EI} \tag{8.7}$$

For emphasis we recap and for convenience we renumber

Load rate: $$w = \frac{dV}{dx} \tag{8.8}$$

Shear: $$V = \frac{dM}{dx} \tag{8.9}$$

Moment: $$M = EI \frac{d\theta}{dx} \tag{8.10}$$

Slope: $$\theta = \frac{dy}{dx} \tag{8.11}$$

The significance of this important chain of equations should not escape the student. Working from Eq. 8.11 to Eq. 8.8, we see that when we have the deflection, taking its derivative with respect to position x gives the slope of the beam. If the slope of the beam θ is known, taking its derivative gives the bending moment M (times $1/EI$). Given the bending moment, its derivative yields the shear V, and the derivative of the shear gives the load rate w. Hence starting with an equation of a beam's deflection and differentiating four times yields successively slope, bending moment, shear, and finally, load rate. This mathematical manipulation must occasionally be carried out. More important to the student, however, is the intimate relation between the variables expressed by these four equations.

In Chapter 6 we noted the integral versions of Eqs. 8.8 and 8.9. Similar relations may be developed for Eqs. 8.10 and 8.11. Thus we may alternately write

Shear: $$V = \int w \, dx \qquad (8.12)$$

Moment: $$M = \int V \, dx \qquad (8.13)$$

Slope: $$\theta = \int \frac{M}{EI} \, dx \qquad (8.14)$$

Deflection: $$y = \int \theta \, dx \qquad (8.15)$$

The order remains the same: load rate to shear to moment to slope to deflection. The integral of one variable yields the following one. Alternately, the derivative or slope of one variable is the instantaneous value of the previous one.

8.4 DEFLECTION BY MULTIPLE INTEGRATION

Equations 8.12 through 8.15 are the basis of this method. It should be recalled from the calculus that each time an integration is executed an arbitrary constant is acquired. This arbitrary constant allows us to make a general mathematical solution fit our particular problem. The conditions that characterize our particular problem are called *boundary conditions*. For example, consider the deflection of a cantilever beam. At the support the beam has zero deflection. This is a boundary condition. Any equation that claims to be a solution to our problem must predict zero deflection at the support; that is, it must satisfy the boundary condition. Otherwise it is not a solution.

Boundary conditions, then, are constraints imposed on the solution by the physical characteristics of the problem. We need one boundary condition for each integration. That allows us to evaluate the one arbitrary constant that accrues at each integration. The following sample problem delineates the method.

EXAMPLE PROBLEM 8.2

Find an equation for the deflection of the beam shown in Fig. 8.2a.

Solution

In order to determine the boundary conditions it is necessary to draw a free body diagram and find the reactions at the support. A coordinate system is also established, Fig. 8.2b.

A mathematical expression for the load rate must be obtained. We note

$$w = 0$$

Then by Eq. 8.12

$$V = \int w\, dx = \int 0\, dx = C_1$$

where C_1 is the first constant. This equation simply says that for any value of x, V is a constant. We force this solution to fit our boundary condition to evaluate C_1. Since C_1 is in terms of shear, it follows that we need a boundary condition for shear; that is, we must determine the numerical value and appropriate sign for shear at some point in the beam. We note that just to the right of the support the shear will be +80 lb. It may be necessary to make a mental shear diagram to see this. Thus our first boundary condition is

$$V_0 = C_1 = 80$$

and

$$V = 80$$

is the equation for shear anywhere in the beam.

FIG. 8.2a

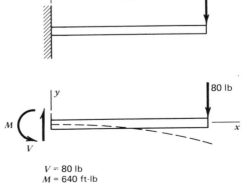

$V = 80$ lb
$M = 640$ ft-lb

FIG. 8.2b Free body diagram and coordinate system.

DEFLECTION BY MULTIPLE INTEGRATION

Next we have

$$M = \int V \, dx = \int 80 \, dx = 80x + C_2$$

This says that the bending moment is a linear function of x. We evaluate C_2 noting the boundary condition at $x=0$

$$M_0 = -640 \text{ in. lb}$$

The negative sign is because the reacting moment produces negative bending in the beam. We evaluate the above equation for $x=0$.

$$M_0 = 80(0) + C_2 = -640$$

$$C_2 = -640$$

and

$$M = 80x - 640$$

From Eq. 8.14

$$\theta = \int \frac{M}{EI} \, dx = \frac{1}{EI} \int (80x - 640) \, dx$$

The parenthesis and dx following the integral sign follow regular algebra rules. Thus

$$\theta = \frac{1}{EI} \left\{ \int 80x \, dx - \int 640 \, dx \right\}$$

$$\theta = \frac{1}{EI} \left\{ \frac{80x^2}{2} - 640x \right\} + C_3$$

To determine a boundary condition for θ, the slope of the beam, we must intuitively describe its shape. Thus we sketch the deflection shown in Fig. 8.2b. Since the support is rigid and will not allow the beam to rotate at the support, we note the boundary condition:

$$\theta_0 = 0$$

Then evaluating the above equation at $x=0$

$$\theta_0 = \frac{1}{EI} \left\{ \frac{80(0)^2}{2} - 640(0) \right\} + C_3 = 0$$

$$C_3 = 0$$

and

$$\theta = \frac{1}{EI}[40x^2 - 640x]$$

Finally we have

$$y = \int \theta \, dx$$

$$= \frac{1}{EI} \int [40x^2 - 640x] \, dx$$

$$= \frac{1}{EI} \left\{ \int 40x^2 \, dx - \int 640x \, dx \right\}$$

$$= \frac{1}{EI} \left\{ 40 \frac{x^3}{3} - 640 \frac{x^2}{2} \right\} + C_4$$

C_4 is evaluated by noting that at $x=0$, $y=0$

$$y_0 = \frac{1}{EI} \left\{ \frac{40}{3}(0)^3 - 320(0)^2 \right\} + C_4 = 0$$

$$C_4 = 0$$

$$y = \frac{1}{EI} \left\{ \frac{40}{3} x^3 - 320 x^2 \right\}$$

The equation may be evaluated for any value of x to find the corresponding value of y, the beam deflection. For instance, at $x = 8$ ft

$$y = \frac{1}{EI} \left\{ \frac{40}{3}(8)^3 - 320(8)^2 \right\} = \frac{-13{,}700}{EI} \text{ ft}^3 \text{ lb}$$

which is the maximum deflection. Since the distance x was put in feet, E and I would have to be converted to feet as well. Substituting in the appropriate values of EI gives a numerical value for the deflection y. Notice it has a negative sign—meaning the deflection is down.

EXAMPLE PROBLEM 8.3

Find the deflection of the beam in Example Problem 8.2 if it is a 1.25-in.-diameter steel rod.

Solution

From Example Problem 8.2 we have

$$y = \frac{13{,}700}{EI} \text{ ft}^3\text{-lb}$$

Further,

$$I = \frac{\pi D^4}{64} = \frac{\pi (1.25)^4}{64} = 0.120 \text{ in.}^4$$

and

$$E = 30 \times 10^6 \text{ psi}$$

Thus

$$y = \frac{-13{,}700 \text{ ft}^3\text{-lb (in.)}^2}{30 \times 10^6 \text{ lb } (0.120 \text{ in.}^4)}$$

$$y = -3.80 \times 10^{-3} \frac{\text{ft}^3}{\text{in.}^2} \left(\frac{12 \text{ in.}}{\text{ft}}\right)^2$$

$$y = -5.47 \times 10^{-1} \text{ ft} = \mathbf{-0.547 \text{ ft}}$$

$$y = \mathbf{-6.56 \text{ in.}}$$

Note that there will frequently be considerable conversion of units associated with evaluation of the deflection.

8.5 DEFLECTION BY GRAPHICAL INTEGRATION

In the previous section beam deflection was obtained by integration. Recall that integration is a summation process. When we express Eq. 8.12 as

$$V = \int w\, dx$$

we are saying that the change in shear V is equal to the area under the load rate curve. When Eq. 8.12 is evaluated with an arbitrary constant it gives the value of shear. When it is evaluated without an arbitrary constant it gives the change in shear. Thus we will also write

$$\Delta V = \int w\, dx$$

as in Chapter 6. In Example Problems 6.4, 6.5, and 6.6 we found the change in shear by finding the area under the load rate curve. This amounts to graphical integration. We call it graphical, although it would be more accurate to call it

200 BEAM DEFLECTION

semi-graphical. There are graphical methods that depend more on drawing lines and less on computation. We also evaluated Eq. 8.13,

$$M = \int V \, dx$$

by graphical integration. In the graphical integration, boundary conditions were accounted for by reproducing concentrated shear and bending moment loads as they occurred. In this section we will handle them in the same manner. In fact, the first two stages of graphical integration, load rate to shear and shear to bending moment, are exactly the same as in Section 6.6.

Evaluation of Eqs. 8.14 and 8.15 may also be done by graphical integration. However, for these two last stages we must identify boundary conditions, as we did for analytical integration. We then force the graphical integration to satisfy these boundary conditions. The following example problem illustrates the method.

EXAMPLE PROBLEM 8.4

Find the maximum deflection of the beam shown in Fig. 8.3a by graphical integration.

Solution

As in Example Problem 8.2 a free body diagram (Fig. 8.3b) is drawn, and V and M are found. An approximate deflection curve is also drawn. Starting with the shear curve at 0, we reproduce the 80-lb upward load taking us to $V = +80$ lb, Fig. 8.3c. From the left end to the right end the load rate is zero, so the area under the load rate curve is zero, and thus there is no change in V.

At the right end the 80-lb down load is reproduced closing the shear curve. Shear is positive across the beam.

The bending moment diagram starts at zero and reproduces the negative moment of 640 ft-lb, Fig. 8.3d. The change from the left end to the right end is the area under the shear curve

$$\Delta M = +80(8) = +640 \text{ ft-lb}$$

This takes us back to zero. Since the shear curve is the zeroth degree, the moment, its integral, will be first degree or linear. Its slope is $+80$, as indicated. The curve closes, and the bending is negative the entire length. At this point a qualitative check may be made by observing that the approximate shape of the beam, as originally drawn, calls for negative bending. We also recall that these first two curves must close to satisfy static equilibrium. As we proceed to the slope curve θ, it will be convenient to plot $EI\theta$ when the beam is uniform in EI. The boundary condition for slope needs to be noted at this point. We observe from the approximate shape that the slope of

FIG. 8.3a

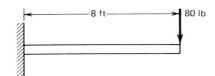

FIG. 8.3b

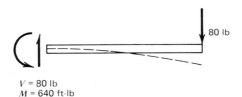

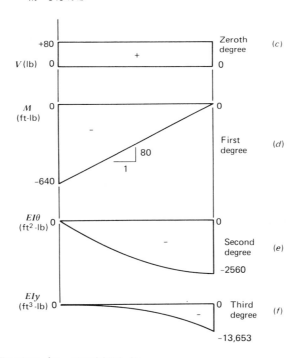

FIG. 8.3c-f Deflection by graphical integration.

the beam at the left end, the fixed end, is zero. Hence the $EI\theta$ diagram starts at zero, Fig. 8.3e. The change in θ is the area under the moment curve

$$\Delta(EI\theta) = \tfrac{1}{2}bh = \tfrac{1}{2}(8 \text{ ft})(-640 \text{ ft-lb}) = -2650 \text{ ft}^2\text{-lb}$$

Thus the $EI\theta$ diagram goes from 0 to -2560 ft^2-lb. The moment curve is first degree, so the slope curve is second degree, which may be concave upward or concave downward. By noting

$$\frac{d}{dx}(EI\theta) = M$$

and that $M=0$ at the right end (this is the slope of the slope curve) we conclude that the desired curve is concave up. The slope of the beam is negative. This can be qualitatively compared with the approximate beam curve. The figure shows a vertical line closing the curve. This is for consistency only. In general, there is no independent check that requires this curve to close.

We also plot EIy for convenience for deflection when EI is constant. We observe that the deflection is zero at the left end, which is the needed boundary condition. So EIy will start at zero, Fig. 8.3f. The change in deflection is the area under the slope curve, so from Table A4.1 for the complement of the area under the parabola

$$\Delta(EIy) = \tfrac{2}{3}bh = \tfrac{2}{3}(8 \text{ ft})(-2560 \text{ ft}^2\text{-lb})$$

$$\Delta(EIy) = -13{,}700 \text{ ft}^2\text{-lb}$$

Thus the deflection ends at $-13{,}700$ ft^3-lb and is negative for the entire beam. Since the $EI\theta$ curve is second degree, the EIy curve will be third degree. The slope of the EIy curve will be zero at the left end corresponding to a zero value for $EI\theta$. Thus the curve is concave downward and agrees with original intuitive approximation of the beam's shape.

We conclude

$$y_{max} = \frac{-13{,}700}{EI} \text{ ft}^3\text{-lb}$$

which agrees with Example Problem 8.2.

8.6 DEFLECTION BY FORMULAS

In Example Problem 8.2, we found the deflection in a beam 8 ft long with a load of 80 lb on the free end. If we replace this specific problem with a more general one—a beam of length L and load P—we would find the deflection to be

$$y = -\frac{PL^3}{3EI}$$

Thus we may find the deflection for any cantilever beam of length L, end load P, modulus of elasticity E, and moment of inertia I simply by evaluating this formula. The type support and the type load must each be the same before this equation may be used.

At this point it is worthwhile to look at the beam deflection formula and observe its behavior. First, we see that deflection is proportional to the load—the greater the load, the greater the deflection. It is proportional to length cubed. Doubling the length increases the deflection by a factor of 8! It is inversely proportional to the modulus of elasticity. A steel beam will deflect only one-third as much as an aluminum one under otherwise identical conditions. Finally, deflection is inversely proportional to the moment of inertia. Remember that the moment of inertia depends on the depth of the beam cubed; so double the depth and cut the deflection by a factor of 8!

The above discussion applies strictly to the cantilever beam under consideration. However, the tendencies indicated may be extrapolated to beams in general.

We can analyze other loading conditions in the manner of Example Problem 8.2. Table A8.1 shows the results of this type of analysis for a number of loading conditions. This type of table is commonly available from a number of sources. The equations may be evaluated readily for a given set of conditions to obtain the deflection at any desired location. When available, equations of this form offer the easiest method of determining the deflection.

EXAMPLE PROBLEM 8.5

A simply supported beam carries a uniform load of 600 lb/ft over its 12-ft span. Find the deflection at the center of the beam if it is steel $W\ 12\times 65$ oriented for minimum deflection.

Solution

From the tables in the Appendix

$$I = 533.4 \text{ in.}^4$$

$$E = 30 \times 10^6 \text{ psi}$$

And from Table A8.1

$$y = \frac{-5wl^4}{384EI}$$

$$y = \frac{-5}{384} \frac{(600 \text{ lb})}{\text{ft}} \frac{(12 \text{ ft})^4}{30 \times 10^6 \text{ lb}} \frac{\text{in.}^2}{(533.4 \text{ in.}^4)} \left(\frac{12 \text{ in.}}{\text{ft}}\right)^3$$

$$y = -0.0175 \text{ in.}$$

8.7 DEFLECTION BY THE METHOD OF SUPERPOSITION

One of the most important and broadly applicable principles in mechanics is superposition. This principle, as applied to deflection, says that the resultant deflection from a combination of loads is the sum of the deflections from each individual load. This is true when the individual deflections are linearly related to the individual loads—as they are in Table A8.1. This principle presents a valuable method of determining beam deflection from a combination of loads.

Suppose we wish to know the maximum deflection of a beam loaded as shown in Fig. 8.4a. This combined loading is not available in Table A8.1. If we separate the loading as shown in Fig. 8.4b and 8.4c, we have loading conditions that are available in the table. We can then calculate

$$y_1 = -\frac{P_1 L^3}{3EI}$$

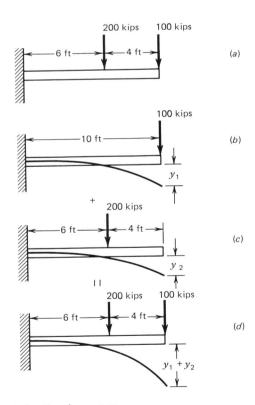

FIG. 8.4a–d Beam deflection by superposition.

and

$$y_2 = \frac{-P_2 a^2(a-3L)}{6EI}$$

The method of superposition says that the resultant deflection from these two individual loads is simply $y_1 + y_2$.

We do not make the calculation at this point, but it is worth noting that in every sense the resultant beam in Fig. 8.4d is the sum of Figs. 8.4b and 8.4c. We can, for example, calculate the reactions at the beam support in b and c. Their sum is the reaction in 8.4d.

EXAMPLE PROBLEM 8.6

The beam shown in Fig. 8.4 is a $W\ 14 \times 78$ of steel. Calculate the maximum deflection of the beam.

Solution

We replace the combined loading of Fig. 8.4a with the two separate loads of Figs. 8.4b and 8.4c. If we visualize superimposing Fig. 8.4c on Fig. 8.4b, we see the result is Fig. 8.4d. We label the deflection in Figs. 8.4b and 8.4c as y_1 and y_2, respectively. From the tables in the Appendix we get the beam properties and from Table A8.1 we get for y_1

$$y_1 = \frac{-PL^3}{3EI} = \frac{-100 \text{ kips } (10 \text{ ft})^3 (\text{in.})^2}{3(30 \times 10^6 \text{ lb})} \cdot \frac{1}{(851.2 \text{ in.}^4)} \times \frac{1000 \text{ lb}}{1 \text{ kips}} \cdot \frac{(12 \text{ in.})^3}{(\text{ft})^3}$$

$$= \frac{-100(10)^3 (1000)(12)^3 \text{ in.}}{3(30 \times 10^6)(851.2)}$$

$y_1 = -2.26$ in.

Similarly for y_2 we get

$$y_2 = \frac{-P(a)^2}{6EI}(a-3L)$$

$$= \frac{-200 \text{ kips } (6 \text{ ft})^2 (\text{in.}^2)}{(6)30 \times 10^3 \text{ kips}} \cdot \frac{(6 \text{ ft} - 3(10) \text{ ft})}{(851.2 \text{ in.}^4)} \times \left(\frac{12 \text{ in.}}{\text{ft}}\right)^3$$

$y_2 = -1.95$ in.

Then by superposition the total deflection is the sum of y_1 and y_2

$$y_t = y_1 + y_2 = -2.26 - 1.95 = -4.20 \text{ in.}$$

8.8 THE MOMENT-AREA METHOD

The moment-area method of determining beam deflection is useful when deflection at a known point is desired. The loading can be of most any pattern, and variation in beam cross section and properties can be routinely handled. The method can also be used for determining the equation of deflection, but it offers no significant advantages for such problems. Its greatest application, then, is where the deflection of one or more points is desired and deflection formula are not readily available.

There are two theorems on which the method is based. This first theorem is simply the application of Eq. 8.14:

$$\theta = \int \frac{M}{EI} dx \qquad (8.14)$$

As we previously indicated, when Eq. 8.14 is integrated between two specific points, the left-hand side of the equation is the change in the angle θ. This principle was used in finding the slope (θ) curve from the moment (M) curve in graphical integration. The theorem may be stated as follows:

Theorem 1. The change in slope between any two points on a beam is equal to the area under the M/EI curve between the same two points.

The theorem is illustrated in Fig. 8.5. The top figure is a general M/EI curve. The bottom is a plot of the deflection, or an elastic curve. (We have not included the slope diagram, which would be used for graphical integration.) As indicated in the figure, the change in slope of the elastic curve (the angle the tangent makes with the horizontal) between points A and B is equal to the area under the M/EI curve between points A and B.

We arrive at the second theorem by considering two points, C and D, that are separated by the infinitesimal distance dx, Fig. 8.6. Tangents to the elastic curve at these two points are drawn in the lower figure. By Theorem 1, the angle between the tangents is

$$d\theta = \frac{M}{EI} dx$$

(This is also Eq. 8.10.) The vertical distance between the two tangents at position

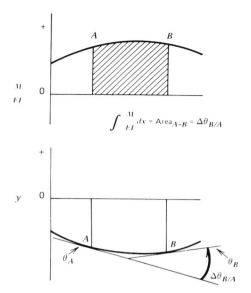

FIG. 8.5 The first moment-area theorem.

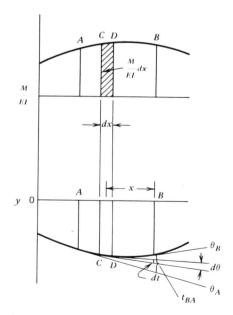

FIG. 8.6 The second moment-area theorem.

208   BEAM DEFLECTION

B is dt. Since the angles are very small, dt may be approximated by the arc length, that is,

$$dt = x\,d\theta$$

where x is the distance from B to the element. Substituting for $d\theta$ from above gives

$$dt = x\left(\frac{M}{EI}\right)dx$$

which is the moment of area of the element $(M/EI)dx$ about B. The deviation of the beam at B from a tangent at A is the summation of the dt's from A to B, or

$$t_{BA} = \int \frac{M}{EI} x\,dx \qquad (8.16)$$

is t_{BA}. The tangential deviation, *not* the deflection of the beam. The integral is the moment of area under the M/EI curve about B. Hence we may write

$$t_{BA} = \bar{x} A_{A-B} \qquad (8.17)$$

where A_{A-B} is the area between A and B under the M/EI curve and $\bar{x}$ is the distance from the centroid of the area to B. In words we have:

Theorem 2. The deviation of a point on a beam from the tangent drawn at a second point is the moment about the first point of the area under the M/EI curve between the two points.

The geometric properties of common areas given in Table A4.1 will be useful in applying the moment area method. Its application is illustrated in the following problems.

EXAMPLE PROBLEM 8.7

Find the slope and deflection of the beam in Fig. 8.7a at its free end.

$$EI = 137{,}000 \text{ ft}^2\text{-lb}$$

Solution

The moment diagram was found for this loading in Example Problem 8.4. An M/EI curve may be found by dividing this curve by EI, Fig. 8.7b. If EI is constant, the division is not necessary. Otherwise it is.

A sketch of the deflected beam is made noting the boundary conditions of

THE MOMENT-AREA METHOD

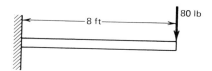

FIG. 8.7a

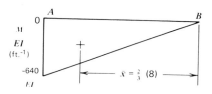

FIG. 8.7b

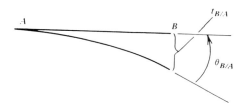

FIG. 8.7c

zero deflection and zero slope at point A, Fig. 8.7c.

We find the slope at the free end, point B, by applying Theorem 1. The change in slope between A and B is the area under the M/EI curve or

$$\theta_{B/A} = \frac{1}{2} \frac{M}{EI} (8)$$

$$\theta_{B/A} = \frac{1}{2} \frac{-640(8)}{137,000} = -0.0187 \text{ rad}$$

Since the slope at A is zero, this change in slope is the actual slope. Since the area is negative, the change in slope is negative. Also note that the angle involved is very small.

We find the deflection by applying Theorem 2. The tangential deviation is the moment of the area about B; thus

$$t_{BA} = A\bar{x} = \frac{1}{2} \frac{(-640)(8)}{137,000} \left(\frac{2}{3}\right)(8)$$

$$t_{BA} = 0.0997 \text{ ft}$$

This is the deviation of the beam at B from the tangent from A. In this case, since (1) the deflection at A is zero and (2) the slope at A is zero, this is also the deflection of the beam. Since cantilever beams have zero slope and deflection at the support, they are especially suitable for solution by the moment-area method.

EXAMPLE PROBLEM 8.8

Find the deflection at the point of loading, point C, for the beam shown in Fig. 8.8a.

Solution

After solving the statics we arrive at the moment diagram shown, Fig. 8.8b. Next the deflected beam $AC'B$ is sketched, Fig. 8.8c. The distance CC' is the desired dimension. We sketch a tangent at A and note that the deflection t_{CA} is the one governed by the second moment-area theorem. If we could find the slope of the beam at A, we could solve the problem from the geometry and the second theorem. We might look to the first theorem here, since it deals with slope. However, we lack a zero slope reference point and must seek elsewhere.

Elsewhere in this case is observing that t_{BA} may be found from the second theorem. Knowing t_{BA}, we can calculate θ_A. The second theorem says that t_{BA}, the deviation of B from the tangent from A, is the moment of the area under the M/EI curve between the two points about B. Thus t_{BA} is found by taking the moment of the two triangular areas in the moment diagram about B:

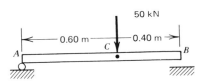

FIG. 8.8a

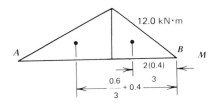

FIG. 8.8b

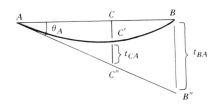

FIG. 8.8c

$$t_{BA} = \bar{x}_1 A_1 + \bar{x}_2 A_2$$

$$= \left[\frac{2(0.40)}{3} \cdot \frac{1}{2} \cdot (0.40)12 + \left(\frac{0.60}{3} + 0.4 \right) \frac{1}{2} (0.60)12 \right] \frac{1}{EI}$$

$$t_{BA} = \frac{2.80}{EI}$$

Since EI is constant, it is not necessary to find the M/EI diagram. Knowing t_{BA}, we calculate θ_A as

$$\theta_A = \frac{t_{BA}}{L} = \frac{2.80}{EI(1.0)} = \frac{2.80}{EI}$$

You may want to say $\tan \theta_A$, but remember we are talking about very small angles where

$$\tan \theta_A = \theta_A$$

Knowing θ_A, we can calculate CC''

$$CC'' = \theta_A (AC) = 0.6 \frac{(2.80)}{EI} = \frac{1.68}{EI}$$

Now we apply the second moment-area theorem to find t_{CA}, which is the moment of the area between A and C about C:

$$t_{CA} = \frac{0.60}{3} \cdot \frac{1}{2} \cdot (0.60) \frac{(12.0)}{EI} = \frac{0.720}{EI}$$

The desired distance CC' is then computed as

$$CC' = CC'' - t_{CA} = \frac{1.680}{EI} - \frac{0.720}{EI} = \mathbf{\frac{0.960}{EI}}$$

Finding deflection by the moment-area method usually involves combinations of the first and second theorems and geometric observations.

8.9 SUMMARY

Curvature:

$$\frac{1}{\rho} = \frac{M}{EI}$$

212 BEAM DEFLECTION

Beam equations:

	Differential Form	**Integral Form**
	Slope	Area
w	$w = \dfrac{dV}{dx}$	$V = \int w\,dx$
V	$V = \dfrac{dM}{dx}$	$M = \int V\,dx$
M	$\dfrac{M}{EI} = \dfrac{d\theta}{dx}$	$EI\theta = \int M\,dx$
θ	$\theta = \dfrac{dy}{dx}$	$y = \int \theta\,dx$
y		

Boundary conditions: One boundary condition is needed for each integration. Beam deflections may be found by:

1. Integration
2. Graphical integration
3. Formula
4. Superposition
5. Moment-area method

Moment-area theorems:

Theorem 1. The change in slope between any two points on a beam is equal to the area under the M/EI curve between the two points.

Theorem 2. The tangential deviation of one point with respect to a second is the moment about the first point of the area under the M/EI curve between the two points.

PROBLEMS

8.1 Check the dimensions for Eq. 8.1.

8.2 A 1 in. thick by 6 in. deep southern pine board is to be bent and set into a curved member by a soaking process. The board is forced onto a form, soaked, and dried. The form has a radius of curvature of 8 ft, Fig. 8.9. What bending moment will be required to set the board and what stress will be developed in it?

8.3 A 25 mm thick by 150 mm deep southern pine board is to be bent and set into a curved member by a soaking process. The board is forced onto a form, soaked, and dried. The form has a radius of curvature of 3 m, Fig. 8.9. What bending moment will be required to set the board and what stress will be developed in it?

8.4 A $\tfrac{3}{8}$-in.-diameter steel rod is to be bent into a circular ring of 400-in.

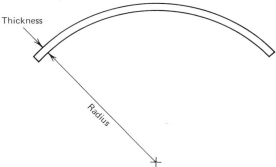

FIG. 8.9 Problems 8.2, 8.3.

diameter. Find the bending moment required to form the ring and the resultant stress.

8.5 A 10-mm-diameter steel rod is to be bent into a circular ring of 10-m diameter. Find the bending moment required to form the ring and the resultant stress.

8.6 A $\frac{1}{4}$-in.-diameter aluminum rod is bent into a circular ring until the maximum stress produced is 25,000 psi. What is the radius of the ring?

8.7 A 6-mm-diameter aluminum rod is bent into a circular ring until the maximum stress produced is 170 MPa. What is the radius of the ring?

8.8 The deflection of beams is given by the following equations. Successively differentiate the equations to find equations for slope, bending moment, shear, and load rate.

(a) $y = \dfrac{1}{EI}\left(100x^2 - 50x^3 + \dfrac{x^5}{20}\right)$

(b) $y = \dfrac{1}{EI}\left(64x^3 - \dfrac{x^6}{40} - \dfrac{32{,}678x}{10}\right)$

(c) $y = \dfrac{1}{EI}(-x^4 + 18x^3 - 180x^2 + 720x)$

In the following problems for each indicated figure:

(a) Write an equation for load rate as a function of beam position.
(b) Write boundary conditions for shear, bending moment, slope, and deflection.

8.9 Fig. 8.10.
8.10 Fig. 8.11.
8.11 Fig. 8.12.

In the following problems for each beam shown, find by successive integration equations for shear, bending moment, slope, and deflection *without* evaluating the constants of integration (C_1, C_2, etc.):

214   BEAM DEFLECTION

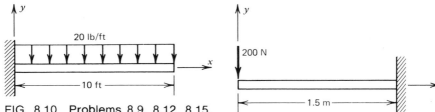

FIG. 8.10 Problems 8.9, 8.12, 8.15, 8.18, 8.25, 8.30, 8.36a.

FIG. 8.11 Problems 8.10, 8.13, 8.16, 8.19, 8.26, 8.31, 8.36b.

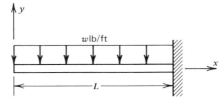

FIG. 8.12 Problems 8.11, 8.14, 8.17, 8.20, 8.36c.

8.12 Fig. 8.10.
8.13 Fig. 8.11.
8.14 Fig. 8.12.

In the following problems for each beam shown:

(a) Find the equation for shear, bending moment, slope, and deflection by successive integration, *including* the evaluation of the constants of integration.

(b) Where the cross section and material are indicated, find the slope and deflection.

8.15 Fig. 8.10—2 by 12 (actual size), lying flat, southern pine.
8.16 Fig. 8.11—100 mm deep by 40 mm wide aluminum.
8.17 Fig. 8.12.

In the following problems for each beam shown, find the maximum deflection by graphical integration:

8.18 Fig. 8.10.
8.19 Fig. 8.11.
8.20 Fig. 8.12.
8.21 to 8.24 Draw shear, bending slope, and deflection beam and calculate maximum deflection for the indicated figure:
Problem 8.21—Fig. 8.13.
Problem 8.22—Fig. 8.13.

Problem 8.23—Fig. 8.15.
Problem 8.24—Fig. 8.16.

8.25 Find the maximum deflection of the beam in Fig. 8.10 by formula and superposition if necessary.

8.26 Find the maximum deflection of the beam in Fig. 8.11 by formula and superposition, if necessary.

8.27 Find the maximum deflection of the beam in Fig. 8.13 by formula and superposition, if necessary.

8.28 Find the maximum deflection of the beam in Fig. 8.14 by formula and superposition, if necessary.

8.29 Find the maximum deflection of the beam in Fig. 8.15 by formula and superposition, if necessary.

8.30 For the beam in Fig. 8.10 using beam formulas find the deflection every foot along the beam. Make a plot of the beam deflection.

8.31 For the beam in Fig. 8.11 find the deflection every 0.3 m along the beam using Table A8.1. Make a plot of the beam deflection.

8.32 For the beam in Fig. 8.14 find the deflection every 0.5 m along the beam using Table A8.1. Make a plot of the beam deflection.

8.33 to 8.34 Find the maximum deflection of the indicated beams by formula and superposition, if appropriate.
Problem 8.33—Fig. 8.17
Problem 8.34—Fig. 8.18

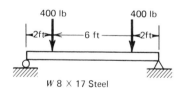

FIG. 8.13 Problems 8.21, 8.27, 8.34a, 8.38a, 8.39a.

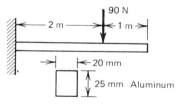

FIG. 8.14 Problems 8.22, 8.28, 8.32, 8.36d.

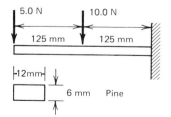

FIG. 8.15 Problems 8.23, 8.29, 8.36e.

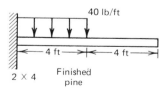

FIG. 8.16 Problem 8.24.

8.35 Find the deflection at the center of the beam by formula and superposition for Fig. 8.19.

8.36 Using the moment-area method find the slope and deflection of the free end of the cantilever beams in:
   (a) Fig. 8.10.
   (b) Fig. 8.11.
   (c) Fig. 8.12.
   (d) Fig. 8.14.
   (e) Fig. 8.15.
   (f) Fig. 8.17.

8.37 Using the moment-are method find the deflection at the center of simply supported beams in:

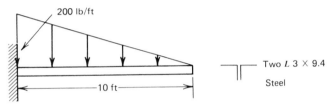

FIG. 8.17   Problems 8.33, 8.36 f.

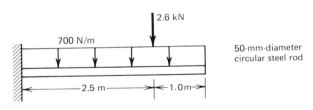

FIG. 8.18   Problem 8.34.

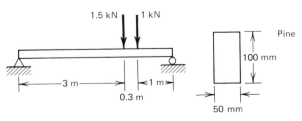

FIG. 8.19   Problems 8.35, 8.37b, 8.38b, 8.39b.

(a) Fig. 8.13.
(b) Fig. 8.19.

8.38 Using the moment-area method find the slope and deflection at the left-hand load in:
(a) Fig. 8.13.
(b) Fig. 8.19.

8.39 Using the moment-area method find the slope and deflection at the right-hand load in:
(a) Fig. 8.13.
(b) Fig. 8.19.

8.40 In analyzing a beam the moment equation was found to be

$$M = 30x^2 - 500$$

The following boundary conditions were also observed.

$$y(0) = 0$$

$$\theta(10) = 0$$

Find the deflection equation for this beam.

8.41 After double integration, the following moment equation was found.

$$M = -200x^2 - 8000x$$

The slope of the beam is observed to be zero when x is zero and the deflection of the beam is zero when x is 8. Find the deflection equation for this beam.

8.42 Find the deflection of the free end of the beam shown by the method of superposition, Fig. 8.20.

$$E = 210 \text{ GPa}$$

$$I = 3.25E6 \text{ mm}^4$$

8.43 Find the deflection of the beam shown (by diagrams) if it is a standard steel I-beam, 10×35.0 (Fig. 8.21).

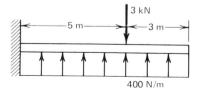

FIG. 8.20 Problem 8.42.

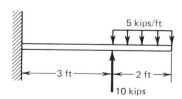

FIG. 8.21 Problem 8.43.

8.44 A beam is observed to have a zero slope 4 m to the right of its left support. The deflection is zero at the right support. Its moment diagram is given in Fig. 8.22. Draw the slope and deflection diagram for this beam, thoroughly labeling all relevant points. Indicate the maximum deflection.

8.45 A moment curve for a beam is as shown in Fig. 8.23. Its slope and deflection are zero 4 ft to the right of the left end. Draw the slope and deflection diagrams for this beam and find the maximum deflection. Thoroughly label all relevant points.

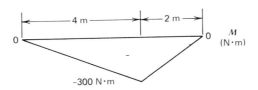

FIG. 8.22 Problem 8.44.

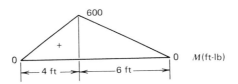

FIG. 8.23 Problem 8.45.

9

Shearing Stresses in Beams

9.1 The shear stress formula
9.2 Shear stress distribution
9.3 Maximum shear stress for common cross sections
9.4 Shear flow
9.5 Summary

We have seen that when a section is passed through a beam, there result three components of force: an axial load, a shear load, and a bending moment. In Chapter 7 we developed a relationship between the bending moment and the normal stress produced. In Chapter 3 the normal stress produced by an axial load was developed. In that chapter we also developed an equation for the average shear stress as a function of the shear load. We commented at that time that although the average shear stress concept was useful, it did not accurately represent the shear stress distribution. In this chapter we will develop a more accurate relation between the shear load and the stress it produces. We will also develop a relation between the shear load and a useful concept known as the shear flow.

The stress equation of this chapter is the most complex of all that we will consider. Thus the student should pay particularly close attention to the development of this equation to assure that he or she thoroughly understands each term in the equation.

9.1 THE SHEAR STRESS FORMULA

Consider a beam as shown in Fig. 9.1. Also shown in the figure are the shear and moment diagrams. The element AB, which has a length dx, will be taken from the

220 SHEARING STRESSES IN BEAMS

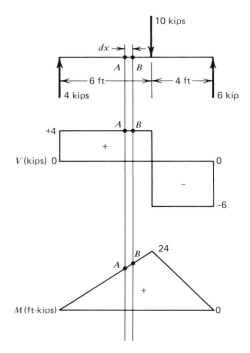

FIG. 9.1

beam. Notice that the shear at A equals the shear at B, while the bending moment at B is greater than that at A. We designate the moment at A as M and that at B as $M+dM$, as shown in Fig. 9.2a. Of course, this results in the stresses at B, σ' being greater than the stresses at A, σ, as shown in Fig. 9.2c. The resulting stress distribution is shown in Fig. 9.2a. Figure 9.2b shows a rectangular cross section these stresses act over. (The method developed is applicable to other cross sections as well.) We examine the area bounded by c, the distance to the extreme fiber, and an arbitrary fiber y_0, designating this area as A'. The area A' contains the element of area dA over which the stresses σ (at section A) and σ' (at section B) act. Figure 9.2c shows the small element dx long, with a cross section A'. The stress at any location, y_1 on the right end is σ'. It is due to $M+dM$ and will be greater than the corresponding stress at the same distance from the neutral axis y_1 on the left end. The stress on the left end is σ, and it is due to M, as shown in Fig. 9.2c. Since σ' is greater than σ, the resultant horizontal force to the left is greater than that to the right. For this element of the beam (Fig. 9.2c) to be in equilibrium, there must be an additional force to the right. This force is produced by the shear stress τ acting over the length dx as indicated in Fig. 9.2c.

To recap, the greater bending moment at B, $M+dM$, produces greater stresses, σ', on this section. The greater stresses produce an unbalanced resultant force to the left which is ultimately balanced by the force due to shear stress, τ.

THE SHEAR STRESS FORMULA

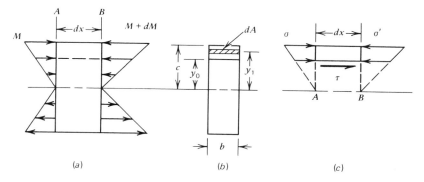

FIG. 9.2a–c Shear stress.

The relation between these variables is found by summing the forces horizontally

$$\sum F_x = 0 = \int_{A'} \sigma \, dA - \int_{A'} \sigma' \, dA + \tau(b \cdot dx)$$

where b is the beam thickness and the area $b \cdot dx$ is the area that τ acts over. From Eq. 7.5:

$$\sigma = \frac{My}{I}$$

and

$$\sigma' = \frac{(M+dM)y}{I}$$

Substituting these into the previous force equation gives

$$\int_{A'} \frac{My}{I} dA - \int_{A'} \frac{(M+dM)y}{I} dA + \tau(b \cdot dx) = 0$$

$$\int_{A'} \frac{My}{I} dA - \int_{A'} \frac{My}{I} dA - \int_{A'} \frac{(dM)y}{I} dA + \tau(b \cdot dx) = 0$$

The first two terms are identical and cancel each other. The third term calls for an integration over the area A'. Neither I nor dM varies as we integrate over the area A', so they may be factored out, giving

$$\frac{-dM}{I} \int_{A'} y \, dA + \tau(b \cdot dx) = 0$$

222 SHEARING STRESSES IN BEAMS

Solving for τ

$$\tau = \frac{1}{Ib}\frac{dM}{dx}\int_{A'} y\, dA$$

From Eq. 6.3

$$\frac{dM}{dx} = V$$

Thus

$$\tau = \frac{V}{Ib}\int_{A'} y\, dA$$

We let

$$Q = \int_{A'} y\, dA \tag{9.1}$$

and then

$$\tau = \frac{VQ}{Ib} \tag{9.2}$$

Examining Eq. 9.1 and referring to Fig. 9.2b, we see that Q is the moment of area bounded by y_0 and c about the neutral axis. Equation 9.2 has the following terms:

τ: The shear stress at y_0.

V: The shear load at the cross section.

Q: The moment of the *partial area* A' bounded by y_0 and c about the neutral axis.

I: The moment of inertia of the entire cross-sectional area about the neutral axis.

b: The thickness of the beam where the stress is being calculated, at y_0 in this case.

Since Q is the moment of area, Eq. 9.2 is often written as

$$\tau = \frac{V(\bar{y}A')}{Ib} \tag{9.3}$$

where A' is the partial area bounded by y_0 and c and $\bar{y}$ is the distance from the neutral axis to the centroid of the partial area A'.

The shear stress defined by Eq. 9.2 is longitudinal, although it is caused by a transverse shear load V. This can be reconciled by recalling that an element with shear stress on one face must have equal shear stress on all four faces, as shown in

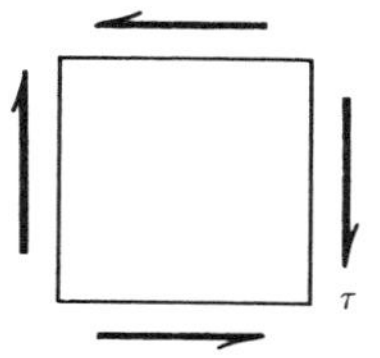

FIG. 9.3 Shear stress element.

Fig. 9.3. Thus the shear load V produces transverse shear stresses resisting V and simultaneously produces longitudinal shear stress along the length of the beam.

It should be noted that although the development of Eq. 9.2 was based on a rectangular cross section, the results are not limited to that.

EXAMPLE PROBLEM 9.1

For the beam shown in Fig. 9.1 which has a 2 in.×8 in. (full size) cross section turned on edge, find the shear stress 2 in. below the top fiber and at the neutral axis for the cross section 3 ft to the right of the left end.

Solution

From the shear diagram in Fig. 9.1 the shear is 4 kips, 3 ft to the right of the left support.

$$V = 4 \text{ kips}$$

I is the moment of inertia of the entire cross section about the neutral axis, Fig. 9.4a.

$$I = \frac{bh^3}{12} = \frac{(2)(8)^3}{12} = 85.3 \text{ in.}^4$$

The width b, 2 in. below the top fiber, is 2 in.:

$$b = 2 \text{ in.}$$

Q is the moment of area about the neutral axis of the area bounded by the extreme fiber and the fiber at which we are finding the shear stress. Thus

$$Q_2 = \bar{y}A' = 3(2 \times 2) = 12.0 \text{ in.}^3$$

Then by Eq. 9.2, the shear stress is

$$\tau_2 = \frac{VQ}{Ib} = \frac{4 \text{ kips }(12.0 \text{ in.}^3)}{85.3 \text{ in.}^4 (2 \text{ in.})} = \mathbf{0.281 \text{ ksi}}$$

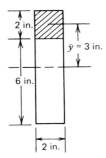

FIG. 9.4a

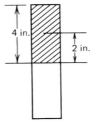

FIG. 9.4b

At the neutral axis only Q changes, Fig. 9.4b

$$Q_{NA} = \bar{y}A' = 2(2 \times 4) = 16.0 \text{ in.}^3$$

$$\tau_{NA} = \frac{4 \text{ kips } (16.0 \text{ in.}^3)}{85.3 \text{ in.}^4 (2 \text{ in.})} = \mathbf{0.375 \text{ ksi}}$$

9.2 SHEAR STRESS DISTRIBUTION

In this section we examine Eq. 9.2 of the previous section to determine how the shear stress varies over the cross section, where the maximum shear stress occurs and where zero values of shear occur. Again we examine a rectangular cross section, although the results may be extended in principle to other common shapes.

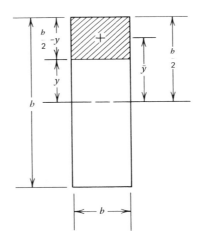

FIG. 9.5 Rectangular cross section.

SHEAR STRESS DISTRIBUTION

Figure 9.5 shows a rectangular cross section whose dimensions are b by h. We ask, what is the shear stress on an arbitrary fiber at a distance y from the neutral axis? To answer the equation we evaluate Eq. 9.2 for this fiber:

$$\tau = \frac{VQ}{Ib} \tag{9.2}$$

V is the shear load on the section and I is the moment of inertia for the entire cross section, so these would be constant for any position y. The width of the fiber under consideration is b. It would change with y for some cross sections, but for a rectangular one it is constant. Thus for this cross section V, I, and b are independent of y, leaving only Q, which defined by

$$Q = \int y \, dA \tag{9.1}$$

This is simply the moment of the area bounded by y and $h/2$ about the neutral axis. Hence

$$Q = \int y \, dA = \bar{y} A'$$

The distance from the neutral axis to the centroid is

$$\bar{y} = \frac{1}{2}\left(\frac{h}{2} + y\right)$$

And the area of the shaded portion is its width times its height

$$A' = b\left(\frac{h}{2} - y\right)$$

So

$$Q = \frac{1}{2}\left(\frac{h}{2} + y\right)(b)\left(\frac{h}{2} - y\right)$$

$$= \frac{b}{2}\left[\left(\frac{h}{2}\right)^2 - (y)^2\right]$$

Then

$$\tau = \frac{V}{Ib} \cdot \frac{b}{2}\left[\left(\frac{h}{2}\right)^2 - y^2\right]$$

We evaluate I for the entire rectangular cross section

$$I = \frac{bh^3}{12}$$

SHEARING STRESSES IN BEAMS

$$\tau = \frac{V(12)}{bh^3} \cdot \frac{1}{2}\left[\left(\frac{h}{2}\right)^2 - y^2\right]$$

$$\tau = \frac{6V}{bh^3}\left[\left(\frac{h}{2}\right)^2 - y^2\right] \quad \tau \text{ Max when } y = 0 \text{ (@ N.A.)} \tag{9.4}$$

Equation 9.4 gives the shear stress for a rectangular cross section. It is of little use in and of itself. However, the conclusions to which it leads us are very important. For a given situation, V, b, and h are constant. Thus the only variable in the equation for the cross section under consideration is y, the distance from the neutral axis to the element under consideration. Notice also that y is squared, and if we were to plot τ versus y we would get a second-degree curve, namely, a parabola. So we see that the shear stress distribution is parabolic. We could find the maximum shear stress by treating Eq. 9.4 as a classical maximum-minimum problem from the calculus—that is, set the derivative equal to zero, find the roots, and substitute them back into the original equation. It is simple enough, however, to find the maximum by inspection in Eq. 9.4. The distance y varies from 0 at the neutral axis to $h/2$ at the extreme fiber. Equation 9.4 is maximum when y is 0, or at the neutral axis. The location is more important than the value in the long run. The maximum value is found by substituting y equal 0 into the equation:

$$\tau_{max} = \frac{6V}{bh^3}\left[\left(\frac{h}{2}\right)^2 - (0)^2\right]$$

$$= \frac{6V}{bh^3}\left(\frac{h}{2}\right)^2$$

$$\tau_{max} = \frac{3}{2}\frac{V}{bh}$$

Note that bh is the area of the cross section, so

$$\tau_{max} = \frac{3}{2}\frac{V}{A} \tag{9.5}$$

This result has been developed only for a rectangular area, but most shapes have the maximum shear stress at the neutral axis.

We further observe that Eq. 9.4 yields zero when y equals $h/2$, that is, at the extreme fiber. We have, therefore, a shear stress distribution that is zero at the extreme fibers, maximum at the neutral axis, and parabolic. This stress distribution is shown in Fig. 9.6. It acts vertically resisting the direct shear and horizontally as required for equilibrium of the element as shown in Fig. 9.3.

SHEAR STRESS DISTRIBUTION

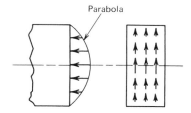

FIG. 9.6 Shear stress distribution for a rectangular cross section.

EXAMPLE PROBLEM 9.2

The cross section of a beam is as shown, Fig. 9.7a. Find the shear stress on the top surface and every 10 mm to the bottom surface. The shear load at the cross section is 12 kN.

Solution

We need to evaluate Eq. 9.2 for the locations A through G. We have the following fixed values:

$$V = 12 \text{ kN}$$

$$b = 30 \text{ mm}$$

$$I = \frac{bh^3}{12} = \frac{30 \text{ mm } (60 \text{ mm})^3}{12} = 5.40E5 \text{ mm}^4$$

Since these are fixed, we need only evaluate Q at each location. First we seek the stress at level A; Q is the moment about the neutral axis of the area bounded by A and the extreme fiber, which "ain't much." It is, in fact, zero. Thus

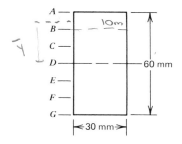

FIG. 9.7a

228 SHEARING STRESSES IN BEAMS

$$\tau_A = \frac{VQ}{Ib} = \frac{12 \text{ kN } (0)}{5.40E5 \text{ mm}^4 \text{ (30 mm)}} = 0$$

Next we seek the stress at level B; Q will be the moment about the neutral axis of the area bounded by B and the extreme fiber, or the top rectangular area, as indicated in the Fig. 9.7b.

$$Q_B = \bar{y}A' = 25 \text{ mm } (10 \text{ mm}) (30 \text{ mm})$$

$$Q_B = 7.50E3 \text{ mm}^3$$

Then

$$\tau_B = \frac{12 \text{ kN } (7.50E3 \text{ mm}^3)}{5.40E5 \text{ mm}^4 \text{ (30 mm)}} = 5.56 \frac{\text{N}}{\text{mm}^2} = \mathbf{5.56 \text{ MPa}}$$

Proceeding in this manner we get

$$Q_C = \bar{y}A' = 20 \text{ mm } (20 \text{ mm}) \, 30 \text{ mm} = 1.20E4 \text{ mm}^3$$

$$\tau_C = \frac{12 \text{kN } (1.20E4 \text{ mm}^3)}{5.40E5 \text{ mm}^4 \text{ (30 mm)}} = \mathbf{8.89 \text{MPa}}$$

$$Q_D = 15 \text{ mm } (30 \text{ mm}) (30 \text{ mm}) = 1.35E4 \text{ mm}^3$$

We save time by proportioning the previous answer.

$$\tau_D = 8.89 \text{ MPa} \left(\frac{1.35}{1.20}\right) = \mathbf{10.0 \text{ MPa}}$$

We can move to point E proceeding as above and as shown in Fig. 9.7c.

$$Q_E = 10 \text{ mm } (40 \text{ mm}) (30 \text{ mm})$$

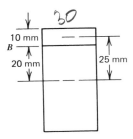

FIG. 9.7b

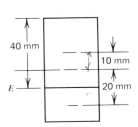

FIG. 9.7c

$$= 1.20E5 \text{ mm}^3$$

It might occur to someone to calculate Q for the area between E and the bottom extreme fiber, rather than the top. If it does, someone is being extremely perceptive and may move to the front of the class. Following someone's lead, we have

$$Q_E = \bar{y}A' = 20 \text{ mm}(20 \text{ mm})(30 \text{ mm}) = 1.20E4 \text{ mm}^3$$

This is, of course, the same value as before. This is because we are taking the moment of area about a centroidal axis. Technically the two moments are of opposite sign, which when added together yield zero, which is a property of centroids. The signs of the moments are of no consequence to us in calculating shear stress. Proceeding, we then get

$$\tau_E = \mathbf{8.89 \text{ MPa}}$$

This is the same value we had for C and raises the question of symmetry. Symmetry does apply in this problem and we also have by symmetry

$$\tau_F = \tau_B = \mathbf{5.56 \text{ MPa}}$$

$$\tau_G = \tau_A = \mathbf{0}$$

Note that the shear stress was zero on the top and bottom surfaces and a maximum value at the neutral axis, as indicated in Section 9.2.

9.3 MAXIMUM SHEAR STRESS FOR COMMON CROSS SECTIONS

In Eq. 9.5 we found that for a rectangular cross section

$$\tau_{max} = \frac{3}{2}\frac{V}{A} \qquad (9.5)$$

In Chapter 3 we found the average shear stress to be

$$\tau_{AV} = \frac{V}{A} \qquad (3.2)$$

Thus we see the actual shear stress varies from zero (at the surface) to maximum (at the neutral axis) which is 50% greater than the average stress.

For a circular cross section we also find the shear stress to be zero at the extreme fiber and maximum at the neutral axis. Its development requires a straightforward (but messy) application of the calculus, yielding

$$\tau_{max} = \frac{4}{3}\frac{V}{A} \qquad (9.6)$$

For an *I*-beam (*S* or *W* shape) the maximum shear stress can be approximated by the equation

$$\tau_{max} = \frac{V}{A_{web}} \tag{9.7}$$

Remember that the web is the vertical connecting center section in the figure in Tables A4.2 and A4.3.

9.4 SHEAR FLOW

Equation 9.2 gives the shear stress caused by a shear load. It has been indicated a number of times that this shear stress acts transversely and longitudinally. An additional useful concept called *shear flow* can be developed from the longitudinal shear stress:

$$\tau = \frac{VQ}{Ib} \tag{9.2}$$

If we multiply by *b*, the width,

$$\tau b = \frac{VQ}{I}$$

which is defined as the shear flow, using the symbol *q*

$$q = \frac{VQ}{I} \tag{9.8}$$

Dimensionally *q* is force per unit length. It can be thought of as the longitudinal shear load per unit length of beam. Its use is illustrated in the following example problem.

EXAMPLE PROBLEM 9.3

A *T*-section (Fig. 9.8) is made by nailing two 2×6's (full size) together as shown. The nails can carry an average shear load of 200 lb. For a vertical shear load of 120 lb, what is the maximum allowable spacing between the nails?

$$\bar{I} = 136 \text{ in.}^4 \qquad \bar{y} = 5 \text{ in.}$$

Solution

From Eq. 9.8

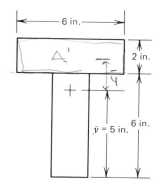

FIG. 9.8

$$q = \frac{VQ}{I}$$

The nail will be in shear where the two 2×6's come together; thus

$$Q = \bar{y}A' = 2(2\times 6) = 24 \text{ in.}^3$$

$$q = \frac{120 \text{ lb } (24 \text{ in.}^3)}{136 \text{ in.}^4} = 21.2 \text{ lb/in.}$$

This is the shear flow longitudinally where the two boards come together. It means that as we move longitudinally along this joint a longitudinal shear load is being accumulated at the rate of 21.2 lb per inch. After 1 in. we have 21.2 lb to be resisted, after 2 in., 42.4 lb, and so forth. The nails that hold these two boards together must resist this shear load. If we have a nail every 1 in., it must resist a shear load of 21.2 lb; if every 2 in., 42.4 lb, and so forth. Thus we write

$$\text{longitudinal shear load} = q \cdot d$$

where d is the distance between the nails. Since our nails can carry 200 lb in shear, we have

$$d = \frac{\text{longitudinal shear load}}{q} = \frac{200 \text{ lb}}{21.2} \frac{\text{in.}}{\text{lb}}$$

$$d = 9.44 \text{ in.}$$

This is the maximum distance between nails for load per nail not to exceed 200 lb.

232 SHEARING STRESSES IN BEAMS

9.5 SUMMARY

Shear stress equation:

$$\tau = \frac{VQ}{Ib}$$

where

$$Q = \int_A y\, dA$$

Q is the moment about the neutral axis of the area bounded by an extreme fiber and the position where shear stress is being determined.

This shear stress is zero at the extreme fibers and is usually maximum at the neutral axis. It acts transversely and longitudinally. For a rectangular cross section:

$$\tau_{max} = \frac{3V}{2A}$$

For circular:

$$\tau_{max} = \frac{4}{3}\frac{V}{A}$$

For an I beam the shear stress may be approximated by:

$$\tau_{max} = \frac{V}{A_{web}}$$

Shear flow:

$$q = \frac{VQ}{I}$$

This last expression is useful for spacing fasteners when making composite beams.

PROBLEMS

9.1 A rectangular 4 in. $\times$ 16 in. beam carries a shear load of 5 kips. Calculate the shear stress on the top and bottom surfaces and at 2-in. intervals across the cross section. Make a plot of stress versus position.

9.2 A rectangular 20 mm $\times$ 100 mm beam carries a shear load of 25 kN. Calculate the shear stress on the top and bottom surfaces and at 20 mm intervals across the cross section. Make a plot of stress versus position.

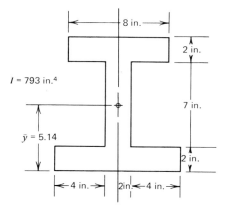

FIG. 9.9 Problem 9.3.

9.3 The cross section of Fig. 9.9 carries a 1000-lb shear load. Calculate the shear stress at:
(a) The bottom surface.
(b) Just below 2.0 in. from the bottom.
(c) Just above 2.0 in. from the bottom.
(d) At the centroid.
(e) Just below 9.0 in. from the bottom.
(f) Just above 9.0 in. from the bottom.
(g) At the top surface.
Plot the above stress distribution.

9.4 A rectangular 100 mm×400 mm beam carries a 4.0-MN shear load. Calculate the shear stress at 50-mm intervals from the neutral axis to the extreme fiber. Make a plot of these stresses against their position across the cross section of the beam.

9.5 A rectangular 4 in.×16 in. beam carries a 10,000-lb shear load. Calculate the shear stress at 1-in. intervals from the neutral axis to the extreme fiber. Make a plot of these stresses against their position across the beam.

9.6 Using the cross section and load of Example Problem 9.3 (Fig. 9.8), calculate the shear stress on the top fiber and every inch to the bottom fiber. (Calculate for just above and just below the joint.) Make a plot of this stress distribution.

9.7 For Fig. 9.10 find the maximum shear and bending stresses in a section at point B.

9.8 For Fig. 9.10 find the maximum shear and bending stresses in the beam.

9.9 For Fig. 9.11 find the shear stress at the neutral axis and 2 in. above it for a section at point B. Compare this with the maximum bending stress at B.

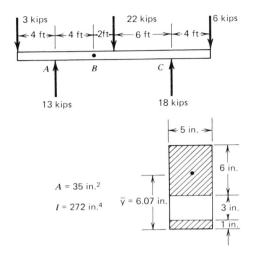

FIG. 9.10 Problems 9.7, 9.8.

9.10 For Fig. 9.11 find the maximum shear and bending stress in the beam.

9.11 For Fig. 9.12 find the maximum shear and bending stresses in the $W\ 8 \times 40$ beam.

9.12 Do Problem 9.11 if the cross section is:
  (a) 4×12 timber (full size).
  (b) A 3-in.-diameter pipe (standard weight).
  (c) 4×12 timber (finished S4S).

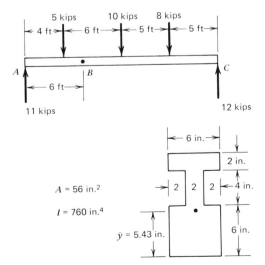

FIG. 9.11 Problems 9.9, 9.10, 9.20.

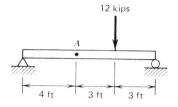

FIG. 9.12 Problems 9.11, 9.12.

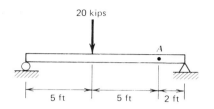

FIG. 9.13 Problem 9.13.

9.13 For Fig. 9.13 find the maximum shear stress in a section at point A and at any location in the $W\ 12\times 36$ beam.

9.14 For Fig. 9.14 find the maximum shear stress in a section at point A and at any location in the $W\ 14\times 38$ beam.

9.15 For Fig. 9.15 find the maximum shear stress in a section at point A and at any location in the $W\ 16\times 50$ beam.

9.16 For the indicated loading and cross section, Fig. 9.16, find the maximum shear stress in a section at point A:

(a) $A-1$ (e) $B-1$ (i) $C-1$ (m) $D-1$
(b) $A-2$ (f) $B-2$ (j) $C-2$ (n) $D-2$
(c) $A-3$ (g) $B-3$ (k) $C-3$ (o) $D-3$
(d) $A-4$ (h) $B-4$ (l) $C-4$ (p) $D-4$

9.17 For the indicated loading and cross section, Fig. 9.16, find the maximum shear stress and the maximum bending stress in the beam:

(a) $A-1$ (e) $B-1$ (i) $C-1$ (m) $D-1$
(b) $A-2$ (f) $B-2$ (j) $C-2$ (n) $D-2$
(c) $A-3$ (g) $B-3$ (k) $D-3$ (o) $D-3$
(d) $A-4$ (h) $B-4$ (l) $C-4$ (p) $D-4$

9.18 For the indicated loading Fig. 9.16, select a beam with a square cross section based on a shear strength of 70 MPa and a factor of safety of 3.

(a) A (c) C
(b) B (d) D

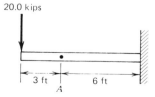

FIG. 9.14 Problem 9.14.

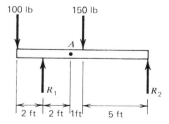

FIG. 9.15 Problem 9.15.

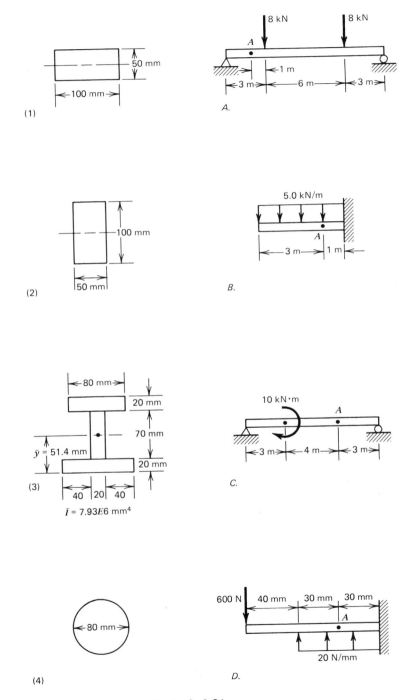

FIG. 9.16 Problems 9.16, 9.17, 9.18, 9.21.

9.19 The beam of Example Problem 9.2 is constructed by glueing three boards together, each having a 20 mm by 30 mm cross section.
  (a) Find the shear flow at the two glued joints.
  (b) If the boards are nailed together instead, and each nail can withstand 8-kN shear load, what is the minimum spacing of the nails?

9.20 Find the shear flow at point B in Fig. 9.11 where the top 2×6 joins the middle 2×4.

9.21 Cross section (3) of Fig. 9.16 has a shear load of 12 kN. Find the shear flow where the top and bottom flange boards join the web. Determine the minimum spacing on the top and bottom for screws that carry a shear load of 1200 N.

10
Putting It All Together: Compound Stress

10.1 Nonaxial loads
10.2 A special case: concrete and masonry structures
10.3 Loads not parallel to the axis
10.4 Superposition of shearing stresses
10.5 Coiled springs
10.6 Summary

Up to now we have learned to calculate and describe the stress distribution that results from a single cause, such as an axial load. But what if the load is not axial? (At this point you may want to examine the fine print on the guarantee that goes with this course.) In any case, it is necessary to examine the fine print that accompanied the stress equations that we have learned. For example, to return to the axial load, as shown in Fig. 10.1, if a load P acts on a straight member, along its longitudinal axis and through the centroid of its cross section, then the resultant stress distribution will be as shown. The maximum stress produced can be calculated from the equation

$$\sigma = \frac{P}{A} \tag{10.1}$$

As indicated in Chapter 3, four characteristics of this situation should be noted:

1. Condition of the load—axial.
2. Condition of the body—straight and homogeneous.

NONAXIAL LOADS 239

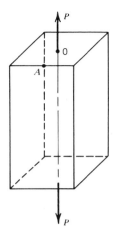

FIG. 10.1a Axial load.

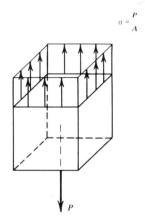

FIG. 10.1b Uniform stress distribution.

3. Stress equation—$\sigma = P/A$.
4. Stress distribution—uniform.

Our applications have emphasized point three and you may have gotten by with mechanically plugging into this stress equation. Bad news! "That ain't gonna work no more!" Successful application of the material in this chapter requires that you thoroughly appreciate all four points above for each stress equation developed thus far. We will postpone our query regarding a nonaxial load while we review these four points for our other normal stress.

BENDING STRESS (Fig. 10.2)

1. Condition of the load—a bending moment about a centroidal or neutral axis of the cross section.
2. Condition of the body—straight, homogeneous, linearly elastic.
3. Stress equation—

$$\sigma = Mc/I \tag{10.2}$$

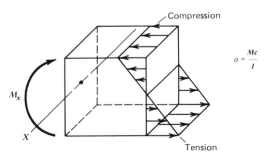

FIG. 10.2 Bending moment and resulting stress distribution.

4. Stress distribution—varies linearly with distance from the neutral axis, where it is zero; maximum tension on one surface and maximum compression on the opposite surface.

Note that the stresses described by Eqs. 10.1 and 10.2 are both normal; that is, they are perpendicular to the surface over which they act.

Suppose that instead of P acting through 0 as shown in Fig. 10.1, it acts through point A on the front surface. It no longer acts through the axis of the member and, therefore, is no longer an axial load. We very cleverly call it a nonaxial load. Since the load is not an axial load, condition 1 for Equation 10.1 is not satisfied, and hence Equation 10.1 does not describe the stress produced by such loading. Equation 10.1 cannot be used directly!

Nonaxial loading (also called eccentric loading—meaning not at the center) is a common occurrence in real problems. In fact, pure axial loading is the ideal case and exists only in the heads of fuzzy-minded professors who write textbooks. A significant problem in materials testing is minimizing the degree of eccentricity in the applied load. Eccentricity always exists to some degree, and you should become sensitive to it as a potential problem. Fortunately, it can be adequately dealt with in most situations, and it turns out the professor's fuzzy-minded ideal has very wide application.

10.1 NONAXIAL LOADS

The nonaxial load problem is solved by applying two principles that will be useful throughout the chapter. The first principle is one from statics. It is illustrated in Fig. 10.3, where a force acts parallel to the longitudinal axis but not along it. In Fig. 10.3b a section has been passed, a free body diagram drawn, and the internal force necessary for equilibrium found. This force does not act through the centroid of the section, and so Equation 10.1 does not apply. Figure 10.3c shows the same free body. This time, however, P is shown acting at 0. Now $\Sigma F = 0$ is satisfied, a necessary condition from Newton's first law. But $\Sigma M = 0$ is not satisfied. This is also a necessary condition; therefore, the free body, as shown, Fig. 10.3c, is not in equilibrium. If a moment $M_x = Pe$ is added as shown in Fig. 10.3d, the second condition is also satisfied. So both Figs. 10.3b and 10.3d show free body diagrams that are in equilibrium. The two conditions are statically equivalent.

What advantage does the analysis in Fig. 10.3d have over that in 10.3b? In 10.3d the load P acts through the centroid. Condition 1 is satisfied, and we may apply Eq. 10.1. Furthermore, M_x is about a centroidal axis and satisfies condition 1 for Eq. 10.2; hence, it also may be applied. This leads us to our first rule, which is:

Rule 1. Replace nonqualifying loads with statically equivalent loads that do satisfy condition 1.

We may alternately state the rule as:

NONAXIAL LOADS 241

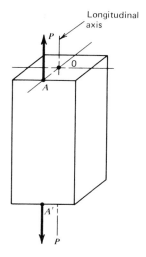

FIG. 10.3a Nonaxial load.

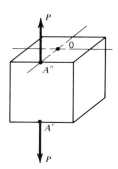

FIG. 10.3b

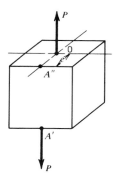

FIG. 10.3c

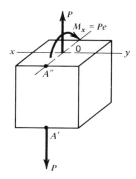

FIG. 10.3d

When finding internal loads use components that satisfy condition 1.

To continue our discussion, loads have been found that satisfy condition 1. Therefore, the stress distribution and stress values from Eqs. 10.1 and 10.2 may be determined. This is illustrated in Fig. 10.4a for the axial load and 10.4b for the bending load.

$$\sigma = \frac{P}{A} \qquad (10.1)$$

$$\sigma = \frac{Mc}{I} \qquad (10.2)$$

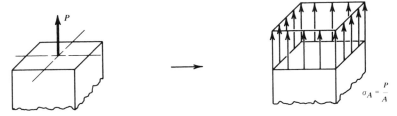

FIG. 10.4a Stress due to axial load.

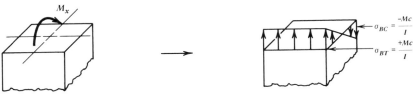

FIG. 10.4b Stress due to bending moment.

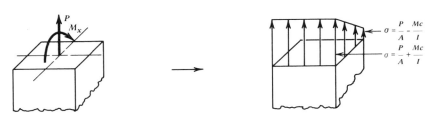

FIG. 10.4c Stress due to combined load.

This gives us the stress distribution and values from each cause. Since the resultant applied load may be obtained by superimposing the bending load on the axial load, it follows that the resultant stress pattern may be obtained by superimposing the stress from the bending load on the stress from the axial load. Hence the bending load of Fig. 10.4b is superimposed on the axial load of Fig. 10.4a, and the resultant stress is obtained by superimposing the bending stress on the axial stress as shown in Fig. 10.4c. In this case the stresses may be added directly as though they were scalar quantities. This is because they are the same type stresses, they have a common direction, and they act over the same area. In general, stresses cannot be combined in a scalar fashion or even as vectors. We will deal with that later. For the time being, note that stresses may be added as though they were scalars when they are of the same type, have the same direction, and act over the same area.

Thus our second rule as illustrated above is:

Rule 2. Find the stress distribution due to each load and superimpose them on each other. If the stresses are of the same type and direction acting over the same area, they may be added as scalars.

The two rules are illustrated in the following problem.

EXAMPLE PROBLEM 10.1

A 400 kip load is applied to the 2 in.×6 in. cast iron machine support as shown in Fig. 10.5a. Find the maximum tensile and compressive stresses.

Solution

We seek a plane of maximum stress and, therefore, maximum loading. All vertical planes will see the same loading in this problem, so we arbitrarily select one and pass a cutting plane as indicated in Fig. 10.5a.

We apply rule 1 and find an equivalent load that satisfies the load conditions, namely loads through and about the centroidal axis. Drawing the free body diagram (Fig. 10.5b) and solving the statics we get

$$P = 400 \text{ kips}$$

$$M = Pe = 400 \text{ kips } (2 \text{ in.})$$

$$M = 800 \text{ in.-kips}$$

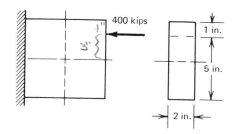

FIG. 10.5a

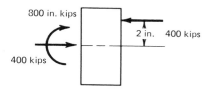

FIG. 10.5b Free body diagram.

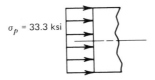

FIG. 10.5c Stress due to axial load.

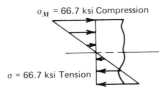

FIG. 10.5d Stress due to bending.

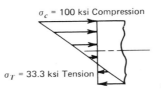

FIG. 10.5e Combined stress.

Next we find the stress distribution for each load and sketch the distribution. The stress due to axial load is (Fig. 10.5c)

$$\sigma_P = \frac{P}{A} = \frac{400 \text{ kips}}{(6 \text{ in.})(2 \text{ in.})} = 33.3 \text{ ksi}$$

The maximum stress due to the bending load is (Fig. 10.5d)

$$\sigma_M = \frac{Mc}{I} = \frac{800 \text{ in.-kips }(3 \text{ in.})(12)}{(2 \text{ in.})(6 \text{ in.})^3} = 66.7 \text{ ksi}$$

The stress distributions are now imposed upon one another, resulting in the stress distribution shown in Fig. 10.5e. The peak values are

$$\sigma_c = -33.3 \text{ ksi} - 66.7 \text{ ksi} = -\mathbf{100 \text{ ksi compression}}$$

$$\sigma_T = -33.3 \text{ ksi} + 66.7 \text{ ksi} = +\mathbf{33.3 \text{ ksi tension}}$$

10.2 A SPECIAL CASE: CONCRETE AND MASONRY STRUCTURES

The above problem can serve to indicate the danger in simply calculating the maximum stress by adding stresses. We illustrate this by asking the question, "If this part were made of gray cast iron, could it carry this load?" The largest stress found was 100 ksi in compression. Gray cast iron has a maximum compressive strength of 120 ksi, and based on this strength, this design appears safe. However, cast iron is much weaker in tension (as are most brittle materials), having an ultimate tensile strength of only 30 ksi. Hence, based on the tensile stress, the design is *unsafe*. This type of consideration must be given anytime the materials used have differing tensile and compressive properties.

Concrete and masonry structures present a special problem, in that without reinforcement, they can carry only very small tensile loads. Hence they must be designed so that no tensile stresses are produced. In Fig. 10.6 the shaded area shown is known as the kern. If a compressive load is applied within this area, the stress will be compressive everywhere. If a compressive load is applied outside this area, tension will be produced. For nonreinforced concrete or masonry structures the load must be applied within the kern.

EXAMPLE PROBLEM 10.2

From Example Problem 10.1 it is obvious that the 400 kip load is applied outside the kern, since a tensile stress of 33.3 ksi is produced on the lower surface. Find the location of P so that no tensile stress is produced—that is, find the upper boundary of the kern.

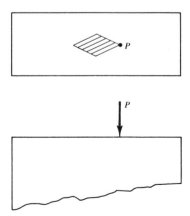

FIG. 10.6

Solution

We assume the force P acts at a distance y from the centroid, Fig. 10.7a.

From the solution of Example Problem 10.1 the stress on the lower element is the difference in the axial load stress and the bending stress, which should equal zero in the limiting case.

$$\sigma = -\frac{P}{A} + \frac{Mc}{I} = 0$$

Substituting general parameters of b and h for Fig. 10.7a, we get

$$\sigma = -\frac{P}{bh} + \frac{Py(12)h}{bh^3(2)} = 0$$

Dividing through by P/bh gives

$$-1 + \frac{6y}{h} = 0$$

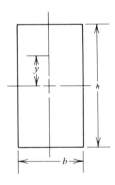

FIG. 10.7a

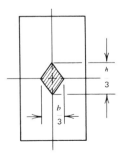

FIG. 10.7b Location of the kern.

Solving for y gives

$$y = \frac{h}{6}$$

This says when $y=h/6$ the stress on the bottom surface will be zero. As we saw in Example Problem 10.1, if the load is applied above this point, tension will be produced. If the load is applied below this level, compressive stresses will be produced on the bottom surface and everywhere else. This will be true as we move down until $y=-h/6$. At this point the stress will be zero on the top surface. As we move below, or outside the point, tension will be produced on the top surface.

Thus if the load is applied on the middle third of the centerline, the stresses will be compressive over the surface. If they are applied outside the middle third, tensile stresses will be produced, and the structure will fail if it is unreinforced masonry. Because of this we very often find these structures with steel reinforcing.

If the same question were asked with regard to the x axis, a similar answer, that is, $x=b/6$, would be obtained. The four points obtained by moving $h/6$ from the center of the section in the y direction and $b/6$ in the x direction constitute boundary points of the kern, Fig. 10.7b. Thus the old rule of thumb that the load should be in the middle third of the structure is supported by the theory. It should also be noted that the solution obtained is perfectly general for rectangular cross sections.

We could have solved Example Problem 10.2 numerically by moving the load toward the center a step at a time until the stresses were all compressive and then perhaps backing up until we found the exact point where the stress on the lower boundary was zero. We avoided this tedious trial-and-error process by generalizing the problem and applying one of the mathematical tools at our disposal—namely, algebra. In doing so we solved not only the problem at hand, but an entire class of problems. We have found the kern for any rectangular cross section. This is another instance in which a general mathematical approach is a superior approach. This will not always be the case, but when it is, it should not be avoided.

10.3 LOADS NOT PARALLEL TO THE AXIS

Another important type of loading is shown in Fig. 10.8a. In this situation the loading is not only not centric, but it also is not parallel to the axis of the member. In Fig. 10.8b, we have replaced the inclined load P with vertical P_y and horizontal P_x, components that are statically equivalent to P. In a second step we transfer these components from point Q to point O, as shown in Fig. 10.8c. As P_x is moved from Q to O it must be accompanied by the moment $M_x = P_x a$ to

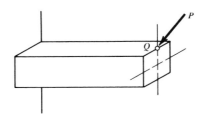

FIG. 10.8a

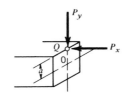

FIG. 10.8b

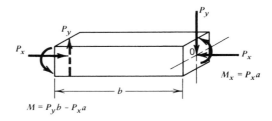

FIG. 10.8c

maintain static equivalency. No moment is developed in the transfer of P_y, since it is transferred along its line of action.

We have applied Rule 1, and all the loads developed at this section satisfy the conditions of previously developed stress equations. Therefore, we can calculate the stresses and determine the stress distribution from each load according to Rule 2. If this were a critical section—one which would likely govern the design—we would do just that. In this loading the supported left end appears more likely to be the critical section. The loads at this section can be found by constructing a free body diagram of the beam and computing the reaction.

When this is done we find the shear and axial loads are of the same magnitude as they were, respectively, on the right end of the beam. The bending moment has changed by the component developed by P_y, the shear load. Since the shear and axial load are constant, the bending moment will determine the critical section. A bending moment diagram will aid in the determination of the critical section.

The stresses due to each load and their distribution may now be determined, as shown in Fig. 10.8d. (The shear stress equation is reviewed in the next section.) The normal stresses due to the axial and bending loads can be combined by superposition, and the critical value is found by scalar addition since we have the same type of stresses (normal) acting on a common area. The shear stress distribution may also be superimposed with the two normal stress distributions.

LOADS NOT PARALLEL TO THE AXIS 249

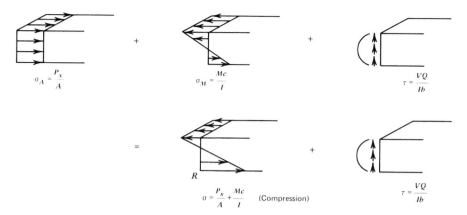

FIG. 10.8*d*

However, its combining effect cannot be determined by simple algebraic addition or even vector addition. The combination of normal and shear stresses is a very complex subject. Notice that the maximum normal stress occurs at the lower surface point. The shear stress is zero at this location. (Once more, knowing stress distribution is as important as knowing the maximum stress.) Therefore, the shear stress will not affect the maximum stress at this critical surface, and we can postpone the combination of shear and normal stresses until a later chapter.

To recap, we have found loads at a critical section meeting the requirements of relevant stress equations. We have calculated stresses and their distributions from those loads and superimposed them. We have found that the maximum normal stress was the algebraic summation of the normal stresses from the axial and bending loads. The shear stress at this extreme surface is zero and, therefore, will not affect the design. It is possible to have the dimensions such that the stress at point R, Fig. 10.8*d*, does not govern the design, in which case the shear stress at other locations would play an important role and the method developed in the next chapter would be required. Usually, if the beam is long and slender, that is, if b is much greater than a, the normal stresses will govern the design.

EXAMPLE PROBLEM 10.3

A 20-kip load is applied to the edge of a 2-in.×1-in. column as shown in Fig. 10.9. Find the maximum normal stress and sketch the stress distribution at the base.

Solution

Replace P with its components at A.

$$P_x = P\cos 45° = 20\cos 45° = 14.14 \text{ kips}$$

$$P_y = P\sin 45° = 20\sin 45° = 14.14 \text{ kips}$$

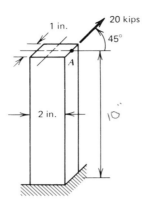

FIG. 10.9a

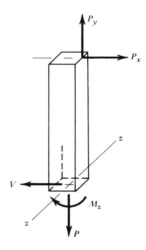

FIG. 10.9b Free body diagram.

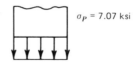

FIG. 10.9c Stress due to axial load.

Complete a free body diagram, Fig. 10.9b, passing a cutting plane through the support.

$$\Sigma F_x = 0, \; V = P_x = \mathbf{14.14 \; kips}$$

$$\Sigma F_y = 0, \; P = P_y = \mathbf{14.14 \; kips}$$

$$\Sigma M_z = 0 = +P_y(1) - P_x(10) - M_z$$

$$M_z = -10 P_x + 1 P_y$$

LOADS NOT PARALLEL TO THE AXIS

$$= -10 \text{ in. } (14.14) + 1 \text{ in. } (14.14)$$

$$M_z = -127.3 \text{ in.-kips}$$

Find the stress and stress distribution due to each load. The stress due to the axial load is

$$\sigma_P = \frac{P}{A} = \frac{14.14 \text{ kips}}{(2 \text{ in.})(1 \text{ in.})} = 7.07 \text{ ksi}$$

with the stress distribution of Fig. 10.9c. The stress due to bending is

$$\sigma_M = \frac{Mc}{I} = \frac{(127.3 \text{ in.-kips})(1 \text{ in.})(12)}{(1 \text{ in.})(2 \text{ in.})^3}$$

$$\sigma_M = 190.9 \text{ ksi}$$

with the stress distribution of Fig. 10.9d.

The stress due to the shear load V will not be calculated, since it is zero where the normal stresses are maximum.

Superimpose the stress distributions and compute the maximum stress

$$\sigma_{max} = \sigma_P + \sigma_M$$

$$\sigma_{max} = 7.07 + 190.9 = \mathbf{198 \text{ ksi } T}$$

$$\sigma = \sigma_P - \sigma_M$$

$$\sigma = 7.07 - 190.9 = \mathbf{184 \text{ ksi } C}$$

The resulting stress distribution is shown in Fig. 10.9e.

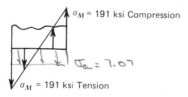

FIG. 10.9d Stress due to bending.

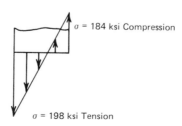

FIG. 10.9e Combined stress.

10.4 SUPERPOSITION OF SHEARING STRESSES

The student will be relieved to know that shear stresses are handled in exactly the same manner as normal stresses—that is, observing rules 1 and 2, which we review here.

> *Rule 1.* Find load components which satisfy the conditions (condition 1) imposed by the stress equations.

> *Rule 2.* Find the stress magnitude and distribution due to each load and superimpose them on the same cross section. If the stresses are of the same type acting over the same area, they may be added as scalars.

We will also review the four characteristics of the two shear stress equations developed previously. From Chapter 9, we have for direct shear as illustrated in Fig. 10.10:

1. Condition of the load—through the centroid and perpendicular to the neutral plane.

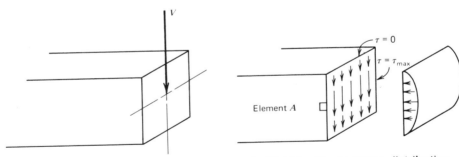

FIG. 10.10*a* Shear load.

FIG. 10.10*b* Shear stress distribution.

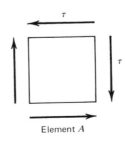

FIG. 10.10*c* Shear stress on element.

2. Condition of the body—straight, homogeneous, and linearly elastic.
3. Stress equation

$$\tau = \frac{V}{It} \int y \, dA \qquad (10.3a)$$

$$\tau_{max} = \frac{3}{2} \frac{V}{A} \text{ for rectangular cross sections} \qquad (10.3b)$$

$$\tau_{max} = \frac{4}{3} \frac{V}{A} \text{ for circular cross sections} \qquad (10.3c)$$

4. Stress distribution—varies depending on the shape of the cross section but generally is second degree. The maximum value usually occurs at the centroid. Free surfaces have zero shear stress. The distribution shown in Fig. 10.10b is for rectangular cross sections only. Also, note that the shear described by the relation is simultaneously acting on the face of the cross section and longitudinally as shown on element A in Fig. 10.10c.

From Chapter 5 we have the conditions for shear stress from torsional loading as follows and as shown in Fig. 10.11:

1. Condition of load—a moment of force (torque) about the longitudinal axis.

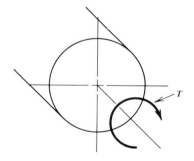

FIG. 10.11a Torsional load.

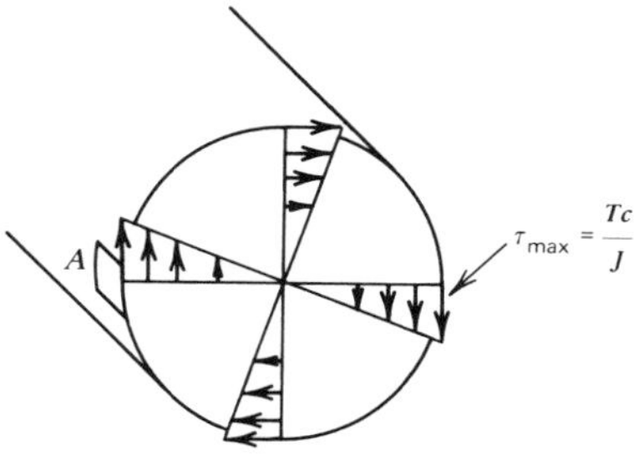

FIG. 10.11b Shear stress distribution.

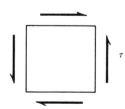

FIG. 10.11c Shear stress on element.

254 PUTTING IT ALL TOGETHER: COMPOUND STRESS

2. Condition of the body—straight, circular cross section, homogeneous, and linearly elastic.

3. Stress equation

$$\tau = \frac{Tc}{J} \tag{10.4}$$

4. Stress distribution—linear from zero at the center to a maximum value on the surface. As with the shear stress from a direct shear load, the shear stress also acts longitudinally as well as on the face of the cross section.

Consider the everyday problem of tightening a bolt with a socket wrench, shown schematically in Fig. 10.12. A force P is applied at the end of the wrench. What stress does this action produce in the extension rod at a section through point A? We solve this problem, as previously, by applying the two rules we have developed. By drawing a free body diagram of the handle with a section passed through point A we find the reactions satisfying condition 1 according to Rule 1, Fig. 10.12b:

$$\Sigma F_y = 0: \quad V = P \quad \text{shear load}$$

$$\Sigma M = 0: \quad T = Pa \quad \text{torque}$$

According to Newton's third law equal but opposite forces act on the shaft as shown in Fig. 10.12c. Applying rule 2 for the shear load V we get

$$\tau_V = \frac{4}{3} \frac{V}{A}$$

and the distribution shown in Fig. 10.12d.

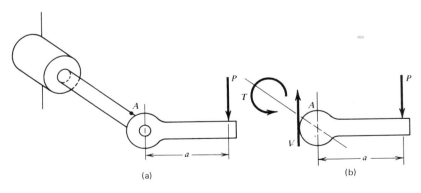

FIG. 10.12 a–b

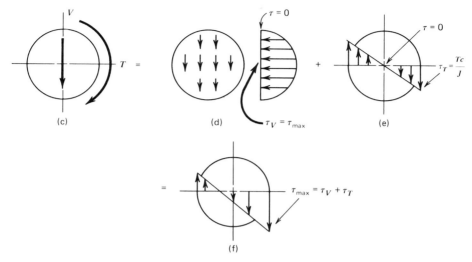

FIG. 10.12c-f

For the torque T we get

$$\tau_T = \frac{Tc}{J}$$ with the distribution shown in Fig. 10.12e.

Then by the method of superposition

$$\tau_{max} = \tau_V + \tau_T = \frac{4}{3}\frac{V}{A} + \frac{Tc}{J}$$

This maximum stress will occur on the right side of the shaft as shown in Figure 10.12f. The stress on the left side will be $\tau_V - \tau_T$. Stresses can be obtained by simple algebraic addition only along this axis. Fortunately that includes the maximum and minimum values.

EXAMPLE PROBLEM 10.4

A 3-in.-diameter gear transmits 5 hp to a short 1-in.-diameter shaft at 300 rpm. Find the maximum shear stress in the shaft.

Solution

The transmitted load is shown schematically in Fig. 10.13a and is found as follows:

$$P = T\omega$$

256 PUTTING IT ALL TOGETHER: COMPOUND STRESS

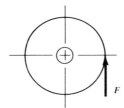

FIG. 10.13a

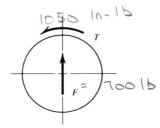

FIG. 10.13b Equivalent qualifying loads.

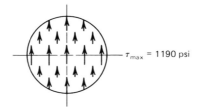

FIG. 10.13c Stress due to shear load.

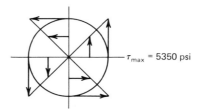

FIG. 10.13d Stress due to torque.

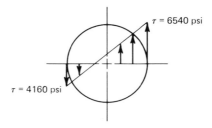

FIG. 10.13e Combined stress.

from which we solve for the torque T

$$T = \frac{P}{\omega} = 5\;\cancel{\text{hp}} \frac{550 \;\cancel{\text{ft-lb}}}{\cancel{\text{s}}\cdot\cancel{\text{hp}}} \left(\frac{\cancel{\text{min}}}{300 \;\cancel{\text{rev}}}\right)\left(\frac{1 \text{ rev}}{2\pi \text{ rad}}\right)\left(\frac{60 \;\cancel{\text{s}}}{\cancel{\text{min}}}\right)\left(\frac{12 \text{ in.}}{\cancel{\text{ft}}}\right)$$

$$= \frac{5(550)(60)(12)}{300(2\pi)} \text{ in.-lb} = \mathbf{1050 \text{ in.-lb}} = T$$

Next we find the transmitted load F which equals the shear load.

$$T = Fr$$

$$F = \frac{T}{r} = \frac{1050}{1.5} = \mathbf{700.3 \text{ lb}} = F$$

The loading is as shown in Fig. 10.13b, and these loads satisfy the restrictions of the stress Eqs. 10.3 and 10.4, so those stresses may now be calculated.

Applying Eq. 10.3 results in the stress distribution shown in Fig. 10.13c and a maximum shear stress of

$$\tau_V = \frac{4}{3}\frac{V}{A} = \frac{4}{3}\frac{(700.3)(4)}{\pi(1)^2}$$

$$\tau_V = 1189 \text{ psi}$$

Applying Eq. 10.4 results in the stress distribution shown in Fig. 10.13d, and the maximum stress is

$$\tau_T = \frac{Tc}{J} = \frac{T\frac{d}{2}}{\left(\frac{\pi d^4}{32}\right)} = \frac{16T}{\pi d^3}$$

$$= \frac{16(1050)}{\pi(1)^3} = 5350 \text{ psi}$$

Superimposing these results gives

$$\tau = \tau_V + \tau_T = 1189 + 5350$$

$$\tau_{\max} = 6539$$

$$\boldsymbol{\tau_{\max} = 6540 \text{ psi}}$$

The stress distribution along the midline is shown in Fig. 10.13e with the maximum shear stress occurring at the right-hand surface.

EXAMPLE PROBLEM 10.5

An 8-mm shaft is rigidly attached to a pinion gear. Figure 10.14a shows one tooth of the gear on the shaft. The tangential component of the load may be treated as acting at a radius of 30 mm. Find the maximum shear stress in the shaft due to the 500-N load.

Solution

We find values of V and T by applying Newton's first law for Fig. 10.14a:

$$V = 500 \text{ N} \rightarrow$$

$$T = 500 \text{ N } (30 \text{ mm})$$

$$T = 15000 \text{ N} \cdot \text{mm} = \mathbf{15.0 \text{ N} \cdot \text{m}}$$

(A bending moment would also occur at this section. We will ignore it for the present.) For convenience in visualization a facing section is shown in Fig. 10.14b. We now calculate the stresses produced by each component of load.

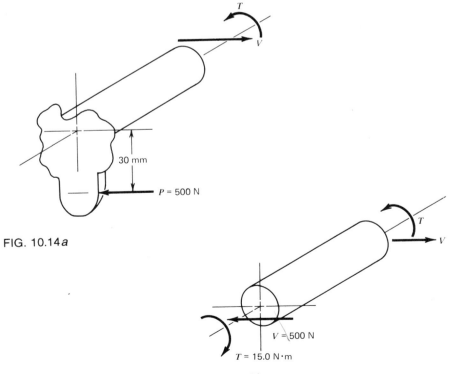

FIG. 10.14a

FIG. 10.14b

For the shear load V:

$$\tau_V = \frac{4}{3}\frac{V}{A} = \frac{4}{3}\frac{500\text{ N}(4)}{\pi(0.008\text{m})^2}$$

$$\tau_V = \mathbf{13.3\text{ MPa}}$$

This stress distribution is sketched in Fig. 10.14c. For the torque T

$$\tau_T = \frac{Tc}{J} = \frac{15.0\text{ N}\cdot\text{m }(0.004\text{m})(32)}{\pi(0.008\text{m})^4}$$

$$\tau_T = \mathbf{149\text{ MPa}}$$

as shown in Fig. 10.14d. These are superimposed to give

$$\tau_{\text{top}} = 149 - 13.3 = \mathbf{136\text{ MPa}}$$

$$\tau_{\text{bottom}} = 149 + 13.3 = \mathbf{162\text{ MPa}}$$

which is the maximum shear stress as shown in Fig. 10.14e.

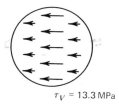

$\tau_V = 13.3$ MPa

FIG. 10.14c Stress due to shear load.

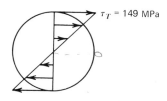

$\tau_T = 149$ MPa

FIG. 10.14d Stress due to torque.

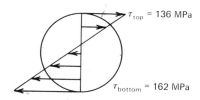

$\tau_{\text{top}} = 136$ MPa

$\tau_{\text{bottom}} = 162$ MPa

FIG. 10.14e Combined stress.

10.5 COILED SPRINGS

Closely coiled helical springs are widely used for absorbing and storing energy. Many automobiles use a coiled spring suspension, especially for the front end. If you are taking notes with a ball-point pen with a retractable point, chances are it contains a coil spring. Coil springs come in many sizes and are made of several materials. While many, perhaps most, are "off the shelf," it is also common to custom-make them. The point to this departure is that the coiled spring represents a direct application of combining shear stresses.

Shown in Fig. 10.15a is a closely coiled helical spring. The wire is circular in cross section. Rectangular cross sections are also common but will not be analyzed here. A vertical section is passed through one turn of the coil, resulting in the free body diagram shown in Fig. 10.15b. Keeping Rule 1 in mind we sum forces:

$$\Sigma F_y = 0: \quad V = F$$

$$\Sigma M = 0: \quad T = \frac{FD}{2}$$

The next step is to apply Rule 2. At this point we will take action that will disturb the student who critically examines the text. We will assume something we know to be false! We will assume that the shear stress due to the direct shear load V is uniform and may be calculated by:

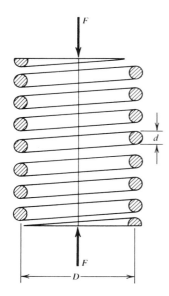

FIG. 10.15a Coiled spring.

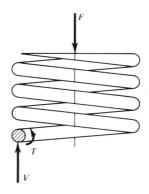

FIG. 10.15b Free body diagram.

COILED SPRINGS

$$\tau_v = \frac{V}{A} = \frac{4F}{\pi d^2} \qquad (10.5)$$

This is illustrated in Fig. 10.16a.

We know from our previous analysis that $\tau_{v_{max}} = (4/3)(V/A)$ and that the distribution is not uniform. So why do we assume a value known to be 33% low? Well, for several reasons.

One reason is that it is customary. An unpleasant fact of life is that we must interact with the world as it is and not as we think it should be. For years, I guess literally centuries, we have put up with the clumsy English units system. In this age of enlightenment we are rapidly moving to the metric system, or more correctly, SI. No matter how convinced we may be of the SI's superiority, we in America will continue to interact with the English system for sometime. And today's student must be conversant in both. So, we do many things because they are customary. That doesn't mean we don't work for change when it's needed (we are changing to the SI). Although, in general, I have observed that things became customary because that was a good way to operate.

A second reason for making assumptions which we know to be incorrect is that it may substantially simplify the problem. In the engineering world we should always use the simplest analysis that produces an acceptable result. This leads to a consideration of the effect of the simplification. If a more rigorous analysis does not lead to significantly different results in the final answer, the rigor is not justified. Meanwhile back at the spring $\cdots$.

Following our patently false assumption we make a more accurate assumption concerning the shear stress due to the torque:

$$\tau = \frac{Tc}{J} = \frac{\left(\frac{FD}{2}\right)\left(\frac{d}{2}\right)}{\left(\frac{\pi d^4}{32}\right)} = \frac{8FD}{\pi d^3} \qquad (10.6)$$

This stress distribution is illustrated in Fig. 10.16b. Superimposing the stress of Eq. 10.5 on that of 10.6 gives a maximum stress of:

$$\tau = \frac{4F}{\pi d^2} + \frac{8FD}{\pi d^3}$$

$$\tau = \frac{8FD}{\pi d^3}\left(\frac{d}{2D} + 1\right) \qquad (10.7)$$

as shown in Fig. 10.16c. Note that if the wire size (d) is small compared to the spring diameter (D) then $d/2D$ is considerably smaller than 1.0 and the direct shear does not have a large effect. Hence we see that our simplifying assumption does not produce significant error in many cases.

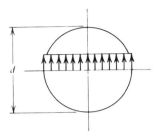

FIG. 10.16a "Uniform" stress due to shear load.

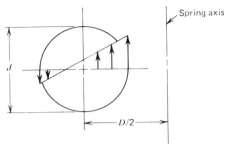

FIG. 10.16c Theoretical resultant stress.

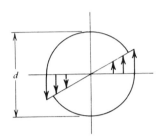

FIG. 10.16b Linearly distributed stress due to torque.

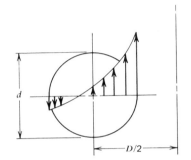

FIG. 10.16d Actual resultant stress.

We can simplify Eq. 10.7 by letting K represent $(d/2D)+1$. This yields:

$$\tau_{max} = K\left(\frac{8FD}{\pi d^3}\right) \qquad (10.8)$$

The bracket in Eq. 10.8 contains the stress due to torsion. Hence at this point the K factor modifies the torsional stress to include the effect of direct shear. Because of the curvature of the spring winding, there will be a further concentration of stress on the inside edge. This more accurate stress distribution is shown in Fig. 10.16d. We now allow K to include this effect as well so we can calculate the maximum stress as the factor K times the stress due to simple torsion. In this form K is known as the Wahl correction factor. Fig. 10.17 is a plot of K versus (D/d). (D/d) is known as the spring index. These values of K come from a solution of the problem using the theory of elasticity—an advanced topic in mechanics.

For small spring indices—springs with a high relative curvature—the stress concentration is high. For larger spring indices it is only slightly greater than 1. Another characteristic of springs should be noted at this point. They are highly stressed compared to other steel parts. Typical ultimate strengths of common steels are from 50 to 150 ksi. Spring steels have strengths up to 300 ksi.

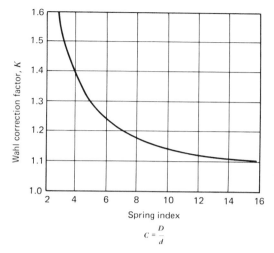

FIG. 10.17 Wahl correction factor.

EXAMPLE PROBLEM 10.6

A coil spring is made of 5-mm wire turned on a 40-mm diameter. Find the theoretical and the actual stress in this spring due to a 500-N load.

Solution

The theoretical load is found by evaluating Eq. 10.7:

$$\tau = \frac{8FD}{\pi d^3}\left(\frac{d}{2D}+1\right) = \frac{8(500\text{ N})(0.040\text{ m})}{\pi(0.005\text{ m})^3}\left(\frac{0.005\text{ m}}{2(0.040\text{ m})}+1\right)$$

$$\tau = 407E6(0.0625+1) = \mathbf{433\text{ MPa}}$$

To calculate the actual stress we need to find the spring index and use Eq. 10.8:

$$C = \frac{D}{d} = \frac{40\text{ mm}}{5\text{ mm}} = \mathbf{8.00}$$

From Fig. 10.17 we read

$$K = 1.17$$

and

$$\tau = K\left(\frac{8FD}{\pi d^3}\right) = 1.17(407E6) = \mathbf{477\text{ MPa}}$$

Thus we see the actual stress is 44 MPa or about 10% greater than the theoretical stress. This higher value has two sources. First, we assumed average stress due to direct shear when it is actually higher. Second, we neglected stress concentrations.

10.6 SUMMARY

All the stress equations are summarized in the following four.
Normal stresses:

$$\sigma = \frac{P}{A}$$

$$\sigma = \frac{Mc}{I}$$

Shear stresses:

$$\tau = \frac{Tc}{J}$$

$$\tau = \frac{VQ}{Ib}$$

We also have the average shear stress equation which is frequently used.

$$\tau_{AV} = \frac{V}{A}$$

By now these basic stress equations need to be thoroughly ingrained. If they are not, work at reproducing them on a blank sheet of paper until they are. Include the load causing the stress, where it must act, and the resulting stress distribution on this paper. Resolve loads into components which satisfy these equations at the cross section of interest. Find the stresses and their distributions. Then superimpose stress distributions to get resultant stress distributions. Stresses of the same type which are in the same direction and act over the same area may be added as scalars. Others cannot.

Coiled spring:

$$\tau_{max} = K \left(\frac{8FD}{\pi d^3} \right)$$

PROBLEMS

10.1 The bracket shown in Fig. 10.18 subjected to $P = 1000$ lb, $Q = 0$. Calculate the stresses at the top and bottom of the bracket at section $A - A$. $I = 10$ in.^{4}, $A = 8$ in.^{2}

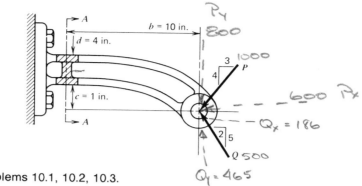

FIG. 10.18 Problems 10.1, 10.2, 10.3.

10.2 The bracket shown in Fig. 10.18 is subjected to $P=0$, $Q=500$ lb. Calculate the stress at the bottom of the bracket at section $A-A$. $I = 8$ in.^{4}, $A = 6$ in.^{2}

10.3 The bracket shown in Fig. 10.18 has the following loads: $P=1000$ lb, $Q=500$ lb. Calculate the stresses at the top and bottom of section $A-A$. Sketch this stress distribution. $I = 10$ in.^{4}, $A = 8$ in.^{2}

10.4 Find the stress at point A in Fig. 10.19.

10.5 Find the stress at point B in Fig. 10.19

10.6 Sketch the stress distribution at the support and locate the fiber with zero stress for Fig. 10.19.

10.7 Find the stress at point A in Fig. 10.20.

10.8 Find the stress at point B in Fig. 10.20.

10.9 Sketch the stress distribution at the support and locate the fiber with zero stress for Fig. 10.20.

10.10 For Fig. 10.21 determine the maximum stress produced in the 3-in.-diameter member when the tension in the rope is 500 lb.

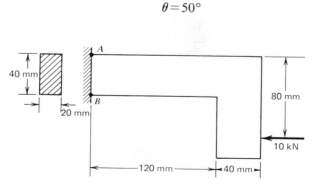

FIG. 10.19 Problems 10.4, 10.5, 10.6.

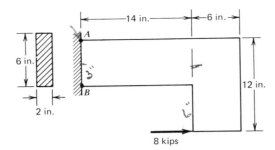

FIG. 10.20 Problems 10.7, 10.8, 10.9.

10.11 In Problem 10.10 let θ vary from 0 to 90° in 30° steps. What is the maximum stress produced and what value of θ corresponds to it? Explain why.

10.12 For Fig. 10.22 determine the maximum stress produced in the 50-mm-diameter member when the tension in the rope is 1200 N.

10.13 In Problem 10.12 let θ vary from 0 to 90° in 30° steps. What is the maximum stress produced and what value of θ corresponds to it? Explain why.

10.14 Find the stress at point A on the 1 in. $\times$ 8 in. (actual size) loaded as shown in Fig. 10.23.

10.15 Find the stress at point B on the 1 in. $\times$ 8 in. (actual size) loaded as shown in Fig. 10.23.

10.16 A 150-mm-diameter post is loaded as shown in Fig. 10.24. Find the stress at point C.

10.17 A 150-mm-diameter post is loaded as shown in Fig. 10.24. Find the stress at point B.

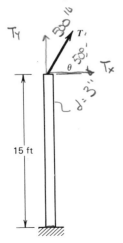

FIG. 10.21 Problems 10.10, 10.11.

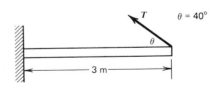

FIG. 10.22 Problems 10.12, 10.13.

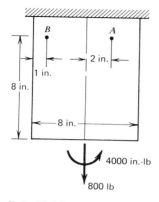

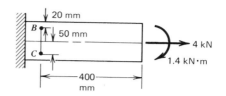

FIG. 10.24 Problems 10.16, 10.17.

FIG. 10.23 Problems 10.14, 10.15.

10.18 A torque wrench similar to Fig. 10.25 has an 8-in. handle (a). A load of 50 lb (P) is applied to the handle. The shaft has a 3/8 in. diameter. Find the maximum shear stress in the shaft due to
   (a) Direct shear.
   (b) Torque.
   (c) Both (a) and (b) combined.

10.19 A torque wrench similar to Fig. 10.25 has a 150-mm handle (a). A load of 220 N (P) is applied to the handle. The shaft has a 10-mm diameter. Find the maximum shear stress in the shaft due to
   (a) Direct shear.
   (b) Torque.
   (c) Both (a) and (b) combined.

10.20 A torque wrench similar to Fig. 10.25 is made of steel with a shear

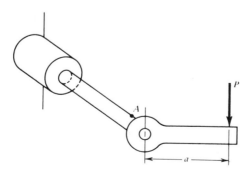

FIG. 10.25 Problems 10.18, 10.19, 10.20, 10.21.

strength of 18 ksi. If the shaft is ½ in. diameter and the handle 6 in. long, what load (P) can it carry? Consider only shear stress.

10.21 A torque wrench similar to Fig. 10.25 is made of steel with a shear strength of 150 MPa. If the shaft is 15 mm diameter and the handle 200 mm long, what load (P) can it carry? Consider only shear stress.

Problems 10.22–10.25 use the following values for Fig. 10.26a.

$P_1 = 150$ lb $P_2 = 120$ lb
$R_1 = 6$ in. $R_2 = 8$ in.
$L_1 = 10$ in. $L_2 = 12$ in.

10.22 For the ¾-in.-diameter shaft shown in Fig. 10.26a between sections A and B, find the combined shear stress on

(a) A top element (Fig. 10.26b).
(b) A right element.
(c) A bottom element.
(d) A left element.

10.23 For the ¾-in.-diameter shaft shown in Fig. 10.26a between sections B and C, find the combined shear stress on

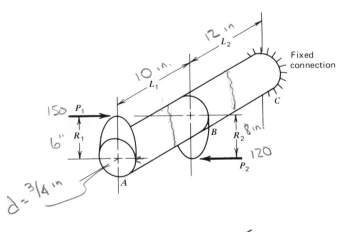

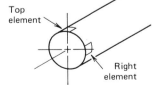

FIG. 10.26a–b Problems 10.22, 10.23, 10.24, 10.25, 10.26, 10.27, 10.28, 10.29.

(a) A top element (Fig. 10.26b).
(b) A right element.
(c) A bottom element.
(d) A left element.

10.24 Based on a shear strength of 16 ksi and a factor of safety of 3, what size shaft is needed between sections A and B, Fig. 10.26a?
(Hint: A trial-and-error solution may be necessary.)

10.25 Based on a shear strength of 16 ksi and a factor of safety of 3, what size shaft is needed between sections B and C, Fig. 10.26a?
(Hint: A trial-and-error solution may be necessary.)

Problems 10.26–10.29 use the following values:
$P_1 = 80$ kN $P_2 = 60$ kN
$R_1 = 300$ mm $R_2 = 500$ mm
$L_1 = 250$ mm $L_2 = 400$ mm

10.26 For the 80-mm-diameter shaft shown in Fig. 10.26a between sections A and B, find the combined shear stress on
(a) A top element (Fig. 10.26b).
(b) A right element
(c) A bottom element.
(d) A left element.

10.27 For the 80-mm-diameter shaft shown in Fig. 10.26a between sections B and C, find the combined shear stress on
(a) A top element (Fig. 10.26b).
(b) A right element.
(c) A bottom element.
(d) A left element.

10.28 Based on a shear strength of 180 MPa and a factor of safety of 3, what size shaft is needed between sections A and B for Fig. 10.26a?
(Hint: A trial-and-error solution may be necessary)

10.29 Based on a shear strength of 180 MPa and a factor of safety of 3, what size shaft is needed between sections B and C, for Fig. 10.26a?
(Hint: A trial-and-error solution may be necessary)

10.30 A $\frac{5}{8}$-in.-diameter shaft is keyed to a 2-in.-diameter gear. It is to transmit 3 hp at 1800 rpm. Find the maximum shear stress on the shaft.

10.31 A 14-mm-diameter shaft is keyed to a 60-mm-diameter gear. It is to transmit 2 kW at 1800 rpm. Find the maximum shear stress on the shaft.

10.32 A 5-in.-diameter gear accepts 10 hp at 3600 rpm. What size shaft is required for the gear based on a shear strength of 24 ksi?

270 PUTTING IT ALL TOGETHER: COMPOUND STRESS

10.33 A 125-mm-diameter gear accepts 6 kW at 2600 rpm. What size shaft is required for the gear based on a shear strength of 300 MPa?

10.34 A coil spring is wound from $\frac{1}{8}$-in.-diameter wire on a 1-in. radius. Find the spring index and stress produced in the spring by a 100-lb load.

10.35 A coil spring is wound from 4-mm-diameter wire on a 30-mm radius. Find the spring index and stress produced in the spring by a 100-N load.

10.36 A steel spring with a shear strength of 110 ksi has a wire diameter of 0.065 in. and spring diameter of $\frac{1}{2}$ in. What load can this spring carry?

10.37 A steel spring with a shear strength of 1.5 GPa has a wire diameter of 1.5 mm and spring diameter of 8 mm. What load can this spring carry?

11

Plane Stress Analysis

11.1 Stress on an inclined plane
11.2 General stress formulas
11.3 Principal stresses
11.4 Maximum shear stress
11.5 Mohr's circle
11.6 Applications of Mohr's circle to failure analysis
11.7 Summary

In the previous chapter we looked at problems in which normal stresses from various causes were combined and where shear stresses from various causes were combined. It was found that when the same type of stresses were acting over the same area they could be simply added. Consideration of those problems should raise the question as to what to do when both shear stresses and normal stresses are present on an element. Alas, they cannot be simply added! They cannot even be added as vectors. (Even though stresses have magnitude and direction, they are not vectors. Mathematically they are described as *tensors*, but only use that term when you need to sound especially precocious.)

In this chapter we will develop methods of handling this important problem. It is necessary for a proper analysis of fairly simple items such as a power transmission shaft which simultaneously has shear stresses produced by torque and a normal stress produced by bending moment. Beyond this problem and more complex ones, it is also necessary for the analysis of failure even in very simply loaded members.

11.1 STRESS ON AN INCLINED PLANE

Consider an axially loaded board as shown in Fig. 11.1 with a 1 in.×0.5 in. cross section. The grain of the wood is at 30° to the direction of loading. The normal stress in the direction of loading is easily calculated as follows:

272 PLANE STRESS ANALYSIS

FIG. 11.1 Axially loaded board.

$$\sigma = \frac{P}{A} = \frac{500 \text{ lb}}{(1 \text{ in.})(0.5 \text{ in.})} = 1000 \text{ psi}$$

In wood the grain represents a plane of weakness. The wood has a relatively low resistance to stresses perpendicular to the grain. Thus we need to know the normal stress in this direction. We develop three methods of handling this problem. The first requires only the application of principles established up to now. The second generalizes that analysis resulting in equations that are broadly applicable. The last results from a graphical interpretation of the general equations. The graphical method will be most useful to us.

To analyze the stress at point A on the board we draw the stress element A, enlarged in Fig. 11.2a to show the stresses acting on it. Also shown on this figure is the line of the grain which makes an angle of 30° with the normal stress. We cut the stress element along this plane, as shown in Fig. 11.2b. On the cut plane there will be a normal stress σ and a shear stress τ, as shown. The original stress, 1000 psi, is also shown. There is no stress on the side parallel to the load.

We do not have a Newton's first law for stresses. So although this partial stress element is in equilibrium, Fig. 11.2b is not a free body diagram. To obtain a free body diagram the stresses acting on the element must be replaced by the associated forces. If the element has a uniform depth (or unity if you prefer) and the inclined area is assigned the value A, then the vertical area will be $A \sin 30°$ and the horizontal area will be $A \cos 30°$, as shown in Fig. 11.2c.

In Fig. 11.2d the stresses are associated with the areas they act over, yielding forces. This is a free body diagram and must satisfy Newton's first law. The force due to the normal stress σ is σA and due to the shear stress τ is τA. Since the vertical area is $A \sin 30°$, the normal force on it will be $(1000)A \sin 30°$. To determine the value of σ, the normal stress on the inclined plane, forces are summed in the normal direction, yielding

$$\Sigma F_N = 0 = +\sigma A - [(1000)A \sin 30°] \sin 30°$$

The last $\sin 30°$ gives the component of the force due to the 1000-psi stress in the direction of the normal stress. The A's may be divided out, yielding

$$\sigma = 1000 \sin^2 30° = \mathbf{250 \text{ psi}}$$

Thus we find there will be a tensile stress of 250 psi perpendicular to the grain attempting to pull the wood apart.

A similar analysis for τ, the shear stress, gives

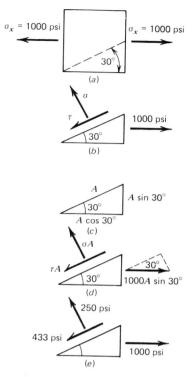

FIG. 11.2a–e Stresses on an inclined plane of a stress element.

$$\Sigma F_T = 0 = -\tau A + [(1000)A \sin 30°] \cos 30°$$

$$\tau = 1000 \sin 30° \cos 30° = \mathbf{433 \text{ psi}}$$

There will also be a 433-psi shear stress attempting to shear along this grain. These stresses are shown on the element in Fig. 11.2e.

This method is further illustrated in Example Problem 11.1.

EXAMPLE PROBLEM 11.1

Find the normal and shear stresses on a plane making a 20° angle with the horizontal, Fig. 11.3a.

Solution

The areas are as shown in Fig. 11.3b. The resulting forces and the free body diagram are shown in Fig. 11.3c. We now sum the forces in the direction of

274 PLANE STRESS ANALYSIS

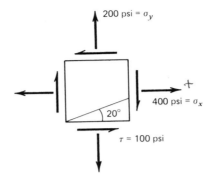

FIG. 11.3a

FIG. 11.3b

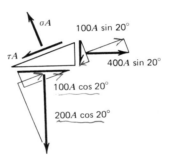

FIG. 11.3c Free body diagram.

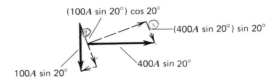

FIG. 11.3d

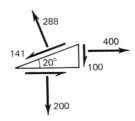

FIG. 11.3e

σ. Figure 11.3d shows the components in the normal direction of the forces acting on the right-hand face.

$$\Sigma F_N = 0 = \sigma A - (400 A \sin 20°)\sin 20° - (100 A \sin 20°)\cos 20°$$
$$- (100 A \cos 20°)\sin 20° - (200 A \cos 20°)\cos 20°$$

The last two terms are the components in the normal direction of the forces acting on the horizontal face.

Dividing out the A and solving for σ we get

$$\sigma = 400 \sin^2 20° + 100 \sin 20° \cos 20° + 100 \cos 20° \sin 20° + 200 \cos^2 20°$$

$\sigma = 288$ psi

Similarly,

$$\Sigma F_T = 0 = -\tau A + (400 A \sin 20°)\cos 20° - (100 A \sin 20°)\sin 20°$$
$$+ (100 A \cos 20°)\cos 20° - (200 A \cos 20°)\sin 20°$$

$$\tau = 400 \sin 20° \cos 20° - 100 \sin^2 20° + 100 \cos^2 20° - 200 \cos 20° \sin 20°$$

$\tau = 141$ psi

The resultant stresses are shown in Fig. 11.3e.

11.2 GENERAL STRESS FORMULA

Without further work the analysis of the previous section would have to be carried out over and over. When that situation presents itself, it suggests a need for a generalized analysis. Beyond eliminating the necessity for repeatedly carrying out the same analysis, the generalized form will allow us to examine the behavior of the stresses on an arbitrary plane, for example, where do maximums occur, what their values are, and similar questions.

Figure 11.4a shows a general stress element. All the stresses shown are in their positive sense for this analysis. Normal stresses are positive when in tension and negative when in compression. Shear stresses are considered positive when up on a right-hand face. This sign system for shear stress differs from the sign system for shear load used previously. An easy way to remember this sign system is by thinking of the primary surfaces as being the right-hand and top surfaces. The positive stresses on these surfaces will then have the sense of positive vectors perpendicular to these surfaces.

We ask, what would be the normal stress on a plane at an angle of θ from the right-hand plane? Note that to arrive at the plane the rotation is counterclockwise. This will continue to be called a positive rotation. A portion of the stress element is drawn in Fig. 11.4b, the new surface being formed at an angle θ from

276 PLANE STRESS ANALYSIS

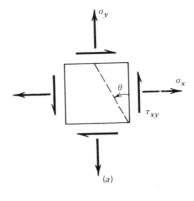

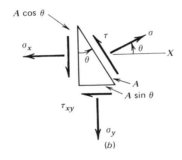

FIG. 11.4a-b Stress element.

the vertical plane. The direction of the normal stress on this plane makes an angle of θ with the original x direction. The normal and shear stresses on this new plane are σ and τ as before. The area of the plane is A, giving $A\sin\theta$ and $A\cos\theta$ as the other two areas as indicated in the figure. Forces may be summed in the directions of σ and τ, yielding explicit equations for each of them. Simpler equations may be obtained by summing forces vertically and horizontally, though they will require algebraic manipulation before yielding explicit values for σ and τ. Also, remember that stresses are not forces and Fig. 11.4b is not a free body diagram until the stresses are associated with their respective areas. However, after doing this, summing the forces horizontally yields

$$\Sigma F_x = 0 = -\sigma_x(A\cos\theta) - \tau A(\sin\theta) - \tau_{xy}(A\sin\theta) + \sigma A(\cos\theta) = 0$$

Dividing through by $A\cos\theta$, noting $\tan\theta = \sin\theta/\cos\theta$, and rearranging gives

$$\sigma - \tau\tan\theta = \sigma_x + \tau_{xy}\tan\theta \qquad (11.1)$$

Summing the forces in the vertical direction gives

$$\Sigma F_y = 0 = -\sigma_y(A\sin\theta) - \tau_{xy}(A\cos\theta) + \tau A(\cos\theta) + \sigma A(\sin\theta)$$

After manipulating as above we get

$$\sigma\tan\theta + \tau = \sigma_y \tan\theta + \tau_{xy} \tag{11.2}$$

Equations 11.1 and 11.2 are two simultaneous equations in the unknowns σ and τ, in terms of the knowns, σ_x, σ_y, τ_{xy}, and θ. The two equations can be solved explicitly for σ and τ. Thus solving and then using numerous trigonometric identities, primarily those for double angles, and after considerable algebraic manipulation, we arrive at the following important results:

$$\sigma = \frac{\sigma_x + \sigma_y}{2} + \frac{\sigma_x - \sigma_y}{2}\cos 2\theta + \tau_{xy}\sin 2\theta \tag{11.3}$$

$$\tau = -\frac{\sigma_x - \sigma_y}{2}\sin 2\theta + \tau_{xy}\cos 2\theta \tag{11.4}$$

These equations allow us to find the normal and shear stresses acting on any plane, defined by the angle θ, given the original state of stresses σ_x, σ_y, and τ_{xy}. However, in using them we must be careful to assign the proper sign to the input variables. Particular care must be taken with τ_{xy} and θ.

EXAMPLE PROBLEM 11.2

Find the normal and shear stresses on a plane making a 20° angle with the horizontal for an element loaded as shown in Fig. 11.3a using Eqs. 11.3 and 11.4.

Solution

From Fig. 11.3a, based on the sign system of Fig. 11.4:

$$\sigma_x = +400 \text{ psi}$$

$$\sigma_y = +200 \text{ psi}$$

$$\tau_{xy} = -100 \text{ psi}$$

The desired plane makes an angle of 20° from the horizontal. From 11.4, the angle θ in Eqs. 11.3 and 11.4 is referenced to the vertical plane, with the counterclockwise being positive. Thus the desired plane is 110° counterclockwise from the vertical

$$\theta = 110°$$

We could also describe it as 70° clockwise or $-70°$.

278   PLANE STRESS ANALYSIS

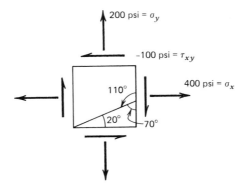

FIG. 11.3a

From Eq. 11.3

$$\sigma = \frac{\sigma_x + \sigma_y}{2} + \frac{\sigma_x - \sigma_y}{2}\cos 2\theta + \tau_{xy}\sin 2\theta$$

$$= \frac{400 + 200}{2} + \frac{(400-200)}{2}\cos 220° + (-100\sin 220°)$$

$$= 300 + 100(-0.766) - 100(-0.643)$$

$\sigma = 288$ psi

And Eq. 11.4

$$\tau = -\frac{\sigma_x - \sigma_y}{2}\sin 2\theta + \tau_{xy}\cos 2\theta$$

$$= -\frac{(400-200)}{2}\sin 220° + (-100\cos 220°)$$

$$= -100(-0.643) - 100(-0.766)$$

$\tau = +141$ psi

These answers, of course, agree with those of Example Problem 11.1 but are obtained with much less effort. The effort was expended in developing Eqs. 11.3 and 11.4, used here. We also note the positive sign associated with τ. This sign has significance concerning the sense of τ. We will not explore it in this text because it is fairly tacky and rarely matters.

11.3 PRINCIPAL STRESSES

Equations 11.3 and 11.4 give the stresses as a function of the angle θ. To predict failure in a mechanical structure requires that we be able to determine the maximum stress that exists anywhere in the structure. We could evaluate Eqs. 11.3 and 11.4 every few degrees, seeking the maximum values of σ and τ. However, the calculus gives us a more direct approach. In the calculus this is known as a maximum-minimum problem. It is based on viewing Eq. 11.3 as

$$\sigma = f(\theta)$$

We could plot this equation on a graph where we label the axes σ and θ (instead of x and y). On this graph the slope of the curve at any point would be

$$\text{slope} = \frac{d\sigma}{d\theta}$$

that is, slope is the geometric meaning of the derivative. Now think about the slope of the curve when the curve reaches a maximum value. Will not the slope be zero?

To apply this to our problem we will take the derivative of σ with respect to θ and set it equal to zero. (σ_x, σ_y, and τ_{xy} are constants in this equation.)

$$\frac{d\sigma}{d\theta} = \frac{d}{d\theta}\left[\frac{\sigma_x + \sigma_y}{2} + \frac{\sigma_x - \sigma_y}{2}\cos 2\theta + \tau_{xy}\sin 2\theta\right]$$

$$= 0 + \frac{\sigma_x - \sigma_y}{2}(-\sin 2\theta)2 + \tau_{xy}(\cos 2\theta)2$$

$$\frac{d\sigma}{d\theta} = -(\sigma_x - \sigma_y)\sin 2\theta + 2\tau_{xy}\cos 2\theta = 0$$

The above equation contains the single unknown θ. If we solve for θ we will know where (what value of θ) the slope is zero, which is where the maximum stress occurs. We must add at this point that minimum values of σ also occur where the slope is zero. So we are getting values of θ corresponding to a maximum or minimum value. At this point we don't know which. We proceed to solve for θ by rearranging

$$(\sigma_x - \sigma_y)\sin 2\theta = 2\tau_{xy}\cos 2\theta$$

$$\frac{\sin 2\theta}{\cos 2\theta} = \frac{2\tau_{xy}}{\sigma_x - \sigma_y}$$

PLANE STRESS ANALYSIS

or

$$\tan 2\theta = \frac{2\tau_{xy}}{\sigma_x - \sigma_y} \tag{11.5a}$$

or

$$\tan 2\theta = \frac{\tau_{xy}}{\left(\dfrac{\sigma_x - \sigma_y}{2}\right)} \tag{11.5b}$$

If the values of τ_{xy}, σ_x, and σ_y are substituted into Eq. 11.5a, two values of 2θ, which are 180° apart, may be obtained. When these values of 2θ are substituted into Eq. 11.3 one will give the maximum value of σ and the other will give the minimum. The resulting two values of σ are known as the principal stresses, and their directions are the principal directions. The principal directions θ will be 90° apart (2θ are 180° apart). The principal stresses are designated σ_1 and σ_2 for the maximum and minimum, respectively.

An equation yielding the principal stresses directly may be obtained by substituting 2θ from Eq. 11.5 into Eq. 11.3. In Fig. 11.5 the angle 2θ is evaluated in terms of a triangle satisfying Eq. 11.5b. The hypotenuse of this triangle may be found by the Pythagorean theorem to be

$$\left[\left(\frac{\sigma_x - \sigma_y}{2}\right)^2 + (\tau_{xy})^2\right]^{1/2}$$

Thus we define

$$\sin 2\theta = \frac{\tau_{xy}}{\left[\left(\dfrac{\sigma_x - \sigma_y}{2}\right)^2 + (\tau_{xy})^2\right]^{1/2}}$$

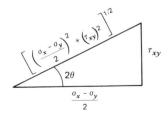

FIG. 11.5 Relation between angles of maximum stress and given variables.

PRINCIPAL STRESSES

and

$$\cos 2\theta = \frac{\left(\frac{\sigma_x - \sigma_y}{2}\right)}{\left[\left(\frac{\sigma_x - \sigma_y}{2}\right)^2 + (\tau_{xy})^2\right]^{1/2}}$$

Then evaluating Eq. 11.3 gives

$$\sigma = \frac{\sigma_x + \sigma_y}{2} + \frac{\sigma_x - \sigma_y}{2} \cos 2\theta + \tau_{xy} \sin 2\theta$$

$$= \frac{\sigma_x + \sigma_y}{2} + \frac{\sigma_x - \sigma_y}{2} \cdot \frac{\left(\frac{\sigma_x - \sigma_y}{2}\right)}{\left[\left(\frac{\sigma_x - \sigma_y}{2}\right)^2 + (\tau_{xy})^2\right]^{1/2}} + \tau_{xy} \frac{\tau_{xy}}{\left[\left(\frac{\sigma_x - \sigma_y}{2}\right)^2 + (\tau_{xy})^2\right]^{1/2}}$$

$$= \frac{\sigma_x + \sigma_y}{2} + \frac{\left(\frac{\sigma_x - \sigma_y}{2}\right)^2 + \tau_{xy}^2}{\left[\left(\frac{\sigma_x - \sigma_y}{2}\right)^2 + (\tau_{xy})^2\right]^{1/2}}$$

$$\sigma_{1,2} = \frac{\sigma_x + \sigma_y}{2} + \sqrt{\left(\frac{\sigma_x - \sigma_y}{2}\right)^2 + \tau_{xy}^2} \qquad (11.6)$$

There are two roots to the radical. One produces the maximum stress, the other the minimum. Equation 11.6 may be used to directly calculate the principal stresses when σ_x, σ_y, and τ_{xy} are given.

What shear stresses accompany these principal stresses? That may be determined by substituting 2θ from Eq. 11.5 into Eq. 11.4. Thus

$$\tau = -\frac{\sigma_x - \sigma_y}{2} \sin 2\theta + \tau_{xy} \cos 2\theta$$

$$= -\frac{\sigma_x - \sigma_y}{2} \frac{\tau_{xy}}{\left[\left(\frac{\sigma_x - \sigma_y}{2}\right)^2 + \tau_{xy}^2\right]^{1/2}} + \tau_{xy} \frac{\left(\frac{\sigma_x - \sigma_y}{2}\right)}{\left[\left(\frac{\sigma_x - \sigma_y}{2}\right)^2 + \tau_{xy}^2\right]^{1/2}}$$

$$\tau = 0 \qquad (11.7)$$

This very important conclusion says that on the planes of principal stresses (maximum and minimum normal stresses) the shear stress is zero.

EXAMPLE PROBLEM 11.3

For the stresses on the element in Fig. 11.3a, find the principal planes and the principal stresses.

Solution

From Fig. 11.3a

$$\sigma_x = +400 \text{ psi}$$

$$\sigma_y = +200 \text{ psi}$$

$$\tau_{xy} = -100 \text{ psi}$$

Equation 11.5:

$$\tan 2\theta = \frac{\tau_{xy}}{\left(\frac{\sigma_x - \sigma_y}{2}\right)} = \frac{-100}{\left(\frac{400-200}{2}\right)}$$

$$\tan 2\theta = -1.00$$

$$2\theta = -45°, \ 135°$$

$$\boldsymbol{\theta = -22.5°, \ 67.5°}$$

$$\sigma = \frac{\sigma_x + \sigma_y}{2} + \frac{\sigma_x - \sigma_y}{2}\cos 2\theta + \tau_{xy}\sin 2\theta$$

$$= \frac{400+200}{2} + \frac{400-200}{2}\cos(-45°) + (-100)\sin(-45°)$$

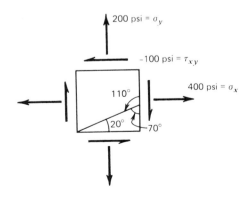

FIG. 11.3a

$$= 300 + 100(+0.707) - 100(-0.707)$$

$\sigma_1 = 441$ psi

$$\sigma_2 = 300 + 100 \cos 135° - 100 \sin 135°$$

$\sigma_2 = 159$ psi

Alternately from Equation 11.6

$$\sigma_{1,2} = \frac{\sigma_x + \sigma_y}{2} \mp \sqrt{\left(\frac{\sigma_x - \sigma_y}{2}\right)^2 + \tau_{xy}^2}$$

$$= \frac{400 + 200}{2} \mp \sqrt{\left(\frac{400 - 200}{2}\right)^2 + (-100)^2}$$

$$\sigma_{1,2} = 300 \mp 141 = \textbf{441 psi, 159 psi}$$

The sketch (Fig. 11.6) shows the original state of stresses and the same state of stress defined in terms of the principal stresses. Note that the principal stresses are 90° apart and the accompanying shear stress is zero.

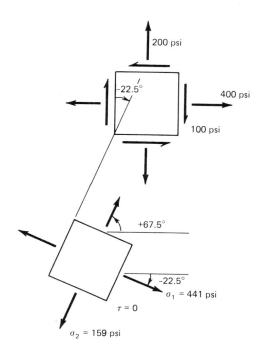

FIG. 11.6

11.4 MAXIMUM SHEAR STRESS

In a manner similar to that used in finding the principal stresses, the maximum shear stress may be found, setting

$$\frac{d\tau}{d\theta} = 0$$

When solved for 2θ this yields

$$\tan 2\theta = -\frac{\left(\dfrac{\sigma_x - \sigma_y}{2}\right)}{\tau_{xy}} \tag{11.8}$$

This equation may be evaluated for 2θ and the value substituted into Eq. 11.4 to determine the maximum value of the shear stress. Or Eq. 11.4 may be evaluated in general as in the preceding article to give

$$\tau_{max} = \sqrt{\left(\frac{\sigma_x - \sigma_y}{2}\right)^2 + \tau_{xy}^2} \tag{11.9}$$

Notice that this equation is the radical from Eq. 11.6. If the normal stress is evaluated by the value of 2θ, from Eq. 11.8 we get

$$\sigma_{\tau_{max}} = \frac{\sigma_x + \sigma_y}{2} \tag{11.10}$$

which is the remaining portion of Eq. 11.6. Thus we find that the maximum shear stress (11.9) is accompanied by a normal stress (11.10), while the maximum normal stresses (principal) are associated with a zero shear stress. The normal stress accompanying the maximum shear stress is equal on all four surfaces.

Comparing Eq. 11.8 with Eq. 11.5 we notice that the right-hand side of one is the negative reciprocal of the other. Noting the trigonometric identity

$$\tan(90 + \phi) = -\frac{1}{\tan \phi}$$

We conclude that

$$2\theta_\tau + 90 = 2\theta_\sigma$$

and

$$\theta_\tau + 45 = \theta_\sigma$$

Hence the plane of maximum shear stress is 45° from the plane of principal stress.

EXAMPLE PROBLEM 11.4

Find the plane of maximum shear stress and the resulting maximum shear stress and associated normal stress for the stress element in Fig. 11.3a.

Solution

Again from Fig. 11.3a

$$\sigma_x = 400 \text{ psi}$$

$$\sigma_y = 200 \text{ psi}$$

$$\tau_{xy} = -100 \text{ psi}$$

Equation 11.8:

$$\tan 2\theta = -\frac{\left(\dfrac{\sigma_x - \sigma_y}{2}\right)}{\tau_{xy}} = -\frac{\left(\dfrac{400-200}{2}\right)}{-100}$$

$$\tan 2\theta = +1.00$$

$$2\theta = 45°, 225°$$

$$\boldsymbol{\theta = 22.5°, 112°}$$

Note that these planes are 90° apart and 45° from the principal planes found in Example Problem 11.3.

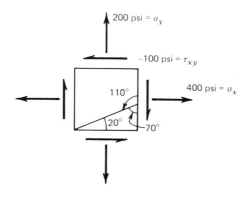

FIG. 11.3a

286 PLANE STRESS ANALYSIS

Equation 11.4:

$$\tau_{max} = -\frac{\sigma_x - \sigma_y}{2}\sin 2\theta + \tau_{xy}\cos 2\theta$$

$$= -\left(\frac{400-200}{2}\right)\sin 45° + (-100)\cos 45°$$

$$\tau_{max} = -100(0.707) - 100(0.707) = \mathbf{141 \text{ psi}}$$

Note that the magnitude of the result is the same whether 45° or 225° is used.

To calculate the normal stress associated with the maximum shear stress use Eq. 11.3:

$$\sigma_{22.5} = \frac{\sigma_x + \sigma_y}{2} + \frac{\sigma_x - \sigma_y}{2}\cos 2\theta + \tau_{xy}\sin 2\theta$$

$$\sigma_{22.5} = \frac{400+200}{2} + \frac{400-200}{2}\cos 45° + (-100)\sin 45°$$

$$= 300 + 100(0.707) - 100(0.707)$$

$$\boldsymbol{\sigma_{22.5} = 300 \text{ psi}}$$

$$\sigma_{112} = \frac{400+200}{2} + \frac{400-200}{2}\cos 225° + (-100)\sin 225°$$

$$= 300 + 100(-0.707) - 100(-0.707)$$

$$\boldsymbol{\sigma_{112} = 300 \text{ psi}}$$

The normal stresses on the two planes are equal.

Alternately, Eqs. 11.9 and 11.10 could have been used:

$$\tau_{max} = \sqrt{\left(\frac{\sigma_x - \sigma_y}{2}\right)^2 + \tau_{xy}^2}$$

$$= \sqrt{\left(\frac{400-200}{2}\right)^2 + (-100)^2} = \sqrt{100^2 + 100^2}$$

$$\tau_{max} = \mathbf{141 \text{ psi}}$$

$$\sigma_{22,112} = \frac{\sigma_x + \sigma_y}{2} = \frac{400+200}{2} = \mathbf{300 \text{ psi}}$$

Figure 11.7 shows the original stress element, the principal stresses, and the

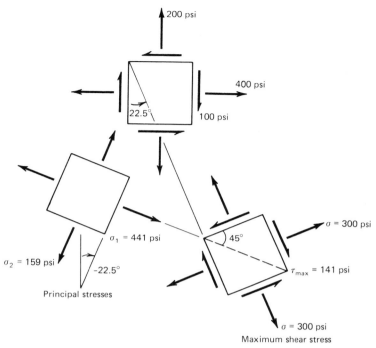

FIG. 11.7 Equivalent states of stress.

maximum shear stresses. These are three different descriptions of the same state of stress.

Note the maximum shear stress is 45° from the principal stresses. Also, notice that the shear stresses accompanying the principal stresses are zero. On the other hand, the normal stresses accompanying the maximum shear stress are not zero in general. However, they are equal on all four faces.

11.5 MOHR'S CIRCLE

The previous ten equations represent the normal and shear stresses on any plane and the locations and values of maximum shear and normal stresses. These equations are accurate, effective, and complete. However, they represent a large amount of detail and are not readily retained unless used frequently. Otto Mohr, the German engineer, discovered that the variables in these equations may be represented as parts on a circle. His method gives a graphical interpretation of these data, but more (or Mohr?) than that they give us the capability to generate the information contained in Eqs. 11.1 through 11.10 without memorizing them.

Recalling Eqs. 11.3 and 11.4, we have

$$\sigma = \frac{\sigma_x + \sigma_y}{2} + \frac{\sigma_x - \sigma_y}{2}\cos 2\theta + \tau_{xy}\sin 2\theta \qquad (11.3)$$

and

$$\tau = -\frac{\sigma_x - \sigma_y}{2}\sin 2\theta + \tau_{xy}\cos 2\theta \qquad (11.4)$$

For a given state of stress, σ_x, σ_y, and τ_{xy} are constants. The variables σ and τ are in terms of a third parameter θ. The parameter θ may be eliminated by combining the two equations. The resulting equation will be that of a circle, known as Mohr's circle. This circle will be extremely valuable in evaluating the stresses acting on an element.

The circle is obtained as follows. The vertical axis is for shear stress and the horizontal axis is for normal stresses, positive being tensile and negative being compressive. Positive signs for the stress element are as shown in Fig. 11.8a. This is exactly the same sign system used in Fig. 11.4 and for Eqs. 11.1 through 11.10.

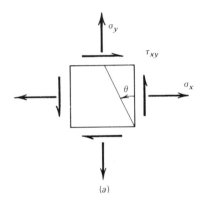

(a)

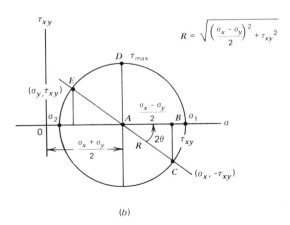

(b)

FIG. 11.8a–b Stress element and Mohr's circle.

Unfortunately, the sign system is not universal. When referring to other texts, check the sign system first. Points on the circle represent the stresses, normal and shear, on a plane. The first point plotted is $(\sigma_x, -\tau_{xy})$, which is point C in Fig. 11.8b. The second is (σ_y, τ_{xy}), point E in Fig. 11.8b. These two points define the circle whose center is on the horizontal or σ axis.

The circle may be analyzed graphically or semigraphically. Graphically the circle is obtained by drawing a straight line from E to C. The center of the circle is where this line intersects the σ axis, point A in Fig. 11.8b. The radius is the distance AC or AE. The circle may easily be drawn now. Significant values, which are discussed below, can be scaled rather than calculated if a purely graphical method is used. By semigraphically we mean the circle is sketched but its actual dimensions are calculated rather than measured, based on the properties of a circle and right triangles. Since we are already familiar with the properties of circles and right triangles, the method becomes very easy to use.

We observe that the center of the circle, A in Fig. 11.8b, must be at $(\sigma_x + \sigma_y)/2$, that is, half-way between points E and C. The distance AB is $(\sigma_x - \sigma_y)/2$ and BC is τ_{xy}. Hence by noting ABC to be a right triangle, we may calculate the hypotenuse, which is the radius of the circle:

$$R = \sqrt{\left(\frac{\sigma_x - \sigma_y}{2}\right)^2 + \tau_{xy}^2}$$

The maximum and minimum values of the normal stress, the principal stresses, are identified as σ_1 and σ_2 and are easily obtained. In fact, we may generate Eq. 11.6, noting σ_1 is OA plus the radius:

$$\sigma_1 = OA + R$$

$$\sigma_1 = \frac{\sigma_x + \sigma_y}{2} + \sqrt{\left(\frac{\sigma_x - \sigma_y}{2}\right)^2 + \tau_{xy}^2} \quad (11.6)$$

The difference in these two terms gives σ_2. Obviously, the shear stress accompanying the principal stresses is zero, since σ_1 and σ_2 are on the horizontal axis.

Point D indicates the maximum shear stress, which is simply the radius of the circle:

$$\tau_{max} = \sqrt{\left(\frac{\sigma_x - \sigma_y}{2}\right)^2 + \tau_{xy}^2}$$

The normal stress accompanying τ_{max} is

$$\sigma_{\tau_{max}} = \frac{\sigma_x + \sigma_y}{2}$$

There is a double-angle relationship between Mohr's circle and real space. However, both angles are positive in the counterclockwise direction. Angles may be calculated using trigonometric relations and the Mohr's circle. The angle θ,

290 PLANE STRESS ANALYSIS

Fig. 11.8a, in real space corresponds to 2θ in Mohr's circle space, Fig. 11.8b. However, the directions of rotation are the same, counterclockwise, as we go from C to σ_1, for example. Point C represents the stresses on the right-hand face, σ_x and $-\tau_{xy}$. Point E represents the stresses on the top face, σ_y and τ_{xy}. Points C and E are 180° apart on Mohr's circle, while the faces they represent are 90° apart in real space, again illustrating the double-angle relationship.

The circle of Fig. 11.8b graphically illustrates several items we have discussed previously. Recall that each point on the circle represents the stresses, σ and τ, on some plane in space. An adjacent plane at 90°, the other side of a square element, will be 180° away on the circle because of the double-angle, 2θ relation. Thus points E and C are 180° apart on the circle, although they represent planes 90° apart in Fig. 11.8a. Any pair of points 180° apart totally describes the stress on an element. Any other pair of points 180° apart totally describes the same state of stress, but in a different coordinate system. The circle consists of all of the possible descriptions. There is no equivalent description that is not represented by points on the circle.

Given the above, it is clear that the greatest σ coordinate for any point on the circle is σ_1. Thus this is the maximum normal or principal stress, and it is accompanied by zero shear stress. It is equally clear that σ_2 in the minimum normal (also principal) stress and that it, too, is accompanied by zero shear stress. We further observe that these two points are 180° apart on the circle, so they are 90° apart on perpendicular planes in real space.

The shear stress, $\tau_{\max}$ is 90° (45° in real space) from σ_1 or σ_2. The minimum shear stress is 180° away (90° in real space) and of equal magnitude. Both the maximum and the minimum shear stresses are accompanied by the same normal stress. Its value is the same as the center of the circle, which is the average normal stress.

It may be helpful to reread the last two paragraphs after you have worked several problems.

EXAMPLE PROBLEM 11.5

For an element loaded as shown in Fig. 11.3a, find the normal and shear stresses on a plane making an angle of 20° with the horizontal using Mohr's circle graphically. Also, find the principal stresses and the maximum shear stress.

Solution

Once more from Fig. 11.3a we have $\sigma_x = +400$ psi, $\sigma_y = +200$ psi, and $\tau_{xy} = -100$ psi. The stresses on the right-hand face are plotted in Fig. 11.9a as (400, 100), since we plot $(\sigma_x, -\tau_{xy})$ $[-(-100) = +100]$, they are labeled point A. The stresses for the top face are plotted as (200, −100) or (σ_y, τ_{xy}) and labeled B. The second point will always use the negative of the sign used for shear in the first point. A line AB connects these two points. As constructed, it intersects the σ axis at the center of the circle C. It breaks AB into two lengths, AC and BC, which are equal to the radius. Placing the

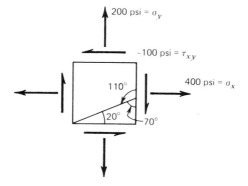

FIG. 11.3a

compass at C with a radius AC the circle is drawn. The circle is now complete, fully describing the given state of stress at any orientation. It only remains to be interpreted.

The stress on a plane at 20° with the horizontal is desired. This plane may also be described as $-70°$ from the right-hand face (Fig. 11.9b). In real space we rotate 70° clockwise (negative) to get the desired plane. For Mohr's circle we double the real space angle but maintain the direction. Thus from A, which represents the right-hand face of the element, we move around the circle clockwise the double angle, 140°.

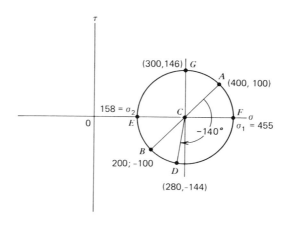

FIG. 11.9a

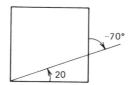

FIG. 11.9b

Drawing this ray gives point D which has the coordinates $(280, 144)$. Thus

$$\sigma_{20°} = 280 \text{ psi}$$

$$\tau_{20°} = 144 \text{ psi}$$

A larger scale would enhance the accuracy of these values. Points E and F give the principal stresses and G the maximum shear stress. Thus

$$\sigma_1 = 455 \text{ psi}$$

$$\sigma_2 = 158 \text{ psi}$$

$$\tau_{max} = 146 \text{ psi}$$

$$\sigma_{\tau_{max}} = 300 \text{ psi}$$

EXAMPLE PROBLEM 11.6

For the stress element of Fig. 11.3a calculate the principal stresses, their location, the maximum shear stresses, and their location by using Mohr's circle.

Solution

Points A and B are plotted (Fig. 11.10) as in Example Problem 11.5. Point C is observed to be halfway between the two, so

$$OC = \frac{400 + 200}{2} = 300$$

CD is one-half the difference between A and B

$$CD = \frac{400 - 200}{2} = 100$$

AD is τ_{xy}

$$AD = 100$$

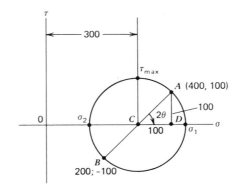

FIG. 11.10

The radius of the circle is the hypotenuse of the triangle ADC

$$R = \sqrt{100^2 + 100^2} = 141$$

The angle 2θ is formed by

$$\tan 2\theta = \frac{100}{100} = 1.00$$

$$2\theta = 45°$$

$$\theta = 22.5° \quad \text{clockwise to } \sigma_1$$

Here we are handling the sign of θ by observing the circle.
Thus

τ_{max} = radius = **141 psi @ 22.5°** ↻ from right-hand surface

$\sigma_{\tau_{max}}$ = **300 psi**

$\sigma_1 = 300 + 141 =$ **441 psi @ 22.5°** ↻ from right-hand surface

$\sigma_2 = 300 - 141 =$ **159 psi**

The sketches of these three descriptions of the same state of stress are given in Fig. 11.11. This figure should be studied and related to the circle above.

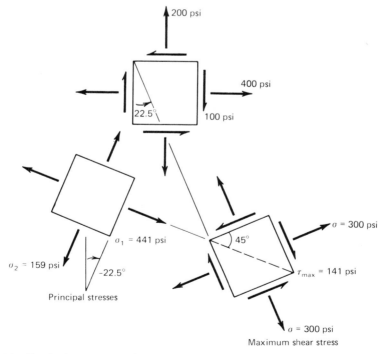

FIG. 11.11 Equivalent states of stress.

11.6 APPLICATIONS OF MOHR'S CIRCLE TO FAILURE ANALYSIS

Mohr's circle (or the other methods we have examined) is very useful for finding the normal or shear stress on a particular plane from combined loading. It can be used to find principal stresses and the maximum shear stresses for complex loading situations. Beyond these applications it is also useful in understanding failure, even in very simple loading situations. Consideration of a circular shaft loaded in pure torsion will illustrate this application.

A shaft under pure torsional loading is shown in Fig. 11.12a. The stress on

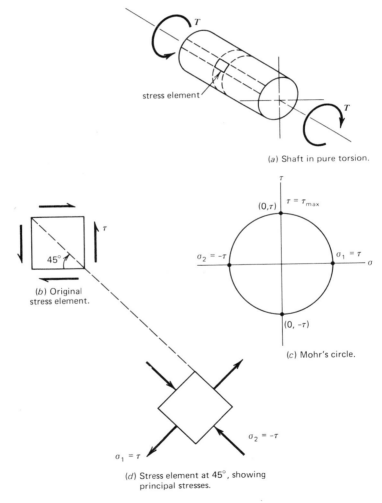

(a) Shaft in pure torsion.

(b) Original stress element.

(c) Mohr's circle.

(d) Stress element at 45°, showing principal stresses.

FIG. 11.12a–d Maximum shear and principal stresses for pure torsion.

the surface element shown is as indicated in Fig. 11.12b. Its magnitude, found by Eq. 5.7, is

$$\tau = \frac{Tc}{J} \tag{5.7}$$

The stress element is in pure shear, that is, there are no normal stresses acting on it in the orientation shown. Mohr's circle is drawn in Fig. 11.12c using the points $(0, \tau)$ and $(0, -\tau)$. This circle has its center at the origin and its radius is τ. The maximum shear stress is the given stress τ. The principle stresses are also equal to τ in magnitude, σ_1 being $+\tau$ and σ_2 being $-\tau$. These stresses occur at 90° to either side of the given shear stresses on the Mohr's circle. Consequently, they occur 45° from the initial planes in real space, as indicated in Fig. 11.12d. Finally Fig. 11.13 shows the element in both the maximum shear stress and the principal stress orientations on the shaft.

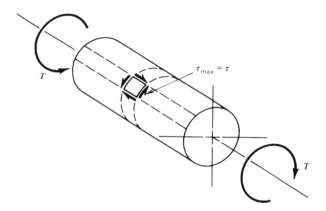

FIG. 11.13a

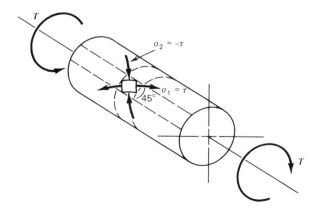

FIG. 11.13b

296 PLANE STRESS ANALYSIS

If the shaft is made of a ductile material it will initially fail by yielding, which occurs because of the maximum shear stress. Thus a ductile shaft, such as mild steel, will fail in the direction of maximum shear stress or in a plane perpendicular to the axis of the shaft, as shown in Fig. 11.14. You can demonstrate this by taking a paper clip and twisting it to failure. (It may take 30 or more turns.) The failure plane will be as shown in Fig. 11.14.

If the shaft is made of a brittle material, it will be weakest in tension. The positive tensile stress σ_1 will initiate failure on a plane at 45° to the axis of the shaft, as shown in Fig. 11.15. You can demonstrate this by twisting a piece of chalk to failure. Be careful not to bend it as you twist it. Note that Fig. 11.15 shows only the plane of failure and not the actual appearance of the failure.

Finally, if the shaft is hollow with a sufficiently thin wall and inadequate stiffness, failure will be by buckling due to the maximum compressive stress σ_2. (Buckling is treated in the next chapter.) This is initiated on a plane that also makes an angle of 45° to the axis of the shaft, but this plane is perpendicular to

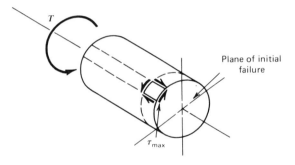

FIG. 11.14 Failure of a ductile shaft in pure torsion.

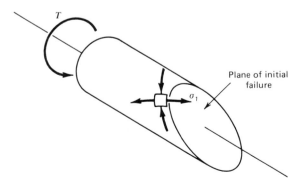

FIG. 11.15 Plane of failure of a brittle shaft in pure torsion.

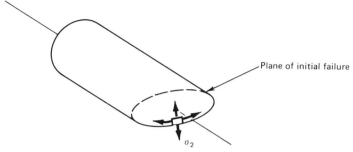

FIG. 11.16 Failure of a hollow tube in pure torsion.

the plane of tensile failure. This plane is shown in Fig. 11.16. This failure can be observed by carefully twisting a soda straw or empty paper tube.

Most load situations are more complex than the pure torsion of this example. However, in many cases, the loading and resultant stresses are analyzed and related to the properties of the material in the manner illustrated here to predict failure.

11.7 SUMMARY

Stresses on an arbitrary plane may be found from the equations

$$\sigma = \frac{\sigma_x + \sigma_y}{2} + \frac{\sigma_x - \sigma_y}{2}\cos 2\theta + \tau_{xy}\sin 2\theta$$

$$\tau = -\frac{\sigma_x - \sigma_y}{2}\sin 2\theta + \tau_{xy}\cos 2\theta$$

Refer to Fig. 11.14 for the sign system. Principal stresses may be found from

$$\sigma_{1,2} = \frac{\sigma_x + \sigma_y}{2} \pm \sqrt{\left(\frac{\sigma_x - \sigma_y}{2}\right)^2 + \tau_{xy}^2}$$

These occur on the principal planes at

$$\tan 2\theta = \frac{\tau_{xy}}{\left(\frac{\sigma_x - \sigma_y}{2}\right)}$$

where σ_1 and σ_2 are 90° apart. The accompanying shear stress is 0. The maximum shear stress may be found from

298 PLANE STRESS ANALYSIS

$$\tau_{max} = \sqrt{\left(\frac{\sigma_x - \sigma_y}{2}\right)^2 + \tau_{xy}^2}$$

occurring at

$$\tan 2\theta = -\frac{\left(\frac{\sigma_x - \sigma_y}{2}\right)}{\tau_{xy}}$$

which is 45° from the principal stresses. The accompanying normal stresses are

$$\sigma = \frac{\sigma_x + \sigma_y}{2}$$

Mohr's circle contains all the above information in a simple graph. It is obtained as follows. Plot $(\sigma_x, -\tau_{xy})$ and (σ_y, τ_{xy}) on a τ(vertical) and σ(horizontal) space. Draw a line between these two points. It will intersect the horizontal axis (σ) at the center of the circle. The distance from the center to either point is the radius. Constructing the circle permits the determination of all important parameters by direct measurement or calculation. There is a double-angle relation between Mohr's circle and real space.

PROBLEMS

11.1 For the stress element shown in Fig. 11.17 find the normal and shear stresses by summation of forces on a plane

(a) 15° counterclockwise from the horizontal.
(b) 30° counterclockwise from the horizontal.
(c) 45° counterclockwise from the horizontal.

11.2 For the stress element shown in Fig. 11.18 find the normal and shear stresses by summation of forces on a plane

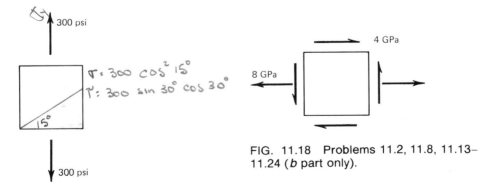

FIG. 11.17 Problems 11.1, 11.7, 11.13–11.24 (*a* part only).

FIG. 11.18 Problems 11.2, 11.8, 11.13–11.24 (*b* part only).

(a) 15° counterclockwise from the horizontal.
(b) 30° counterclockwise from the horizontal.
(c) 45° counterclockwise from the horizontal.

11.3 For the stress element shown in Fig. 11.19 find the normal and shear stresses by summation of forces on a plane
(a) 15° counterclockwise from the horizontal.
(b) 30° counterclockwise from the horizontal.
(c) 45° counterclockwise from the horizontal.

11.4 For the stress element shown in Fig. 11.20 find the normal and shear stresses by summation of forces on a plane
(a) 20° counterclockwise from the horizontal.
(b) 40° counterclockwise from the horizontal.
(c) 60° counterclockwise from the horizontal.

11.5 For the stress element shown in Fig. 11.21 find the normal and shear stresses by summation of forces on a plane
(a) 20° counterclockwise from the horizontal.
(b) 40° counterclockwise from the horizontal.
(c) 60° counterclockwise from the horizontal.

11.6 For the stress element shown in Fig. 11.22 find the normal and shear stresses by summation of forces on a plane
(a) 20° counterclockwise from the horizontal.
(b) 40° counterclockwise from the horizontal.
(c) 60° counterclockwise from the horizontal.

11.7 For the stress element shown in Fig. 11.17 find the normal and shear stresses, using Eqs. 11.3 and 11.4, on a plane
(a) 15° counterclockwise from the horizontal.

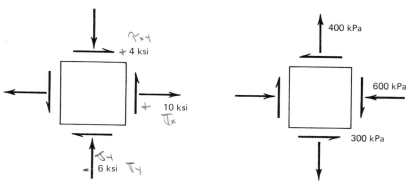

FIG. 11.19 Problems 11.3, 11.9, 11.13–11.24 (c part only).

FIG. 11.20 Problems 11.4, 11.10, 11.13–11.24 (d part only).

300 PLANE STRESS ANALYSIS

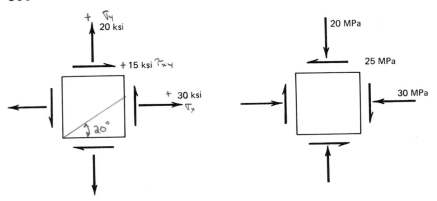

FIG. 11.21 Problems 11.5, 11.11, 11.13–11.24 (*e* part only).

FIG. 11.22 Problems 11.6, 11.12, 11.13–11.24 (*f* part only).

    (b) 30° counterclockwise from the horizontal.
    (c) 15° counterclockwise from the vertical.
    (d) 30° counterclockwise from the vertical.
    (e) 15° clockwise from the vertical.

11.8 For the stress element shown in Fig. 11.18 find the normal and shear stress, using Eqs. 11.3 and 11.4, on a plane
    (a) 15° counterclockwise from the horizontal.
    (b) 30° counterclockwise from the horizontal.
    (c) 15° counterclockwise from the vertical.
    (d) 30° counterclockwise from the vertical.
    (e) 15° clockwise from the vertical.

11.9 For the stress element shown in Fig. 11.19 find the normal and shear stress, using Eqs. 11.3 and 11.4, on a plane
    (a) 15° counterclockwise from the horizontal.
    (b) 30° counterclockwise from the horizontal.
    (c) 15° counterclockwise from the vertical.
    (d) 30° counterclockwise from the vertical.
    (e) 15° clockwise from the vertical.

11.10 For the stress element shown in Fig. 11.20 find the normal and shear stress, using Eqs. 11.3 and 11.4, on a plane
    (a) 20° counterclockwise from the horizontal.
    (b) 40° counterclockwise from the horizontal.
    (c) 20° counterclockwise from the vertical.
    (d) 40° counterclockwise from the vertical.
    (e) 20° clockwise from the vertical.

PROBLEMS   301

11.11 For the stress element shown in Fig. 11.21 find the normal and shear stress, using Eqs. 11.3 and 11.4, on a plane
   (a) 20° counterclockwise from the horizontal.
   (b) 40° counterclockwise from the horizontal.
   (c) 20° counterclockwise from the vertical.
   (d) 40° counterclockwise from the vertical.
   (e) 20° clockwise from the vertical.

11.12 For the stress element shown in Fig. 11.22 find the normal and shear stress, using Eqs. 11.3 and 11.4, on a plane
   (a) 20° counterclockwise from the horizontal.
   (b) 40° counterclockwise from the horizontal.
   (c) 20° counterclockwise from the vertical.
   (d) 40° counterclockwise from the vertical.
   (e) 20° clockwise from the vertical.

11.13 Find the principal planes for the stress elements shown in the following figures from Equation 11.5:
   (a) Fig. 11.17   (d) Fig. 11.20
   (b) Fig. 11.18   (e) Fig. 11.21
   (c) Fig. 11.19   (f) Fig. 11.22

11.14 From the results of Problem 11.13 and Eq. 11.3 calculate the principal stresses and sketch the element showing them for
   (a) Fig. 11.17   (d) Fig. 11.20
   (b) Fig. 11.18   (e) Fig. 11.21
   (c) Fig. 11.19   (f) Fig. 11.22

11.15 Using Eq. 11.6 calculate the principal stresses for
   (a) Fig. 11.17   (d) Fig. 11.20
   (b) Fig. 11.18   (e) Fig. 11.21
   (c) Fig. 11.19   (f) Fig. 11.22

11.16 For the indicated figure calculate and plot the normal stresses every 15° from 0 to 180°. Calculate the principal stresses and their locations. Show these on the plot also.
   (a) Fig. 11.17   (d) Fig. 11.20
   (b) Fig. 11.18   (e) Fig. 11.21
   (c) Fig. 11.19   (f) Fig. 11.22

11.17 Find the planes of maximum shear stress for the stress element shown in
   (a) Fig. 11.17   (d) Fig. 11.20
   (b) Fig. 11.18   (e) Fig. 11.21
   (c) Fig. 11.19   (f) Fig. 11.22

302 PLANE STRESS ANALYSIS

11.18 From the results of Problem 11.17 and Eqs. 11.3 and 11.4, calculate the maximum shear stress and the normal stress accompanying it for the stress element shown in

(a) Fig. 11.17
(b) Fig. 11.18
(c) Fig. 11.19
(d) Fig. 11.20
(e) Fig. 11.21
(f) Fig. 11.22

11.19 Using Eqs. 11.9 and 11.10 find the maximum shear stress and the associated normal stress for

(a) Fig. 11.17
(b) Fig. 11.18
(c) Fig. 11.19
(d) Fig. 11.20
(e) Fig. 11.21
(f) Fig. 11.22

11.20 Calculate and plot the shear stress every 15° from 0 to 180° for the following stress elements. Calculate the maximum shear stress and its location. Show this on the plot also.

(a) Fig. 11.17
(b) Fig. 11.18
(c) Fig. 11.19
(d) Fig. 11.20
(e) Fig. 11.21
(f) Fig. 11.22

11.21 Using Mohr's circle, *graphically* determine the principal stresses, the maximum shear stress and the accompanying normal stress for the stress element shown in these figures:

(a) Fig. 11.17
(b) Fig. 11.18
(c) Fig. 11.19
(d) Fig. 11.20
(e) Fig. 11.21
(f) Fig. 11.22

11.22 Using Mohr's circle *calculate* the maximum shear stress, the accompanying normal stress, the principal stresses, and all relevant angles and sketch the appropriate stress elements showing these stresses and their orientation for the stress element shown in these figures:

(a) Fig. 11.17
(b) Fig. 11.18
(c) Fig. 11.19
(d) Fig. 11.20
(e) Fig. 11.21
(f) Fig. 11.22

11.23 Using Mohr's circle, *graphically* find the stresses on the indicated plane for the indicated figure

(a) Fig. 11.17—30° counterclockwise from the horizontal.
(b) Fig. 11.18—30° clockwise from the horizontal.
(c) Fig. 11.19—30° counterclockwise from the vertical.
(d) Fig. 11.20—30° clockwise from the vertical.
(e) Fig. 11.21—45° clockwise from the horizontal.
(f) Fig. 11.22—45° clockwise from the vertical.

11.24 Do Problem 11.23 *calculating* (using Mohr's circle) the indicated values.

11.25 Find the maximum shear stress and the principal stresses at point A in Problem 10.4.

11.26 Find the maximum shear stress and the principal stresses at point B in Problem 10.4.

11.27 Find the maximum shear stress and the principal stresses at point A in Problem 10.7.

11.28 Find the maximum shear stress and the principal stresses at point B in Problem 10.7.

11.29 For each element in Problem 10.22 find the maximum shear stress and the principal stresses.

11.30 For each element in Problem 10.23 find the maximum shear stress and the principal stresses.

11.31 For each element in Problem 10.26 find the maximum shear stress and the principal stresses.

11.32 For each element in Problem 10.27 find the maximum shear stress and the principal stresses.

12

Columns

12.1 Buckling
12.2 Euler's equation
12.3 Using Euler's equation
12.4 End conditions for columns
12.5 Short columns and long columns
12.6 Idealized design of columns
12.7 Design formulas
12.8 Summary

12.1 BUCKLING

The buckling of structures is not something we observe everyday—fortunately. The reason we rarely see buckling failures is that when they do occur they are usually catastrophic. Catastrophic failures are unpleasant and limit one's opportunity to exercise his design talent. We, therefore, expend considerable effort assuring that buckling does not occur.

While buckling is not observed frequently in structures, we are nonetheless quite familiar with the phenomenon. The futility of pushing on a rope is documented in legend, fairy tale, and even a few ethnic jokes. We know that the cable that can carry tons in tension cannot carry pounds in compression. That is a form of buckling. We can place a fairly large load perpendicular to the page you are reading—in excess of a hundred pounds. The paper will not be crushed. We can cut this page from the book and pull on it with 10 or 20 lb (if we grip it carefully). However, if we push on the same paper, we find that it folds quickly at the slightest load. We can dramatically change the capacity of this sheet of paper to carry a compressive load if we roll it into a cylinder. If we do that it will now support its own weight (it would not before) and some small load besides.

These are examples of members that buckle under very low loads. For other members the load required for buckling can be more substantial. And we can have controlled buckling where catastrophic failure does not occur. Energy is simply stored elastically and then restored at the appropriate time. Large amounts of energy can be stored in this manner. A pole vaulter's pole is an example of such a device in which controlled buckling is a favorable event. Buckling can also be used to advantage by absorbing large amounts of energy as it occurs. The accordian-type buckling of the sheet metal and frame of an automobile takes much of the energy of a collision that otherwise would go to the vehicle occupant.

Thus we see that buckling can be elastic or plastic. It also can be catastrophic or advantageous. Buckling can occur in most types of loading, and it shows up in many forms. If we are addressing the compressive near axial loading of members, we frequently call the members columns. In this book we are only concerned with the buckling of columns.

12.2 EULER'S EQUATION (THAT'S "OILER," PADNUH!)

Consider a column as shown in Fig. 12.1a. This column is initially straight. It carries a compressive perfectly axial load P just sufficient to initiate buckling. This load is called the critical load. The end conditions are very important to this analysis. The column is pinned on both ends. We call it "pinned-pinned" (once again demonstrating our profound grasp of the language). By pinned we normally mean in statics that only rotations about the pin are allowed. We mean that here, plus, we will allow a small movement vertically at one of the supports. Otherwise we could not get the load to the member. Thus Fig. 12.1b represents the loaded and slightly buckled beam. As in statics, no motion to the left or right is allowed by these supports.

Establishing a coordinate system of positive y to the right and positive x downward, we move to any arbitrary distance x from the top support, pass a cutting plane, and draw a free body diagram of a portion of the column in equilibrium, as in Fig. 12.1c. In addition to the compressive load P, the cut

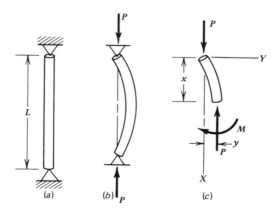

FIG. 12.1a–c Buckling of a simply supported column.

section will have a moment equal to the force P times y, the distance the section has buckled away from its original center line. If we sum the moments of force about the origin, we find that the moment at the section is equal to P times y:

$$M = Py \tag{a}$$

This is a vector equation, and as we have seen a vector sign system is not adequate to fully describe internal loads. In this case the bending moment is what we have called negative bending, in that it bends the beam "above" ($+y$) the x axis. Thus we write

$$M = -Py \tag{b}$$

where P will carry the magnitude of the critical load. From Chapter 8 we learned that the bending moment was proportional to the curvature. The curvature is approximated by the second derivative for small deflections, and we have

$$\frac{M}{EI} = \frac{d^2y}{dx^2} \tag{8.4}$$

whose sign system fits the problem at hand. Making this substitition yields

$$\frac{d^2y}{dx^2} = -\frac{P}{EI} y \tag{12.1}$$

We transform the equation to the form

$$\frac{d^2y}{dx^2} + \frac{P}{EI} y = 0 \tag{c}$$

Our analysis of the problem has led us to an equation describing it. All that remains is the solution of the equation. When studying beam deflections we were able to solve second-degree differential equations (heavy stuff) by integrating them twice. That won't work here because of the y on the right side of Eq. 12.1. (We cannot integrate $\int y\, dx$ unless we know how y is related to x, which is, of course, what we are after to start with.) The solution of equations like this is formally studied in courses on differential equations. Without getting into that, we will simply state that it has a solution of the form:

$$y = A \sin \sqrt{\frac{P}{EI}}\, x + B \cos \sqrt{\frac{P}{EI}}\, x \tag{d}$$

which is the general equation of a sine wave. Note that the deflected column in Fig. 12.1b resembles a portion of a sine wave. To show that equation (d) is *the* solution is quite an effort; however, it may be shown to be *a* solution quite easily. If we find its second derivative and substitute the appropriate values into equation (c), it will be satisfied. Any value or function for y that satisfies the equation is *a* solution to the equation. This exercise is left to the reader.

The proposed solution has two arbitrary constants, A and B. As we saw when we studied deflection, the values of arbitrary constants depend on the boundary conditions of the problem. They make the general solution fit the specific problem. In this case there is no deflection in the y direction at either end. Thus the boundary conditions are

$$y_0 = 0 \tag{e}$$

$$y_L = 0 \tag{f}$$

Imposing these conditions leads to

$$B = 0$$

and

$$A \sin \sqrt{\frac{P}{EI}} (L) = 0 \tag{g}$$

If $A = 0$, then equation (d) becomes

$$y = 0 \tag{h}$$

for all values of x. This is a trivial solution, since it means there is no buckling. The alternate way of satisfying equation (g) is for

$$\sin \sqrt{\frac{P}{EI}} (L) = 0 \tag{i}$$

The sine is equal to zero when the argument (angle) is $0, 180°, 360°, \ldots, n(180°)$ or in radian measure $0, \pi, 2\pi, \ldots, n\pi$. Zero degrees (or radians) is a trivial solution (requiring a zero load, so no problem). The first nontrivial solution is

$$\sqrt{\frac{P}{EI}} (L) = \pi \tag{j}$$

Solving this for P we have

$$P_{CR} = \frac{\pi^2 EI}{L^2} \tag{12.2}$$

The subscript CR denotes critical, and we call P_{CR} the critical load, that is, the load at which buckling will be initiated. Loads below this value will not produce buckling in perfectly straight, axially loaded columns. The equation is known as Euler's equation. It is pronounced "oilers."

12.3 USING EULER'S EQUATION

Now that we've got it (Euler's equation), let's see what it is good for. First of all, let's examine just what it says about the critical or limiting load for buckling. It says the critical load is proportional to the modulus of elasticity or stiffness of the material. That means a steel column can carry three times the load of an identical aluminum column. Note that it is *stiffness* and not strength that governs the load-carrying capability in buckling. By heat treating steel, its strength can be greatly increased while its stiffness is unchanged. Thus although heat treating can increase a member's resistance to stress, it cannot increase a column's resistance to buckling. Next we see that the critical load is proportional to the moment of inertia. The smaller the moment of inertia, the smaller the load required for buckling. If the end conditions are the same in every direction, then buckling will occur in the direction of minimum moment of inertia. Thus, a 2×4 buckles in the 2-in. direction and not the 4-in. one. If the cross section does not have symmetry, such as an angle iron, the buckling may not be parallel to a side but in some third direction. In Tables A4.6 and A4.7 the cross-sectional properties of angle iron are given. As shown in Fig. 12.2 a third axis, Z-Z, is given, about which the moment of inertia is less than it is about either the X-X or Y-Y axes. When buckling occurs in these members its direction will be perpendicular to the Z-Z axis, the axis of minimum moment of inertia. Accordingly, Tables A4.6 and A4.7 give these properties as well.

Lastly, the Euler formula indicates that the critical load varies indirectly with the length squared. Thus doubling the length of a column will reduce the load it

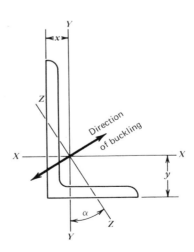

FIG. 12.2

can carry without buckling by a factor of four. Of course, changing the length has no effect on the direct stress.

Recall now that the radius of gyration is defined as the square root of the moment of inertia divided by the area. Or

$$I = k^2 A \tag{h}$$

Substituting into Eq. 12.2 gives

$$P_{CR} = \frac{\pi^2 E k^2 A}{L^2} \tag{12.3}$$

We divide both sides by A and move k^2 to the denominator, yielding

$$\frac{P_{CR}}{A} = \frac{\pi^2 E}{\left(\dfrac{L}{k}\right)^2} \tag{12.4}$$

Equations 12.2 through 12.4 are all referred to as Euler's equation. The form of Eq. 12.4 is probably the most common. It gives the load over the area on the left-hand side, which may be thought of as the critical stress; that is, it is the value of normal stress at which buckling will occur. The denominator on the right-hand side, L/k, is known as the slenderness ratio. This is just the length divided by the radius of gyration (with the two having the same units). It is a governing parameter for columns, as we shall see.

EXAMPLE PROBLEM 12.1

An 8-ft-long 2×4 (actual size) is to be used as a column, Fig. 12.3. Calculate the slenderness ratio that will govern buckling if the supports are uniform in every direction.

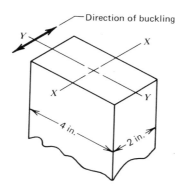

FIG. 12.3

Solution

We find the moment of inertia about the X and Y axes.

$$I_x = \frac{bh^3}{12} = \frac{2(4)^3}{12} = 10.7 \text{ in.}^4$$

$$I_y = \frac{bh^3}{12} = \frac{4(2)^3}{12} = 2.67 \text{ in.}^4$$

From Eq. 12.2 we see that the minimum moment of inertia will govern buckling. Thus we use

$$I_y = 2.67 \text{ in.}^4$$

Then

$$k_y = \sqrt{\frac{I_y}{A}} = \sqrt{\frac{2.67 \text{ in.}^4}{2 \text{ in. }(4 \text{ in.})}} = 0.577 \text{ in.}$$

Note that a minimum radius of gyration corresponds to a minimum moment of inertia. Finally,

$$\frac{L}{k_y} = \frac{8 \text{ ft}}{0.577 \text{ in.}}\left(\frac{12 \text{ in.}}{\text{ft}}\right) = 166$$

For comparison purposes we calculate the alternate slenderness ratio:

$$k_x = \sqrt{\frac{10.7}{8}} = 1.15 \text{ in.}$$

$$\frac{L}{k_x} = \frac{8 \text{ ft}}{1.15 \text{ in.}}\left(\frac{12 \text{ in.}}{\text{ft}}\right) = 83.5$$

Thus we see that minimum moment of inertia, minimum radius of gyration, and maximum slenderness ratio all go together and correspond to the limit on the critical load.

EXAMPLE PROBLEM 12.2

Find the critical load for the column of Example Problem 12.1 if it is simply supported (pinned ends) and of southern pine. What compressive stress would the critical load produce?

Solution

From Eq. 12.4 we have

$$P_{CR} = \frac{\pi^2 EA}{\left(\dfrac{L}{k}\right)^2}$$

From Example Problem 12.1 the slenderness ratio was found to be

$$\frac{L}{k} = 166$$

From Table A3.2 we get a modulus of elasticity of

$$E = 1.6 \times 10^6 \text{ psi}$$

Then

$$P_{CR} = \frac{\pi^2 (1.6 \times 10^6 \text{ lb})(8 \text{ in.}^2)}{\text{in.}^2 (166^2)} = \mathbf{4580 \text{ lb}}$$

The stress produced would be

$$\sigma_{CR} = \frac{P_{CR}}{A} = \frac{4580 \text{ lb}}{8 \text{ in.}^2} = \mathbf{573 \text{ psi}}$$

This wood has a compressive strength of 8400 psi. In this case the load it can carry without buckling is much less than that required for a compressive failure.

12.4 END CONDITIONS FOR COLUMNS

Equation 12.2 indicates that the critical load is a function of the modulus of elasticity, the length and the cross-sectional moment of inertia. A fourth factor, not revealed by this equation, is the end conditions. In Fig. 12.1 the column was pinned on each end. In Fig. 12.4 we show a similar column that is fixed on each end; that is, the ends will not be permitted to rotate. If we analyze this column as we did the pinned-pinned column in Section 12.2 we will find the following result:

$$P_{CR} = \frac{4\pi^2 EI}{L^2} \tag{k}$$

which says the critical load for a column with fixed ends is four times what it is for one with pinned ends. We can rewrite this as

$$P_{CR} = \frac{\pi^2 EI}{(L/2)^2} \tag{l}$$

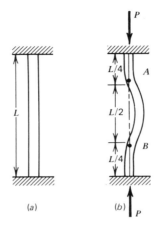

FIG. 12.4a–b Buckling of a column with fixed ends.

If we now introduce an equivalent length as

$$L_E = \tfrac{1}{2}L \qquad \text{(m)}$$

we can express the previous equation as

$$P_{CR} = \frac{\pi^2 EI}{L_E^2} \qquad (12.5)$$

This is the same form as Eq. (12.2)

We might have arrived at this conclusion intuitively by noticing that the middle portion of the column in Fig. 12.4b is similar in shape to the entire length of the column in Fig. 12.1b. Other end conditions are shown in Fig. 12.5. Their equivalent lengths are as indicated. Equation (m) is generalized as

$$L_E = KL \qquad (12.6)$$

with values for K indicated in Figs. 12.1, 12.4, and 12.5 and Table 12.1. It should be noted that these values of K are theoretical. In practice, slightly larger, more conservative values are used. Also note that we are using a lowercase k for radius of gyration and a capital K for the equivalent length factor.

Table 12.1 Equivalent length factors, $L_E = KL$

End conditions	K
Fixed-fixed	0.5
Fixed-pinned	0.707
Pinned-pinned	1.0
Fixed-free	2.0

END CONDITIONS FOR COLUMNS 313

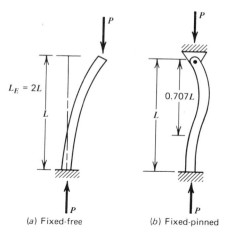

FIG. 12.5a-b Additional buckling modes.

Noting that the critical load varies inversely with the square of the equivalent length, we see that a column with fixed-fixed end conditions can carry four times the load of pinned-pinned; fixed-pinned can carry twice as much, and fixed-free can carry only one-fourth as much.

Equation 12.4 is rewritten as

$$\frac{P_{CR}}{A} = \frac{\pi^2 E}{\left(\dfrac{L_E}{k}\right)^2} \tag{12.6}$$

EXAMPLE PROBLEM 12.3

The 2×4 of Example Problem 12.2 has both ends fixed. What is its critical load?

Solution

From Example Problem 12.2

$$k = 0.577 \text{ in.}$$

$$E = 1.6 \times 10^6 \text{ psi}$$

$$L = 96 \text{ in.}$$

From Table 12.1, $K = 0.5$. The equivalent length is

$$L_E = KL = 0.5(96 \text{ in.}) = 48 \text{ in.}$$

And the equivalent slenderness ratio is

$$\frac{L_E}{k} = \frac{48}{0.577} = 83.2$$

Lastly, we find the critical load

$$P_{CR} = \frac{\pi^2 EA}{\left(\frac{L_E}{k}\right)^2} = \frac{\pi^2 (1.6 \times 10^6 \text{ lb}) \, 8 \text{ in.}^2}{\text{in.}^2 \, (83.2)^2}$$

$$P_{CR} = 18{,}300 \text{ lb}$$

This is four times the 4580 lb the same column could carry with pinned-pinne end conditions.

EXAMPLE PROBLEM 12.4

A 50 mm × 150 mm timber is pinned along the X axis as shown in Fig. 12.6a on both ends. If it is 4 m long and of southern pine, find its critical loads.

Solution

For the support shown the pin allows rotation in the 150 mm, or y, direction (about the x axis). However, it does not allow rotation in the 50-mm, or x, direction (about the y axis). Consequently, we must examine both modes of buckling. The critical load for the column will be the smaller of the two.

Buckling in the 150-mm Direction (Fig. 12.6b)

The moment of inertia is about the x axis, so b is parallel to this axis. Thus $b = 50$ mm

$$I_x = \frac{bh^3}{12} = \frac{50 \text{ mm} \, (150 \text{ mm})^3}{12} = 1.406E7 \text{ mm}^4$$

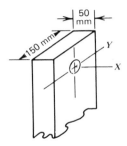

FIG. 12.6a

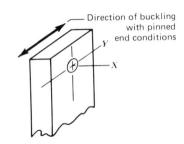

FIG. 12.6b

The radius of gyration is

$$k_x = \sqrt{\frac{I_x}{A}} = \sqrt{\frac{1.406E7 \text{ mm}^4}{50 \text{ mm }(150 \text{ mm})}} = \sqrt{1.875E3 \text{ mm}^2} = 43.3 \text{ mm}$$

For this direction the end conditions are pinned-pinned, so $K=1$ and the equivalent length is the actual length:

$$L_E = KL = 1 \,(4 \text{ m}) = 4 \text{ m}$$

The slenderness ratio is

$$\frac{L_E}{k_x} = \frac{4 \text{ m}}{0.0433 \text{ m}} = 92.4$$

and the critical load is

$$P_{CR} = \frac{\pi^2 EA}{\left(\frac{L_E}{k}\right)^2} = \frac{\pi^2 (11.0E9 \text{ N})\,(0.150 \text{ m})\,(0.050 \text{ m})}{\text{m}^2 \,(92.4)^2}$$

$$P_{CR} = 95.4 \text{ kN} \qquad \text{for the 150-mm direction}$$

Buckling on the 50-mm Direction (Fig. 12.6c)

$$I_y = \frac{bh^3}{12} = \frac{150(50)^3}{12} = 1.562E6 \text{ mm}^4$$

$$k_y = \sqrt{\frac{I_y}{A}} = \sqrt{\frac{1.562E6 \text{ mm}^4}{50 \text{ mm }(150 \text{ mm})}} = 14.4 \text{ mm}$$

For this direction the end conditions are fixed-fixed, so $K=0.5$ and the equivalent length is

$$L_E = KL = 0.5 \,(4 \text{ in.}) = 2.00 \text{ m}$$

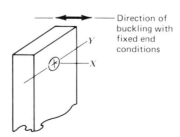

FIG. 12.6c

The equivalent slenderness ratio is

$$\frac{L_E}{k} = \frac{2 \text{ m}}{0.0144 \text{ m}} = 139$$

$$P_{CR} = \frac{\pi^2 EA}{\left(\dfrac{L_E}{k}\right)^2} = \frac{\pi^2 (11.0E9 \text{ N}) \, (0.150 \text{ m}) \, (0.050 \text{ m})}{\text{m}^2 \, (139)^2}$$

$$P_{CR} = 42.1 \text{ kN}$$

Thus the critical load for the column is 42.1 kN, and buckling will occur first in the 50-mm direction. Notice that in the critical load calculation the numbers are identical except for the slenderness ratios. We could have selected the failure mode by comparing slenderness ratios, saving one calculation.

12.5 SHORT COLUMNS AND LONG COLUMNS

In understanding column behavior it is useful to make a plot of Equation 12.6. The line AB in Fig. 12.7 is such a plot.

We see from this plot that for small slenderness ratios the critical stress is very high. For very high slenderness ratios the stress is very low. If we draw an additional line CD on this plot, we can illustrate some very important properties of columns. The line CD represents the yield strength of the material. If the critical stress is greater than the yield strength (points on AD), the column will yield prior to buckling. On the other hand, if the critical stress is less than the

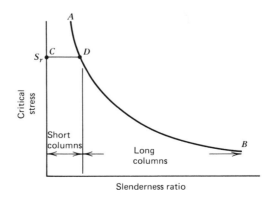

FIG. 12.7 Stress versus slenderness ratio.

yield strength, the column will buckle prior to yielding. Thus the line CDB represents failure of a column and the limiting compressive stress. From C to D the column will fail because of yielding and from D to B because of buckling. At D yielding and buckling are equally likely. We call a column governed by yield strength a short column. This will theoretically be from C to D. We call a column governed by Euler's equation a long column. This will theoretically be from D to B (and beyond).

We have hinted that the characterization of long and short columns is theoretical. This is true, especially as we approach point D. Column tests are difficult to conduct. Members are supposed to be perfectly straight and perfectly homogeneous. The load is perfectly axial and the end conditions approach the idealized state. Of course, these conditions are never absolutely realized, which leads to variation in the results. Near point D, Fig. 12.7, test results tend to fall under the theoretical curve. This undercutting is shown as the cross-hatched area in Fig. 12.8. The cross-hatching is bounded by points E and F. The region from E to F represents intermediate columns. The region C to E represents short columns. These columns are governed by the yield strength. The region beyond F represents long columns which are governed by Euler's formula.

There are a number of design approaches for intermediate columns, most of which are empirical. By empirical we mean we have an equation or procedure that fits the test results without it necessarily having a theoretical basis. Some of these formulas take into account some eccentricity of the load as well. For the purpose of this text—developing an understanding of the buckling phenomenon—we now return to the idealized analysis of Fig. 12.7, where columns are either long or short. We assume they will fail on the basis of yield (or ultimate) strength or by buckling according to Euler's equation. We leave the various empirical methods to courses in structures or machine design. You should note that the method presented here *will not* be valid for columns with eccentric loading or columns in the intermediate domain.

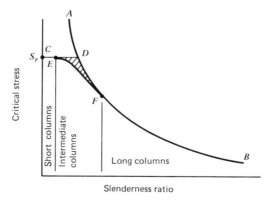

FIG. 12.8 Intermediate columns.

EXAMPLE PROBLEM 12.5

A mild steel column ($S_y = 35$ ksi, $E = 29,000$ ksi) is to carry 20 kips. Assuming the ends to be fixed, determine the factor of safety for this 12 ft $S\ 8 \times 18.4$.

Solution
From Table A4.2

$$A = 5.34 \text{ in.}^2$$
$$r_x = 3.26 \text{ in.} = k_x$$
$$r_y = 0.84 \text{ in.} = k_y$$

Because the end conditions are the same in either direction, the critical stress will be limited by the smaller radius of gyration. From Table 12.1 the equivalent length factor, K, is 0.5. Then the equivalent length is

$$L_E = KL = 0.5\ (12 \text{ ft}) \frac{12 \text{ in.}}{\text{ft}} = 72.0 \text{ in.}$$

and the equivalent slenderness ratio is

$$\frac{L_E}{k} = \frac{72.0 \text{ in.}}{0.84 \text{ in.}} = 85.7$$

Then by Eq. 12.6 the critical stress is

$$\sigma_{CR} = \frac{\pi^2 E}{(L/k)^2} = \frac{\pi^2 (29,000 \text{ kips})}{\text{in.}^2\ (85.7)^2} = \mathbf{39.0 \text{ ksi}}$$

This is the stress the column can carry without buckling.

$$S_y = 35.0 \text{ ksi}$$

This is the stress the column can carry without yielding. Since it is less than σ_{CR}, it will limit the load, and the allowable load is

$$P_{AL} = S_y A = 35.0\ (5.34 \text{ in.}^2) = 187 \text{ kips}$$

The factor of safety N will be the allowable load divided by the actual load.

$$N = \frac{P_{\text{allowable}}}{P_{\text{actual}}} = \frac{187 \text{ kips}}{20 \text{ kips}} = \mathbf{9.34}$$

12.6 IDEALIZED DESIGN OF COLUMNS

From the previous articles we have Euler's equation in two forms:

$$P_{CR} = \frac{\pi^2 EI}{L_E^2} \qquad (12.5)$$

$$\frac{P_{CR}}{A} = \frac{\pi^2 E}{\left(\dfrac{L_E}{k}\right)^2} \qquad (12.6)$$

Equation 12.6 is the more functional form of the equation as we seek to understand the buckling phenomenon. It is in terms of the slenderness ratio which we find to be a governing parameter for columns. Most of the empirical methods are in terms of the slenderness ratio as well. This equation also is useful for analysis, giving a critical stress that can be readily compared to the yield strength of the material to determine which governs. The governing stress can then be compared with the applied stress to determine the adequacy of the column.

If our problem is not analysis but design, Eq. 12.6 is awkward. In design problems we seek to find a structural member that will carry the required load. Equation 12.6 has its geometric factors, A and k, separated. For design purposes, Eq. 12.5 is much more functional. We substitute in the problem parameters, solve for I, and select an appropriate member from tables or geometric formulas. Finally, the column must be checked against its yield or ultimate strength.

EXAMPLE PROBLEM 12.6

Select a steel, equal leg, angle member for a column to carry a 10-kip load, with a safety factor of 3. The column is to be 14 ft long with each end rigidly fixed.

$$S_y = 35 \text{ ksi}, \qquad E = 29{,}000 \text{ ksi}$$

Solution

From Eq. 12.5

$$P_{CR} = \frac{\pi^2 EI}{L_E^2}$$

Solving for I:

$$I = \frac{P_{CR} L_E^2}{\pi^2 E}$$

In the past we have applied the factor of safety to the strength of the material. Since buckling does not depend on the strength of the material, that will not do here. If we apply the factor of safety to the load we will take care of strength *and* buckling. Thus $P_{\text{allowable}} = NP_{\text{actual}}$

$$P_{AL} = 3(10) = 30 \text{ kips}$$

For fixed-fixed ends K is 0.5 and the equivalent length is

$$L_E = 0.5 \ (14 \text{ ft})\left(\frac{12 \text{ in.}}{\text{ft}}\right) = 84.0 \text{ in.}$$

Then

$$I = \frac{30 \text{ kips } (84 \text{ in.})^2 \ (\text{in.}^2)}{\pi^2 (29{,}000) \text{ kips}} = \mathbf{0.740 \text{ in.}^4}$$

Going to Table A4.6 to select a number, note that r (our k) will be minimum about the Z-Z axis. This will require computing $I_z = Ak^2$, since only the radius of gyration is given in the table. We select the following candidates:

L 5×10.3	$3.03 \ (0.99)^2 = 2.97 \text{ in.}^4$
L 4×6.6	$1.94 \ (0.80)^2 = 1.24$
L 3.5×5.8	$1.69 \ (0.69)^2 = 0.805$
L 3×4.9	$1.44 \ (0.59)^2 = 0.501$ too small

The smallest member with sufficient I is the **L 3.5×5.8**.

Checking for direct stress

$$A = \frac{P}{S_y} = \frac{30 \text{ kip } (\text{in.})^2}{35 \text{ kips}} = 0.857 \text{ in.}^2$$

The cross-sectional area of the specific angle iron is 1.69 in.² which is greater than required for yielding. So buckling governs. Use the **L 3.5×5.8**.

12.7 DESIGN FORMULAS

The foregoing discussion has emphasized the concepts of buckling and how a member in compression may fail. It draws out the significance of such terms as slenderness ratio, effective length, end conditions, short columns, intermediate columns, and long columns. It emphasizes the relation between buckling and the applied load, the cross-sectional geometry, and the material properties. Having done all that, is it too much to ask that it also *actually* be a practical method for designing columns? In many cases, yes!

In the previous discussion we treated the problem in an idealized fashion, and we presumed to understand some things that are not so well understood. Beams usually are not "perfectly straight," loads are not "perfectly centered," and end conditions cannot be perfectly described as "pinned" or "fixed." Alas, the "real world" of columns, like so many "real worlds," is not perfect.

To account for these imperfections numerous empirical equations have been developed for various applications. By empirical we mean equations that are based on experimental and/or field data. They describe our laboratory or field experience to some degree but may not lend themselves to theoretical development.

One of the more common of these is the J. B. Johnson formula. It can be written as

$$\frac{P_{CR}}{A} = a - b\left(\frac{L}{k}\right)^2$$

where the constants a and b depend on the material properties and other factors. An equation that takes into account the eccentricity of the load is known as the *secant formula*:

$$\frac{P}{A} = \frac{S_y}{1 + \left(\frac{ec}{k^2}\right)\sec\left[\frac{L}{k}\sqrt{\frac{P}{4AE}}\right]}$$

In this formula e is the eccentricity of the load, that is, its distance from the centroid. The distance from the neutral axis to the extreme fiber is c as it was in bending.

In addition to these fairly general forms, very specific forms have been specified by trade organizations concerned with steel, aluminum, wood, and other products. Some of these and others are also contained in building codes which are legally binding.

12.8 SUMMARY

Euler's equation is the basic governing equation for columns:

$$P_{CR} = \frac{\pi^2 EI}{L_E^2}$$

Note that the load a column can carry without buckling, the critical load, depends on the column length, moment of inertia of its cross section, end conditions, and the modulus of elasticity. It does not depend on the strength of the material. In terms of critical stress, Euler's equation is

$$\sigma_{CR} = \frac{P_{CR}}{A} = \frac{\pi^2 E}{\left(\dfrac{L_E}{k}\right)^2}$$

where (L_E/k) is the slenderness ratio. The equivalent length is

$$L_E = KL$$

where K depends on the end conditions.

For *short* columns

$$\sigma_{CR} > \sigma_{yield}$$

and

$$\sigma_{yield} \text{ governs}$$

For *long* columns

$$\sigma_{CR} < \sigma_{yield}$$

and

$$\sigma_{CR} \text{ governs}$$

Intermediate columns are transitional regions between short and long columns where empirical equations govern.

The above equations emphasize the phenomenon of buckling and its related parameters under ideal conditions. Empirical equations prescribed by trade organizations and codes will supersede them.

PROBLEMS

For problems 12.1 through 12.17 assume the end effects apply in every direction.

12.1 Find the slenderness ratios (in both directions if applicable) for the following members which are 12 ft long and pinned on each end.
  (a) 2×4 (actual size).
  (b) 4×12 (S4S).
  (c) S 8×23.0.
  (d) 3-in.-diameter standard steel pipe.
  (e) 3-in.-diameter solid shaft.
  (f) L 4×18.5.
  (g) L 5×19.8.

12.2 Find the slenderness ratios (in both directions if applicable) for the following members which are 4 m long and pinned on each end.
- (a) 50 mm × 100 mm.
- (b) 100 mm × 300 mm.
- (c) 80-mm-diameter solid shaft.
- (d) A pipe with 80 mm O.D. and 60 mm I.D.
- (e) Cross section of Fig. 12.9.

12.3 Find the critical loads for the above columns of Problem 12.1 if they are
- (a–b) Southern pine.
- (c–g) Steel.

12.4 Find the critical loads for the above columns of Problem 12.2 if they are:
- (a–b) Southern pine.
- (c–e) Steel.

12.5 Do Problem 12.3a if the end conditions are
- (a) Fixed-fixed.
- (b) Fixed-free.
- (c) Fixed-pinned.

12.6 Do Problem 12.4a if the end conditions are
- (a) Fixed-fixed.
- (b) Fixed-free.
- (c) Fixed-pinned.

12.7 Do Problem 12.3b if the end conditions are
- (a) Fixed-fixed.
- (b) Fixed-free.
- (c) Fixed-pinned.

12.8 Do Problem 12.4b if the end conditions are
- (a) Fixed-fixed.

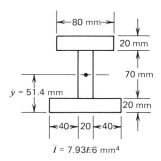

FIG. 12.9 Problems 12.2e, 12.4e, 12.14.

324 COLUMNS

    (b) Fixed-free.

    (c) Fixed-pinned.

12.9 Do Problem 12.3c if the end conditions are

    (a) Fixed-fixed.

    (b) Fixed-free.

    (c) Fixed-pinned.

12.10 Do Problem 12.4c if the end conditions are

    (a) Fixed-fixed.

    (b) Fixed-free.

    (c) Fixed-pinned.

12.11 Do Problem 12.3d if the end conditions are

    (a) Fixed-fixed.

    (b) Fixed-free.

    (c) Fixed-pinned.

12.12 Do Problem 12.4d if the end conditions are

    (a) Fixed-fixed.

    (b) Fixed-free.

    (c) Fixed-pinned.

12.13 Do Problem 12.3e if the end conditions are

    (a) Fixed-fixed.

    (b) Fixed-free.

    (c) Fixed-pinned.

12.14 Do Problem 12.4e if the end conditions are

    (a) Fixed-fixed.

    (b) Fixed-free.

    (c) Fixed-pinned.

12.15 Do Problem 12.3f if the end conditions are

    (a) Fixed-fixed.

    (b) Fixed-free.

    (c) Fixed-pinned.

12.16 The following members are to be used as columns. In each case calculate the load it can carry based on its yield strength and based on Euler's formula. Indicate which one will govern the design.

    (a) 4 in. $\times$ 4 in. (S4S), 4-ft long, southern pine, fixed-fixed.

    (b) 6-in. pipe, 18-in. long, steel, fixed-fixed.

    (c) $L\ 8 \times 51.0$, 6-ft long, steel, fixed-free.

    (d) Do (c) if the material is aluminum.

12.17 The following members are to be used as columns. In each case calculate the load it can carry based on its yield strength and based on Euler's formula. Indicate which one will govern the design.

(a) 100 mm × 100 mm, 1.40-m long, southern pine, fixed-fixed.

(b) 150 mm (125-mm I.D.) pipe, 500-mm long, steel, fixed-fixed.

(c) Do (b) if the material is aluminum.

12.18 Member BD is to be a solid steel rod ($S_y = 35$ ksi), Fig. 12.10. Find the diameter required. Check both buckling and yielding.

12.19 A flag pole sitter weighs 200 lb. He performs on top of a 30-ft pole. Using a factor of safety of 4, select an appropriate steel pipe for this flag pole.

12.20 A flag pole sitter weighs 900 N. He performs on top of a 10-m pole. Using a factor of safety of 4, select an appropriate steel pipe for this flag pole if I.D. = 0.80 O.D.

12.21 The truss in Fig. 12.11 is to be made of southern pine. Each member is to be square. Assume each joint to be pinned. Identify the members where buckling is of concern. Select the size of each indicated member.

(a) AB (f) EF

(b) AE (g) CD

(c) BE (h) CF

(d) BC (i) FD

(e) EC

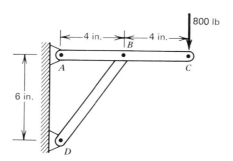

FIG. 12.10 Problem 12.18.

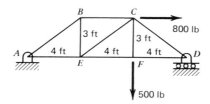

FIG. 12.11 Problem 12.21.

326 COLUMNS

12.22 The truss in Fig. 12.12 is to be made of aluminum. Each member has a square cross section. Assume each joint to be pinned. Identify the members where buckling is of concern. Select the size of each indicated member.

(a) AB
(b) AE
(c) BE
(d) BC
(e) EC
(f) EF
(g) CD
(h) CF
(i) FD

12.23 Identify all compression members in Fig. 12.13. Select a solid steel rod for each compression member.

12.24 Identify all compression members in Fig. 12.14. Select a solid steel rod for each compression member.

12.25 Identify all compression members in Fig. 12.15. Select a square bronze column for each member in compression.

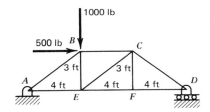

FIG. 12.12   Problem 12.22.

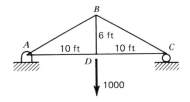

FIG. 12.13   Problem 12.23.

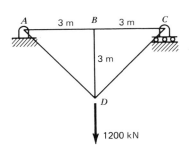

FIG. 12.14   Problem 12.24.

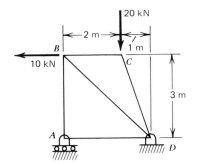

FIG. 12.15   Problem 12.25.

13

Miscellaneous Applications

13.1 Cylindrical pressure vessels
13.2 Spherical pressure vessels
13.3 Bolted and riveted connections
13.4 Weldments
13.5 Stress concentration
13.6 Fatigue
13.7 Summary

The material of the previous chapters constitutes the basics of strength of materials. In this chapter we look at some additional applications of this material as well as some limitations. These are just a few of the great many applications, but they illustrate how the principles may be applied in a few important ways. Additional applications can be found in texts on machine design and structural analysis and design. These texts will also amplify further the limitations of these principles as they are applied.

13.1 CYLINDRICAL PRESSURE VESSELS

The results of the analysis of pressure vessels are important in themselves because of the wide use of pressure vessels. They are important to us beyond that, in that the analysis indicates how the principles may be applied to new problems.

First, we will consider a cylindrical tank with plane ends as shown in Fig. 13.1a. It is important to note that these results are not limited to tanks with plane ends, but this condition may make the analysis clearer for those not familiar with fluid mechanics. The internal pressure in the tank is p. A cutting plane is passed through the tank perpendicular to the longitudinal axis, resulting in Fig. 13.1b. Also shown in this figure is the pressure acting on the flat end of the tank and the

328 MISCELLANEOUS APPLICATIONS

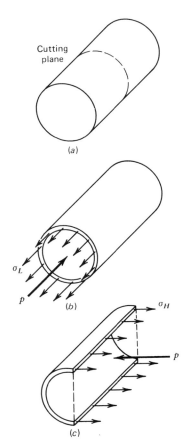

FIG. 13.1a–c Cylindrical pressure vessel.

stress acting on the cross section of the tank. For a thin shell (thickness t less than 0.1 times the radius R) any variation of this stress with the radial direction is ignored. This stress is in the longitudinal direction and is, therefore, called the longitudinal stress, σ_L.

We now sum the forces in the longitudinal direction. The force due to pressure will be the pressure times the area it acts over, or

$$pA = p\pi R^2$$

where R is the radius of the tank.

The longitudinal stress acts on the area of the thin ring of thickness t. Thus it produces a force of

$$\sigma_L A = \sigma_L (2\pi R t)$$

Since these are the only forces acting longitudinally, they must be equal:

$$\sigma_L(2\pi R t) = p\pi R^2$$

and

$$\sigma_L = p\left(\frac{R}{2t}\right) \tag{13.1}$$

We see that the longitudinal stress is equal to the internal pressure times the ratio of the radius to twice the thickness. Since we have required in our analysis that R be at least 10 times t, σ_L will be at least 5 times p.

We originally indicated that the end of the cylinder was flat. That was so that we could calculate the force due to pressure as simply pressure times area. Based on this, Eq. 13.1 would be valid only for tanks with flat ends. However, it turns out that the force in the longitudinal direction, because of pressure, will be the same regardless of the shape of the end. Regardless of the shape, the force will be the pressure times the *projected* area. The projected area is the area of a flat end, and Eq. 13.1 is valid for any cylindrical tank.

A second cutting plane is passed perpendicular to the longitudinal axis and a third through the longitudinal axis resulting in the "C"-shaped section of Fig. 13.1c. With this section we can examine the stresses acting in the circumferential direction. These are labeled σ_H in the figure and are called hoop stresses. Now we sum forces in the horizontal direction. The force due to internal pressure will be the pressure times the projected area. Thus

$$pA = p(2RL)$$

where L is the length of the section.

The force due to the hoop stress is

$$2\sigma_H A = 2\sigma_H(tL)$$

Setting the two equal and noting $A = tL$ gives

$$2\sigma_H(tL) = p(2RL)$$

Solving for the hoop stress gives

$$\sigma_H = p\left(\frac{R}{t}\right) \tag{13.2}$$

This is twice the longitudinal stress and, therefore, will govern the design. As there are no shear stresses on these planes, the hoop and longitudinal stresses are the principal stresses, and the hoop stress, being the greater, will govern the design. Ductile failure of a cylindrical pressure vessel will often reveal this stress pattern by showing a longitudinal split in the vessel as the hoop stress pulls it apart.

The maximum normal stress in a cylindrical tank will be the ratio of the radius to wall thickness times the pressure. This will be at least 10 times the pressure, typically greater. This stress will govern the design. These equations do not apply to "thick" pressure vessels in which the stress varies across the thickness of the vessel wall. Analysis of the stresses in thick vessels is a more advanced topic.

13.2 SPHERICAL PRESSURE VESSELS

In Fig. 13.2 a section of a spherical pressure vessel is shown where a cutting plane has been passed through its center. The normal stress σ and the internal pressure p are indicated. The force due to the internal pressure is the pressure times the projected area.

$$pA = p(\pi R^2)$$

The force due to the stress σ is the stress times the area of the thin ring it acts over:

$$\sigma A = \sigma(2\pi R t)$$

Summing the forces horizontally sets the two equal:

$$\sigma(2\pi R t) = p(\pi R^2)$$

Solving for σ gives

$$\sigma = p\left(\frac{R}{2t}\right) \tag{13.3}$$

This is the same as the longitudinal stress given by Eq. 13.1 but only half of the hoop stress given by Eq. 13.2. Hence the maximum stress produced in a spherical vessel is only one-half of the maximum stress produced in a cylindrical one. A spherical vessel is inherently more efficient than a cylindrical one.

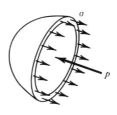

FIG. 13.2 Spherical pressure vessel.

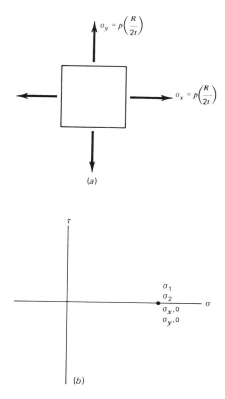

FIG. 13.3a-b Mohr's circle for a spherical pressure vessel.

Plotting Mohr's circle for this pressure vessel produces a very interesting result. Although a vertical cutting plane was used for Fig. 13.2, its selection was arbitrary. Any other cutting plane through the center would produce a similar result. Thus we have equal normal stresses on perpendicular planes and no shear stress, Fig. 13.3a. To plot Mohr's circle, we plot $(\sigma_x, 0)$ and $(\sigma_y, 0)$. Since σ_x and σ_y are both defined by Eq. 13.3, these two points lie on top of each other, Fig. 13.3b. In fact, the Mohr's circle for this state of stress is this point! Further, the maximum stress, the minimum stress, and the stress on any plane are equal to this value. The shear stress on any plane is zero.

EXAMPLE PROBLEM 13.1

A cylindrical pressure vessel is to be 10 ft long and 4 ft in diameter. Find its thickness based on a yield strength of 36 ksi, if it is to carry 180 psi with a factor of safety of 4.

Solution

Equation 13.2 governs

$$\sigma_H = p\left(\frac{R}{t}\right)$$

Solving for t gives

$$t = \frac{p}{\sigma_H}(R)$$

The allowable stress is

$$(\sigma_H)_{AL} = \frac{36 \text{ ksi}}{4} = 9.00 \text{ ksi}$$

Then

$$t = \frac{180 \text{ psi}}{9000 \text{ psi}}(24 \text{ in.}) = \mathbf{0.480 \text{ in.}}$$

13.3 BOLTED AND RIVETED CONNECTIONS

We will discuss bolts and rivets interchangeably in this chapter, since the design procedures for them are common. In Chapter 3 we examined simple riveted connections in shear. There, we looked only at the shear of the rivet. We now consider other factors in the design of a complete connection. In all of our considerations, we assume a logical but somewhat idealistic single mode of failure. Actual failures are often more complex than those assumed. Nonetheless, the methods of analysis and design presented here do give workable results.

In this analysis we also ignore the transfer of load by friction forces. The friction load can play a significant role, and in fact with high-strength bolts that are highly torqued, the entire load can be transferred by friction.

Figure 13.4 shows a single rivet in shear; this joint is called single riveted lap joint. In Chapter 3, when we investigated the failure of this connection in shear, we found that the load P must be supported by the cross-sectional area of the rivet, designated as DE in Fig. 13.4. That is only one of several possible modes of failure that must be evaluated for proper design of a riveted connection. Other modes are as follows.

There can be a bearing failure over the surface ABC (Fig. 13.4). This failure can occur in the rivet or in the plate. Only the weaker material needs to be evaluated, and that is normally the plate. Although the failure actually occurs on the curved surface ABC, the design is based on the projected rectangular plane area AC.

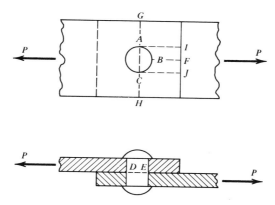

FIG. 13.4 Riveted connection.

A tensile failure of the plate can also occur. In this case the area involved is GA and CH. There are also stress concentrations around the hole, which we will ignore for the present.

The last modes of failure we will consider are tearing out of the end of the plate, area BF (Fig. 13.4), or shearing out the end of the plate, areas AI and CJ. Although we can make similar calculations for either of these two failures, they are usually handled by following a rule of thumb which says to make the center of the rivet at least two diameters from the edge of the plate. Following this rule of thumb eliminates the need for calculating stresses associated with tear out or end plate shear.

Thus in a bolted or riveted connection we need to examine the fastener shear, the plate or fastener bearing stress, and the plate tensile stress. We have discussed these for a single rivet. If additional rivets are involved, a similar analysis may be used as long as the load is through the centroid of the rivet pattern. The problem is more complicated when the load is not through the centroid of the rivet pattern. That problem is left for more advanced texts. In analyzing the failure of a connection involving several rivets we assume the entire structure moves as a rigid body except at the regions of failure. This will be illustrated in Example Problem 13.2. When three plates are used, additional calculations may be needed, but they will be variations of the above.

Typical properties of materials to be used in the problems are:
Plates:

        Bearing strength   100 ksi (690 MPa)
        Tensile strength   70 ksi (483 MPa)

Rivets:

        Shear strength   50 ksi (345 MPa)

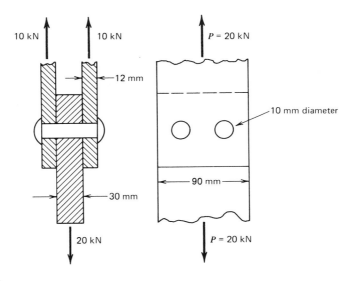

FIG. 13.5a

EXAMPLE PROBLEM 13.2

The joint shown in Fig. 13.5a carries 20 kN. Find all the design stresses for analyzing this joint.

Solution

Shear

Fig. 13.5a shows one-half of a double riveted butt joint. The rivet is in double shear, which means the shear load is only one-half the total load, as shown in Fig. 13.5b. Since there are two rivets, the load per rivet is one-half the 10 kN load per plate, or 5 kN. The average shear stress in the rivet is calculated

$$\tau_{AV} = \frac{V}{A} = \frac{5 \text{ kN}(4)}{\pi(10 \text{ mm})^2} = 63.7 \text{ MPa}$$

Bearing Stresses

Two different bearing stresses must be analyzed: those on the center plate and those on the outside plates. Although it is the half cylindrical surface that is in bearing, calculations are based on the projected area, the shaded rectangular plane in Fig. 13.5c. Again, since there are two rivets, the bearing load on the outside plates will be 5 kN per rivet and 10 kN per rivet on the center plate. For the outside plate and rivet:

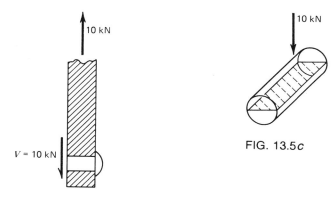

FIG. 13.5b

FIG. 13.5c

$$\sigma_{OB} = \frac{P}{A} = \frac{5 \text{ kN}}{10 \text{ mm } (12 \text{ mm})} = \mathbf{41.7 \text{ MPa}}$$

For the center plate:

$$\sigma_{CB} = \frac{P}{A} = \frac{10 \text{ kN}}{10 \text{ mm } (30 \text{ mm})} = \mathbf{33.3 \text{ MPa}}$$

σ_{OB} being the greater will govern if all the materials are the same or if the center plate is the strongest material (including the rivet).

Tensile Stresses

Again two situations must be evaluated, the center and the outside plates. The outside plates carry 10 kN. The minimum area in tension is the 12×90 cross section minus the two 10-mm holes.

$$\sigma_{OT} = \frac{P}{A} = \frac{10 \text{ kN}}{12 \text{ mm } (90 - 2(10)) \text{ mm}} = \mathbf{11.9 \text{ MPa}}$$

For the inside plate we have

$$\sigma_{CT} = \frac{P}{A} = \frac{20 \text{ kN}}{30 \text{ mm } (90 - 2(10)) \text{ mm}} = \mathbf{9.52 \text{ MPa}}$$

σ_{OT} will govern unless the center plate has a lower strength.

13.4 WELDMENTS

Welds are one of the most common methods of joining metals. They are widely used in pipelines, building structures, and all sorts of mass- and custom-produced

336 MISCELLANEOUS APPLICATIONS

products. Even works of art take advantage of this modern technology. In spite of the very widespread use of this process in very sophisticated industries, the routine design of weldments remains simplistic and conservative. The conservatism seems well founded. The great industrialist and inventor, R. G. LeTourneau, once said he could teach monkeys to weld. As far as I know he was never successful in that endeavor, but I have seen weldments that suggest he may have been.

Weld designs are usually based on the "throat" area. The definition of the "throat" varies with the type of weld, but it is generally the minimum cross section in a weld. Figure 13.6 shows a butt weld. For a butt weld the throat area is the area of the plate being welded. Thus the area would be h times the length of the weld. The rounded reinforcement shown in Fig. 13.6 adds strength in static loading but is not included in the design height h. This reinforcement is ground off for fatigue loading, since a stress concentration is developed at the corner where the reinforcement intersects the plate. This stress concentration is detrimental in fatigue loading but is of no consequence in static loading. The same throat area is used when the load is in shear. Thus we have for tensile or compressive loads

$$\sigma = \frac{P}{hL} \qquad (13.4)$$

where L is the length of the weld and

$$\tau = \frac{P_s}{hL} \qquad (13.5)$$

for shear loads where P_s is a shear load. The shear load P_s is perpendicular to the paper in Fig. 13.6.

Fig. 13.7a shows a lap joint with a fillet weld. The size of the weld refers to the dimension w. The design of fillet welds is also based on the throat area, and it is assumed to fail in shear regardless of the direction of loading. The weld, shown in detail in Fig. 13.7b, looks like a 45° right triangle in cross section. Consequently, the throat depth will be

$$h = w \cos 45° = 0.707w$$

The shear stress will be as given by Eq. 13.5, that is

$$\tau = \frac{P}{hL}$$

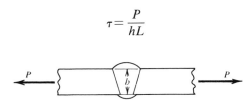

FIG. 13.6 Butt weld.

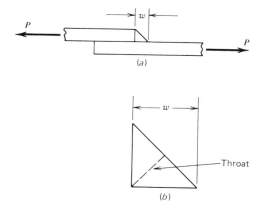

FIG. 13.7a–b Lap joint with filleted welds.

or

$$\tau = \frac{P}{0.707wL} \tag{13.6}$$

Typical allowable strengths for welds in structural steels are 20,000 psi (140 MPa) for tensile stress and 13,500 psi (93.1 MPa) for shear stress.

As in the case of bolted or riveted points, these loads must be through the centroid of the total weld area. When they are not, they have a tendency to twist the connection, and much greater stresses are generated. Many applications of weld design are governed by code. Of course, these codes should be observed when they apply.

EXAMPLE PROBLEM 13.3

Find the average shear stress in the 8-mm fillet weld shown in Fig. 13.8.

Solution

The total length of weld is

$$L = (100 + 80 + 100) = 280 \text{ mm}$$

The throat depth is

$$h = 0.707 \, (8) = 5.66 \text{ mm}$$

and the total shear area is

$$A = hL = 5.66(280) = 1580 \text{ mm}^2$$

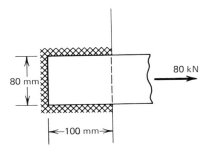

FIG. 13.8

The average shear stress is

$$\tau = \frac{P}{hL} = \frac{80\ E3\ \text{N}}{1580\ \text{mm}^2} = 50.5\ \text{MPa}$$

Note that the load is through the centroid of the weld and has no tendency to twist it.

13.5 STRESS CONCENTRATION

All the stress equations developed thus far have assumed that there were no irregularities in the cross section in the vicinity of the section being analyzed. It was also assumed that the section was well removed from concentrated loads. If these conditions are not met, the stresses produced will be higher than those predicted by the stress equations, and the stress distribution will be different from what we have assumed or developed. Figure 13.9a shows a flat plate with a hole. From the method of Chapter 3 we calculate the stress from

$$\sigma = \frac{P}{A} \qquad (3.1)$$

and assume the stress distribution to be uniform as shown in Fig. 13.9b. The actual stress distribution is as shown in Fig. 13.9c. Clearly, it is not uniform, and the maximum stress can be two to three times as large as the average stress. This higher stress is the result of the sudden change in cross section. These rapid changes, called stress risers, are to be avoided in critical sections of machines and structures. Gradual changes in shape that minimize the effect of stress risers are desirable when changes in geometry are necessary.

The ratio of the maximum stress to the average stress is usually designated by the letter K, called the *stress concentration factor*. We write

$$K = \frac{\sigma_{\max}}{\sigma_{AV}}$$

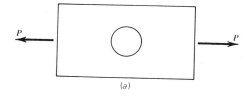
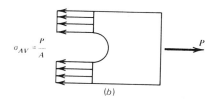
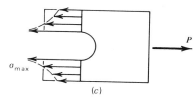

FIG. 13.9a–c The effect of stress concentration on axial stresses.

or more commonly

$$\sigma_{max} = K\sigma_{AV} \qquad (13.7)$$

Or we may modify Eq. 3.1 to read

$$\sigma_{max} = K\frac{P}{A} \qquad (13.8)$$

In a similar manner the other stress equations we have developed may be written as

$$\tau_{max} = K\frac{V}{A} \qquad (13.9)$$

$$\tau_{max} = K\frac{Tc}{J} \qquad (13.10)$$

$$\sigma_{max} = K\frac{Mc}{I} \qquad (13.11)$$

We use Eqs. 13.8 through 13.11 just as we have the ones on which they are based.

340 MISCELLANEOUS APPLICATIONS

Generally the nominal stress calculation is based on the net area, but not always, so care must be taken. The stress concentration factor K can be determined from the theory of elasticity for very simple shapes. As the problem becomes more complex we must rely on experimental methods, primarily photoelasticity. This is a technique using polarized light and plastic models that respond to stresses by varying the transmission of light. The result is very colorful patterns indicating the direction and intensity of the principal stresses. More recently, stress concentration factors have been calculated using numerical methods on the digital computer.

Most machine design texts include an assortment of stress concentration factors similar to those shown in Fig. 13.10. In using these charts to determine

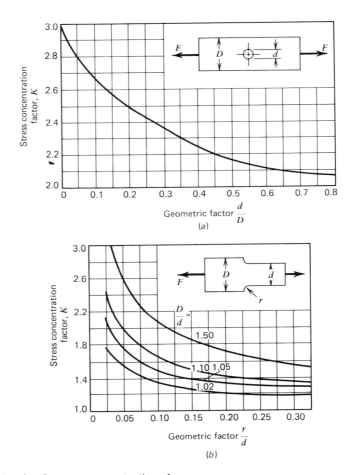

FIG. 13.10 *a–b* Stress concentration factor for a flat bar. (From R. E. Peterson, Design Factors for Stress Concentration, *Machine Design*, Vol. 23, 1951. Used by permission.)

stress concentration factors care must be taken to apply them correctly. Make sure the loading is that described in your problem. Be sure you are using the correct area. And finally, be certain as to whether the chart is for a flat piece or a round one.

It should be noted from Fig. 13.10 that when the change in shape is very abrupt, characterized by relatively small holes and fillets, the stress concentrations are the highest. These abrupt changes should generally be avoided.

Now it's time for a bad news–good news story. The bad news is that stress risers increase the stress beyond the nominal value. The good news is that sometimes it does not matter, and in a few cases it is actually helpful. Specific rules about when and how to use them are best left to texts on machine design. However, a couple of generalities can be made:

1. When designing with ductile materials and static loads, stress risers may be ignored.
2. When designing with repeated loads, (fatigue) stress concentrations are used, although materials differ in their sensitivity to them.
3. When using brittle materials with stress concentrations in critical sections under tensile loading... stay out from under it!

EXAMPLE PROBLEM 13.4

A $\frac{1}{4}$-in.-thick plate is 2 in. wide and has a $\frac{3}{4}$-in. hole cut in the center of it, Fig. 13.11. It carries a 300-lb load. Find the maximum stress in the plate.

Solution

The net area is

$$A = (2 - 0.75)(0.25) = \mathbf{0.312 \text{ in.}^2}$$

The nominal stress is

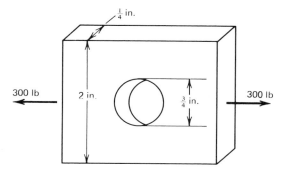

FIG. 13.11

$$\sigma = \frac{P}{A} = \frac{300 \text{ lb}}{0.312 \text{ in.}^2} = \mathbf{960 \text{ psi}}$$

To read Fig. 13.8 we need d/D

$$\frac{d}{D} = \frac{\frac{3}{4}}{2} = 0.375$$

From Fig. 13.10a we get

$$K = 2.27$$

Hence

$$\sigma_{max} = K\sigma_{AV} = 2.27(960) = \mathbf{2180 \text{ psi}}$$

13.6 FATIGUE

If you are still with us after having started umpteem chapters ago, you may be feeling something known as fatigue. Well, surprisingly, materials get tired too—in a sense. When a material gets tired we call it fatigue also. We can take a mild steel with an ultimate strength of say 60,000 psi (414 MPa) and a yield strength of say 40,000 psi (276 MPa). If we repeatedly load this material to maybe 35,000 psi (241 MPa), below both the ultimate and yield strengths, it will fail after several hundred thousand cycles. This failure from cyclic loading is known as fatigue.

Fatigue failure is a failure that leaves its own characteristic signature and usually can be readily identified by materials experts. It is usually initiated at a point of stress concentration. The stress concentration may be due to an abrupt change in geometry, as discussed in the previous article or due to a material flaw or even a surface scratch. A minute crack is initiated at one of the stress concentration locations. As the load is cycled, the crack grows, reducing the area available to carry the load. Suddenly the remaining portion fails, usually in a brittle fashion. Because this failure is sudden and without warning, it is frequently dangerous.

The above type of failure has been studied extensively. It does not yield readily to analysis and shows considerable statistical scatter. Several methods have been developed to design against fatigue failure, but extensive testing is usually recommended as well. Fatigue analysis is needed in most machinery and equipment subject to dynamic loading such as internal combustion engines, electric motors, washing machines, airplanes, and helicopters—especially helicopters!

The most common laboratory test for fatigue is the Moore rotating beam test. The load configuration is as shown in Fig. 13.12. This loading results in a constant bending moment of magnitude Pa everywhere between the two central loads. All the loads are applied through bearings that allow the beam (often polished) to be continuously rotated. For each rotation of the beam the stress on

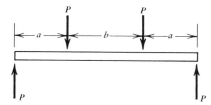

FIG. 13.12 Load configuration for Moore fatigue test.

a given element on the beam cycles from plus the maximum stress to minus the maximum stress as indicated in Fig. 13.13. The maximum stress is set at a particular level. The beam is rotated until it fails; the number of cycles required for failure is recorded. This test is repeated on another specimen at a different stress level until enough data are collected to plot a *S-N* diagram. The *S-N* diagram is a plot of stress level (S) versus the number of cycles (N) required for failure. Fig. 13.14 is typical of such plots. On the vertical axis the stress is plotted.

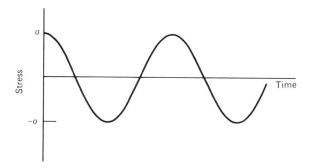

FIG. 13.13 Load cycle in fatigue test.

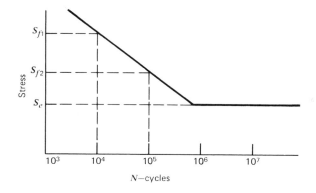

FIG. 13.14 *S–N* diagram.

On the horizontal axis the number of cycles to failure is plotted. A logarithmic scale is usually used for the number of cycles and frequently for the stress.

From the S-N diagram the fatigue strength may be determined. For instance, S_{f1} is the fatigue strength corresponding to 10,000 cycles in Fig. 13.14. Fatigue strength is always for a given number of cycles. A second fatigue strength, S_{f2} is observed for 100,000 cycles, and so on. A last point S_e is known as the endurance limit. If the stress is below the endurance limit, failure will not occur regardless of the number of cycles. It is common to assume that a specimen that has not failed after 1,000,000 cycles (or 10,000,000 in some cases) will not fail. For steels the endurance limit is often estimated as one-half the ultimate strength.

13.6 SUMMARY

Cylindrical pressure vessels:

$$\sigma_L = p\left(\frac{R}{2t}\right)$$

$$\sigma_H = p\left(\frac{R}{t}\right)$$

Spherical pressure vessels:

$$\sigma = p\left(\frac{R}{2t}\right)$$

Bolted and rivet connections when the load is through the centroid of the pattern design for:

1. Shear in rivets.
2. Tensile failure in plates.
3. Bearing failure in plates or rivets.

Weldments:

$$\sigma = \frac{P}{hL}$$

$$\tau = \frac{P_s}{hL}$$

based on the throat or minimum area.

Stress concentration:

$$\sigma = K\frac{P}{A}$$

$$\sigma = K\frac{Mc}{I}$$

$$\tau = K\frac{V}{A}$$

$$\tau = K\frac{Tc}{J}$$

where K is the stress concentration factor. Stress concentration factors are obtained from data like Fig. 13.10.

Fatigue:

   Fatigue strength—the strength of a part corresponding to a given number of cycles of the applied load.

   Endurance limit—the strength of a part corresponding to an unlimited number of cycles.

   S-N diagram—a plot of fatigue strength versus the corresponding number of cycles.

PROBLEMS

13.1 A cylindrical pressure tank is 6 ft in diameter and $\frac{3}{4}$ in. thick. If the pressure is 600 psi
   (a) Find the hoop stress.
   (b) Find the longitudinal stress.
   (c) Plot Mohr's circle for the stress element.
   (d) Find the maximum shear stress in the cylinder.

13.2 A cylindrical pressure tank is 2 m diameter and 20 mm thick. If the pressure is 4.00 MPa
   (a) Find the hoop stress.
   (b) Find the longitudinal stress.
   (c) Plot Mohr's circle for the stress element.
   (d) Find the maximum shear stress in the cylinder.

13.3 A cylindrical pressure vessel is 3 ft in diameter and $\frac{1}{2}$ in. thick. Based on an allowable stress of 40,000 psi, what pressure can it hold?

13.4 A cylindrical pressure vessel is 1.2 m in diameter and 12 mm thick. Based on an allowable stress of 270 MPa, what pressure can it hold?

13.5 Find the maximum normal stress in a spherical pressure vessel that is 5 ft in diameter and 0.80 in. thick if the pressure is 800 psi.

13.6 Find the maximum normal stress in a spherical pressure vessel that is 1.5 m in diameter and 20 mm thick if the pressure is 5.0 MPa.

13.7 Design a spherical pressure vessel to hold 180 ft^3 of gas at 500 psi based on a yield strength of 20 ksi.

13.8 Design a spherical pressure vessel to hold 2 m³ of gas at 3.5 MPa based on a yield strength of 140 MPa.

13.9 What is the normal stress in a balloon inflated to 8 psi when it is 6 in. in diameter and 0.065 in. thick?

13.10 A cylinder (Fig. 13.15) is 150 mm in diameter and 8 mm thick. It carries a compressive longitudinal load of 40 kN.
(a) At what internal pressure will the longitudinal stress be zero?
(b) What will be the corresponding hoop stress?

13.11 A cylinder (FIg. 13.15) is 8 in. in diameter and 0.30 in. thick. It carries a compressive longitudinal load of 8000 lb.
(a) At what pressure will the longitudinal stress be zero?
(b) What will be the corresponding hoop stress?

13.12 The connection shown in Fig. 13.4 is 2 in. wide. The rivet is $\frac{1}{2}$ in. in diameter and the plates are $\frac{1}{2}$ in. thick. Based on the properties given in Section 13.3, find the load this connection can carry.

13.13 The connection shown in Fig. 13.4 is 20 mm wide. The rivet is 12 mm in

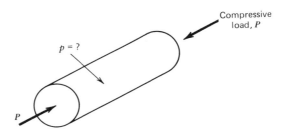

FIG. 13.15 Problems 13.10, 13.11.

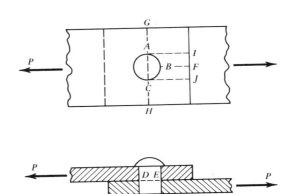

FIG. 13.4 Problems 13.12, 13.13.

diameter and the plates are 12 mm thick. Based on the properties given in Section 13.3, find the load this connection can carry.

13.14 The joint shown in Fig. 13.16 consists of a single set of $\frac{1}{4}$-in. rivets. The external plates are $\frac{1}{4}$ in. thick and the internal ones are $\frac{3}{8}$ in. thick. The plates are $\frac{3}{4}$ in. wide. If the load on this joint is 1000 lb find:

(a) The shear stress in the rivets.

(b) The tensile stress in the exterior plates.

(c) The tensile stress in the interior plates.

(d) The bearing stress in the exterior plates.

(e) The bearing stress in the interior plates.

13.15 Do Problem 13.14 if there are two sets of rivets and the plates are $1\frac{1}{4}$ in. wide.

13.16 Do Problem 13.14 if there are three sets of rivets and the plates are $1\frac{3}{4}$ in. wide.

13.17 The joint shown in Fig. 13.16 consists of a single set of 6-mm rivets. The external plates are 6 mm thick and the internal ones are 10 mm thick. The plates are 20 mm wide. If the load on this joint is 5000 N find:

(a) The shear stress in the rivets.

(b) The tensile stress in the exterior plates.

(c) The tensile stress in the interior plates.

(d) The bearing stress in the exterior plates.

(e) The bearing stress in the interior plates.

13.18 Do Problem 13.17 if there are two sets of rivets and the plates are 30 mm wide.

13.19 Do Problem 13.17 if there are three sets of rivets and the plates are 40 mm wide.

13.20 Find the load the joint described in Problem 13.14 can carry based on the properties given in Section 13.3.

13.21 Find the load the joint described in Problem 13.17 can carry based on the properties given in Section 13.3.

13.22 The simple lap joint in Fig. 13.17 has 1-in. rivets, and the load is 12,000

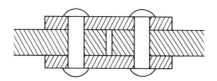

FIG. 13.16 Problems 13.14, 13.15, 13.16, 13.17, 13.18, 13.19, 13.20, 13.21.

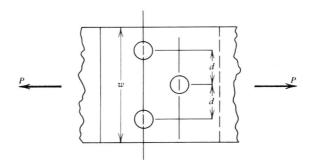

FIG. 13.17 Problems 13.22, 13.23, 13.24, 13.25.

lb. The plates are $\frac{1}{2}$ in. thick. Make the relevant stress calculations for the design of this joint based on the following dimensions:

$$d = 2.5 \text{ in.}$$
$$w = 8.0 \text{ in.}$$

13.23 The simple lap joint in Fig. 13.17 has 25-mm rivets and the load is 60 kN. The plates are 12 mm thick. Make the relevant stress calculations for the design of this joint based on the following dimensions:

$$d = 60 \text{ mm}$$
$$w = 200 \text{ mm}$$

13.24 Given the geometry of Problem 13.22, the material properties of Section 13.3, and a factor of safety of 4, what load can be can be carried by the joint of Fig. 13.17?

13.25 Given the geometry of Problem 13.23, the material properties of Section 13.3, and a factor of safety of 5, what load can be carried by the joint of Fig. 13.17?

13.26 For the lap joint shown in Fig. 13.7 what minimum weld size (w) is required if the weld is to be 2 in. long, the allowable stress is 13 ksi, and the load is 2000 lb?

13.27 For the lap joint shown in Fig. 13.7, what minimum weld size (w) is required if the weld is to be 80 mm long, the allowable stress is 90 MPa, and the load is 12 kN?

13.28 What size (w) fillet weld is needed for the weld design of Example Problem 13.3 if an allowable stress of 90 MPa is used?

13.29 For the joint shown in Fig. 13.18 find the average shear stress in the $\frac{1}{4}$-in. weld if A is 3 in. and the load is 800 lb.

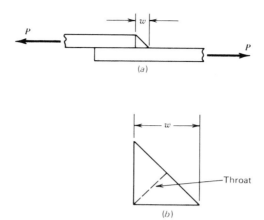

FIG. 13.7 Problems 13.26, 13.27.

13.30 For the joint shown in Fig. 13.18 find the average shear stress in the 10-mm weld if A is 80 mm and the load is 4 kN.

13.31 If the load in Fig. 13.18 is 12,000 lb and the maximum value of A is 6 in., design an appropriate weld (length and width) for this connection based on an allowable stress of 14,000 psi.

13.32 If the load in Fig. 13.18 is 60 kN and the maximum value of A is 200 mm, design an appropriate weld (length and width) for this connection based on an allowable stress of 100 MPa.

13.33 If the strap in Fig. 13.18 is 2 in. wide, $\frac{3}{8}$ in. thick, and overlaps the $\frac{3}{8}$-in.-thick plate by 3 in., what is the maximum load P that can be carried based on an allowable stress of 14 ksi and a safety factor of 3?

13.34 A 5-in.-wide band is $\frac{1}{2}$ in. thick. It has a 1.5-in. hole cut in its center. It is

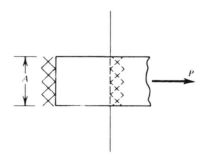

FIG. 13.18 Problems 13.29, 13.30, 13.31, 13.32, 13.33.

to carry 2000 lb in tension. Find the maximum stress in the band, including the effect of stress concentration.

13.35 A 150-mm-wide band is 40 mm thick. It has a 80-mm hole cut in its center. It is to carry 15 kN in tension. Find the maximum stress in the band, including the effect of stress concentration.

13.36 A 5-in.-wide band is $\frac{1}{2}$ in. thick. It uniformly tapers to a 3-in. band using 1-in.-radius fillets. Find the maximum stress including stress concentration in the band if it is to carry 3000 lb in tension.

13.37 A 150-mm-wide band is 40 mm thick. It uniformly tapers to a 100-mm band using 25-mm radius fillets. Find the maximum stress including stress concentration in the band if it is to carry 25 kN in tension.

13.38 A 4-in.-wide steel strap has a 1-in. hole in it. How thick must this strap be to carry 1500 lb in tension based on an allowable stress of 20 ksi?

13.39 A 80-mm-wide steel strap has a 20-mm hole in it. How thick must this strap be to carry 6 kN in tension based on an allowable stress of 140 MPa?

14

Statically Indeterminate Members

14.1 Axially loaded members
14.2 Beams
14.3 Beams by superposition
14.4 Summary

Up to this point we have dealt with a very convenient type of problem—those whose external loads could be found by the methods of statics alone. Needing a name for all God's children we call these creatures "statically determinate structures," meaning that from the principles of *statics* alone we can *determine* the loads on these structures. Fortunately, many real problems are statically determinate, and our effort is not in vain. An additional large class of problems may be treated as though they were statically determinant without introducing appreciable error in the analysis. (Considerable judgement–read experience—no, read, you've done it wrong enough times to know better—is required to know when loads may be safely neglected.) Alas, there are those problems that are not statically determinate, nor can they reasonably be approximated as such. These are called statically indeterminate structures.

Statically indeterminate members occur anytime the number of unknowns present exceeds the number of independent, nontrivial statics equations available. Thus such problems cannot be solved by the methods of statics alone. From algebra we know that we must have as many equations as there are unknowns for a problem to be solvable. Additional equations—beyond those available from statics—are obtained by considering the geometric constraints of the problem and the force-deflection relations that exist. Thus the force-deflection relations of Chapters 3, 5, and 8 will be very important.

For any two-dimensional free body diagram there will be, at most, three statics equations. Therefore, anytime our free body diagram shows more than

352 STATICALLY INDETERMINATE MEMBERS

three unknowns we are potentially dealing with a statically indeterminate problem. As we shall see in the following sections there are problems where three or fewer unknowns result in statically indeterminate problems.

14.1 AXIALLY LOADED MEMBERS

The steel reinforced column shown in Fig. 14.1a is an axially loaded structure that presents a statically indeterminate problem. In this case we wish to know what portion of the load is carried by the concrete and what is carried by the steel. A free body diagram is drawn in Fig. 14.1b after passing a cutting plane through an arbitrary section of the column. Summing forces vertically we get

$$\Sigma F y = 0 = -20 \; kN + F_s + F_c \tag{a}$$

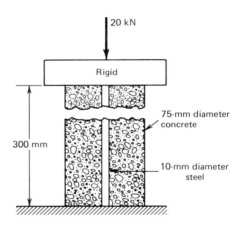

FIG. 14.1a Axially loaded member.

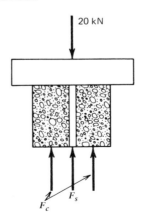

FIG. 14.1b Free body diagram.

where F_s is the force in the steel and F_c is the force in the concrete. The remaining two statics equations only tell us that zero equals zero. We call these equations trivial, since they offer no useful information. We are left with the single equation (a) that contains two unknowns, which, according to my abacus, is one too many.

What is needed is a second equation. The second equation is generated from geometric constraints and force-deflection relationships.

First, the geometric constraints. We note that the cap is rigid and will translate in the direction of the 20-kN load. Thus there is a compression of the column δ that will be common for the concrete and the steel, as shown in Fig. 14.1c, and we may write

$$\delta_s = \delta_c \tag{b}$$

This equation represents the geometric constraints. It is generated by recognizing the deflection that takes place in the structure. This will be greatly aided if a detail sketch of the deflection is drawn. Now we have two equations, (a) and (b), but four unknowns. (So far we have taken one step backward!)

But on to the deflection equations. Equation 3.5 relates axial load to axial deflection. Thus, we may write for the steel reinforcing rod

$$\delta_s = \frac{F_s l}{A_s E_s} \tag{c}$$

and for the concrete

$$\delta_c = \frac{F_c l}{A_c E_c} \tag{d}$$

Equations (c) and (d) may be substituted into equation (b), yielding

$$\frac{F_s l}{A_s E_s} = \frac{F_c l}{A_c E_c} \tag{e}$$

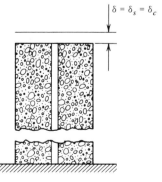

FIG. 14.1c Deflection diagram.

354 STATICALLY INDETERMINATE MEMBERS

Now we have two equations, (a) and (e), and two unknowns, F_s and F_c, and once more it's algebra from here.

Our problem here was to find F_s and F_c. Of course, once they are found, we may compute the actual deflection of the column and the stress in the concrete and steel.

The above principles are basic to problems of this type. As their complexity increases, it will be necessary to apply the principle of superposition. The same approach is also useful for many problems involving stresses caused by temperature changes in a structure. This will be demonstrated in Example Problem 14.2.

EXAMPLE PROBLEM 14.1

For the problem in Fig. 14.1 find the force and stress in each member and the total deflection.

Solution

From the previous analysis we have

$$F_c + F_s = 20 \qquad (a)$$

and

$$\frac{F_s l}{A_s E_s} = \frac{F_c l}{A_c E_c} \qquad (e)$$

Solving for F_s

$$F_s = \frac{A_s E_s}{A_c E_c} F_c$$

$$A_s = \frac{\pi D_s^2}{4} = \frac{\pi (10 \text{ mm})^2}{4} = 78.5 \text{ mm}^2$$

$$A_c = \frac{\pi D_c^2}{4} - A_s = \frac{\pi (75 \text{ mm})^2}{4} - 78.5 = 4339 \text{ mm}^2$$

Then

$$F_s = \frac{78.5(207E9)}{4340(14E9)} F_c = 0.267 F_c$$

Substituting into equation (a) gives

$$F_c + 0.267 F_c = 20$$

$$F_c = \frac{20}{1.267} = \textbf{15.8 kN}$$

$$F_s = 4.22 \text{ kN}$$

then

$$\sigma_s = \frac{F_s}{A_s} = \frac{4.22 \text{ kN}}{78.5 \text{ mm}^2} = 53.8 \text{ MPa}$$

$$\sigma_c = \frac{F_c}{A_c} = \frac{15.8 \text{ kN}}{4340 \text{ mm}^2} = 3.64 \text{ MPa}$$

and

$$\delta = \frac{F_s l}{A_s E_s} = \frac{4.22 \text{ kN } 300 \text{ mm}(\text{mm})^2}{78.5 \text{ mm}^2 \, 207E3 \text{ N}} = 77.9 \, \mu\text{m}$$

EXAMPLE PROBLEM 14.2

The concentric pipe assembly shown in Fig. 14.2a is firmly attached at both ends. Find the stress generated in each pipe if the assembly is heated 200°F.

$$A_s = 5.581 \text{ in.}^2 \qquad A_B = 1.075 \qquad L = 60 \text{ in.}$$

Solution

First we compute the change in each member that would occur due to change in temperature if it were free, Fig. 14.2b. From Eq. 3.9, we have

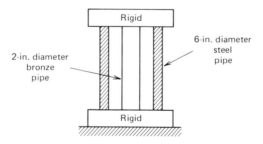

FIG. 14.2a

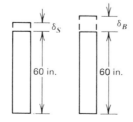

FIG. 14.2b Change in length due to temperature change.

356 STATICALLY INDETERMINATE MEMBERS

$$\varepsilon_s = \alpha(\Delta T)$$

from which

$$\delta_s = \alpha L(\Delta T)$$

$$= 6.5 \frac{E-6}{°F} (60 \text{ in.}) (200°F)$$

$$\delta_s = 7.80E - 2 \text{ in.}$$

Similarly,

$$\delta_B = 10 \frac{E-6}{°F} (60 \text{ in.}) (200°F) = 1.20E - 1 \text{ in.}$$

The bronze wants to expand more than the steel (Fig. 14.2c) by the amount

$$\delta = \delta_B - \delta_s = (1.20E - 1) - (7.80E - 2) = 4.20E - 2 \text{ in.}$$

Hence it will stretch the steel an additional amount δ_{s-p}, generating a tensile force P_s. The steel, on the other hand, will pull the bronze back by the amount δ_{B-p}, generating a compressive force. Because of the rigid cap, the final length of the two pipes must be equal. Thus, the stretching of the steel plus the compression of the bronze will close the gap created by the differences in deformation due to the temperature change and we may write

$$\delta = \delta_{s-p} + \delta_{B-p}$$

The deflection of each member due to its axial load is found by Eq. 3.5, giving

$$\delta = \frac{P_s L}{A_s E_s} + \frac{P_B L}{A_B E_B}$$

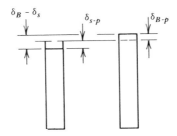

FIG. 14.2c Deflection diagram.

AXIALLY LOADED MEMBERS 357

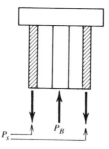

FIG. 14.2d Free body diagram.

The forces P_s and P_B are unknown in the above equation. We pass a cutting plane through the assembly and draw a free body diagram, Fig. 14.2d. There being no external forces, we conclude

$$P_B = P_s = P$$

substituting into the above equation

$$\delta = \frac{PL}{A_s E_s} + \frac{PL}{A_B E_B} = PL\left(\frac{1}{A_s E_s} + \frac{1}{A_B E_B}\right)$$

$$P = \frac{\delta}{L\left(\dfrac{1}{A_s E_s} + \dfrac{1}{A_B E_B}\right)}$$

$$P = \frac{4.20E-2 \text{ in.}}{60 \text{ in.}\left[\dfrac{1 \text{ in.}^2}{5.581 \text{ in.}^2 (30E6)\text{lb}} + \dfrac{1 \text{ in.}^2}{1.075 \text{ in.}^2 (16E6)\text{lb}}\right]}$$

P = 10,900 lb

This is the load in each pipe. From it we calculate the following stresses.

$$\sigma_s = \frac{10,900}{5.581} = \textbf{1960 psi}$$

$$\sigma_B = \frac{10,900}{1.075} = \textbf{10,200 psi}$$

Note that the final deformation of the whole structure is greater than that for the steel alone, but less than that of the bronze alone.

14.2 BEAMS

Beams show up in at least two ways in statically indeterminate problems. The first is just as another elastic element in a general indeterminate problem. It will be illustrated in Example Problem 14.3. The second deals specifically with the indeterminate beam, and it will be handled by the method of superposition in the next section.

EXAMPLE PROBLEM 14.3

For Fig. 14.3a find the load carried by the steel cable with a cross section of 8 mm^2. The beam is steel. $I = 4.17E6$ mm^4.

Solution

We draw a free body diagram, Fig. 14.3b, from which we get two equations and three unknowns, P_c, V, and M. The force equation is

$$\Sigma F_y = 0 = V + P_c - 1200$$

An additional equation is obtained by observing that the deflection of the beam and the cable are the same, Fig. 14.3c.

$$\delta_b = \delta_c$$

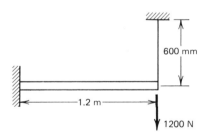

FIG. 14.3a

FIG. 14.3b Free body diagram.

FIG. 14.3c Deflection diagram.

From Chapter 3 the deflection of the cable due to the load on it, P_c, is

$$\delta_c = \frac{PL}{AE} = \frac{P_c(600 \text{ mm}) \text{ mm}^2}{8 \text{ mm}^2 (207E3 \text{ N})} = 3.62E-4 \, P_c$$

From Table A8.1 the deflection of the beam due to the *net* load on it, P_b, is

$$\delta_b = \frac{P_b L^3}{3EI} = \frac{P_b(1200 \text{ mm})^3 \text{ mm}^2}{3(207E3 \text{ N})(4.17E6 \text{ mm}^4)} = 6.67E-4$$

Thus

$$(3.62E-4) \, P_c = (6.67E-4) \, P_b$$

from which

$$P_c = 1.84 \, P_b$$

P_b is the same as V so we can substitute into the first equation, giving

$$P_b + 1.84 \, P_b = 1200$$

$$P_b = \frac{1200}{2.84} = \mathbf{422 \text{ N}}$$

This is the net force on the right end of the beam and the shear on the left end. For force in the cable is

$$P_c = 1.84 \, P_b = \mathbf{777 \text{ N}}$$

14.3 BEAMS BY SUPERPOSITION

The problem of a statically indeterminate beam is illustrated in Fig. 14.4. This beam has three supports. A free body diagram (Fig. 14.4b) shows three unknowns, R_A, R_B, and R_C. We can write two nontrivial equations. Thus there is one excess unknown, and the problem is said to be indeterminate to the first degree. Alternately, we may say there is one redundant support. To solve a problem of this type, we approach it in a manner similar to those previously considered in this chapter, plus we use the method of superposition. We'll think of one of the reactions, R_B in this case, as simply another external load. We will then say, based on the method of superposition, that the solution for the unknowns in Fig. 14.4b is the sum of the solutions in Fig. 14.4c and 14.4d. That is, if we add Fig. 14.4d to Fig. 14.4c, we get Fig. 14.4b. Therefore,

$$R_A = R_{A1} + R_{A2}$$

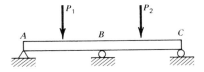

FIG. 14.4a

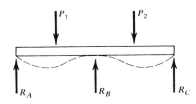

FIG. 14.4b

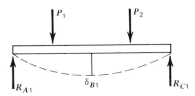

FIG. 14.4c

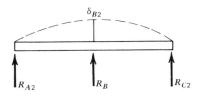

FIG. 14.4d

$$R_C = R_{C1} + R_{C2}$$

Now we still have an excessive unknown. It is handled by imposing the geometric constraints of the problem on the solution. We note that the deflection at the support at B, Fig. 14.4b, is zero. Therefore, the deflection produced by the loading of Fig. 14.4c when added to that produced by Fig. 14.4d must produce a net deflection of zero for point B.

$$\delta_{B1} + \delta_{B2} = 0$$

Using equations from Table A8.1 allows us to relate the deflection to the load and consequently the loads to each other. This generates the additional

EXAMPLE PROBLEM 14.4

The beam of Fig. 14.4 is 16 ft long with the supports and loads equally spaced at 4-ft intervals. The beam is a steel S 8×23.0 which has an I of 64.2 in.^{4} Find the reactions at supports A, B, and C if the loads P_1 and P_2 are each equal to 10 kips.

Solution

Referring to Fig. 14.4, as noted in the preceding discussion

$$-\delta_{B1} = \delta_{B2}$$

From Table A8.1

$$\delta_{B1} = \frac{Pa}{24EI}(3L^2 - 4a^2)$$

$$= \frac{10 \text{ kips (4 ft)}}{24EI}\left[3(16 \text{ ft})^2 - 4(4 \text{ ft})^2\right]$$

$$\delta_{B1} = \frac{1173}{EI} \text{ kips-ft}^3$$

and

$$\delta_{B2} = \frac{PL^3}{48EI} = \frac{R_B(16 \text{ ft})^3}{48EI} = \frac{85.3}{EI} R_B \text{ ft}^3$$

Setting these equal since they are of opposite sign

$$\frac{85.3}{EI} R_B \text{ ft}^3 = \frac{1173 \text{ kips-ft}^3}{EI}$$

$$R_B = \frac{1173}{85.3} = \mathbf{13.75 \text{ kips}}$$

Now that R_B has been found, we may use either the free body diagrams Fig. 14.4c and d or return to Fig. 14.4b. In either case the problem has been reduced to one that is statically determinate. For Fig. 14.4b

$$\Sigma M_A = 0 = -4(10) + 8(13.75) - 12(10) + 16 R_c$$

$$R_c = \frac{40 - 110 + 120}{16} = \mathbf{3.12 \text{ kips}}$$

362 STATICALLY INDETERMINATE MEMBERS

$$\Sigma F_y = 0 = R_A - 10 + 13.75 - 10 + 3.12$$

$$R_A = 3.12 \text{ kips}$$

The symmetric answer should have been anticipated.

14.4 SUMMARY

When the number of unknown forces exceeds the available statics equations, additional equations must be found. These are developed by observing the geometric constraints of the problem and using force-deflection relations. These principles coupled with the principle of superposition allow their solution.

Briefly we can summarize the general method of solution as follows:

1. Draw a free body diagram of the structure.
2. Write equilibrium equations and determine the degree of indeterminancy (number of excess unknowns).
3. Make a deflection sketch or sketches to develop a deflection equation for each excessive unknown. The principle of superposition may be necessary as well as additional free body diagrams.
4. Relate deflection to loads using force-deflection equations.
5. There should now be as many equations as there are unknowns. Carry out the algebra.

PROBLEMS

14.1 The 60-in.-long, 6-in.-diameter concrete column in Fig. 14.5 has two $\frac{1}{2}$-in. steel reinforcing rods.

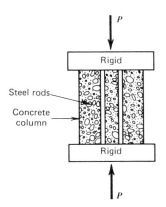

FIG. 14.5 Problems 14.1, 14.2, 14.3, 14.4, 14.15, 14.16, 14.17, 14.18.

(a) Find the loads carried by the concrete and the load carried by the steel if the applied load P is 2 kips.

(b) Find the stress in the concrete and in the steel rods.

(c) Calculate the deflection in each to see if they are equal.

14.2 Do Problem 14.1 if there are four 1-in.-diameter rods.

14.3 The 1.5-m-long, 150-mm-diameter concrete column in Fig. 14.5 has two 12-mm steel reinforcing rods.

(a) Find the load carried by the concrete and the load carried by the steel if the applied load P is 20 kN.

(b) Find the stress in the concrete and in the steel rods.

(c) Calculate the deflection in each to see if they are equal.

14.4 Do Problem 14.3 if there are four 25-mm-diameter rods.

14.5 The bronze rods in Fig. 14.6 are $\frac{1}{2}$ in. in diameter; the aluminum one is $\frac{3}{4}$ in. All are 48 in. long. The weight is 3000 lb. Find

(a) The load in each rod.

(b) The stress in each rod.

(c) The strain in each rod.

(d) The total elongation of each rod.

14.6 The bronze rods in Fig. 14.6 are $\frac{3}{4}$ in. in diameter; the aluminum one is 1 in. All are 36 in. long. The weight is 8000 lb. Find

(a) The load in each rod.

(b) The stress in each rod.

(c) The strain in each rod.

(d) The total elongation of each rod.

14.7 The bronze rods in Fig. 14.6 are 10 mm in diameter; the aluminum one is 15 mm. All are 1.2 m long. The weight is 12 kN. Find

(a) the load in each rod.

(b) The stress in each rod.

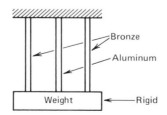

FIG. 14.6 Problems 14.5, 14.6, 14.7, 14.8, 14.19, 14.20, 14.21, 14.22.

364 STATICALLY INDETERMINATE MEMBERS

  (c) The strain in each rod.
  (d) The total elongation of each rod.

14.8 The bronze rods in Fig. 14.6, are 18 mm in diameter; the aluminum one is 25 mm. All are 2.0 m long. The weight is 30 kN. Find
  (a) The load in each rod.
  (b) The stress in each rod.
  (c) The strain in each rod.
  (d) The total elongation of each rod.

14.9 A 2-in.-diameter steel pipe carries a total load of 12,000 lb on the collar shown in Fig. 14.7. Find the reactions at the top and bottom of the support.

14.10 The two aluminum rods of Fig. 14.8 are exactly 18 in. long and have a cross section of 1.5 in.^{2} The bronze one has a cross section of 2.5 in.^{2} and is slightly short, leaving a gap of 0.008 in.
  (a) What force P is required to close the gap?
  (b) How will an additional 50,000 lb be carried in the members?
  (c) What will be the final length of the aluminum rods?

14.11 The two aluminum rods of Fig. 14.8 are exactly 30 in. long and have a cross section of 2.0 in.^{2} The bronze one has a cross section of 3.0 in.^{2} and is slightly short, leaving a gap of 0.010 in.
  (a) What force P is required to close the gap?
  (b) How will an additional 30,000 lb be carried in the members?
  (c) What will be the final length of the aluminum rods?

14.12 The two aluminum rods of Fig. 14.8 are exactly 500 mm long and have a cross section of 900 mm^2. The bronze one has a cross section of 1500 mm^2 and is slightly short, leaving a gap of 0.200 mm.

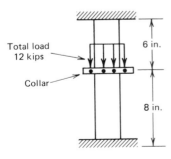

FIG. 14.7 Problem 14.9.

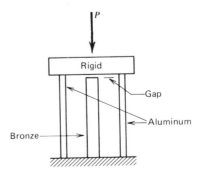

FIG. 14.8 Problems 14.10, 14.11, 14.12, 14.13, 14.23, 14.24.

(a) What force P is required to close the gap?
(b) How will an additional 2.0 MN be carried in the members?
(c) What will be the final length of the aluminum rods?

14.13 The two aluminum rods of Fig. 14.8 are exactly 800 mm long and have a cross section of 1500 mm². The bronze one has a cross section of 3000 mm² and is slightly short, leaving a gap of 0.300 mm.

(a) What force P is required to close the gap?
(b) How will an additional 3.0 MN be carried in the members?
(c) What will be the final length of the aluminum rods?

14.14 For Fig. 14.9 and the following conditions, find the load and elongation of each of the cables.

	a	b	c	d	P	Cable area	Material
(a)	5 ft	3 ft	2 ft	5 ft	1.5 kips	0.1 in.²	Aluminum
(b)	8 ft	4 ft	2 ft	14 ft	2.0 kips	0.2 in.²	Aluminum
(c)	2 m	4 m	2 m	4 m	10 kN	50 mm²	Aluminum
(d)	2 m	1 m	1 m	4 m	20 kN	80 mm²	Aluminum

14.15 The 60-in.-long, 6-in.-diameter concrete column in Fig. 14.5 has two $\tfrac{1}{2}$-in. diameter steel reinforcing rods. There is no external load on the column but with the ends attached to both the concrete and steel, it is heated 400°F. Find the stress in the steel rods and the concrete.

14.16 The 2-m-long, 150-mm-diameter concrete column in Fig. 14.5 has two 12-mm diameter steel reinforcing rods. There is no external load on the column but with the ends attached to both the concrete and the steel, it is heated 300°C. Find the stress in the steel rods and the concrete.

14.17 If the column in Problem 14.15 also has a compressive load of 2 kips, find the stress in the steel and the concrete.

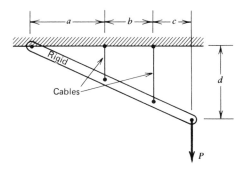

FIG. 14.9 Problem 14.14.

14.18 If the column in Problem 14.16 also has a compressive load of 20 kN, find the stress in the steel and the concrete.

14.19 Do Problem 14.5 if, in addition, the assembly is heated 300°F.

14.20 Do Problem 14.6, if, in addition, the assembly is heated 400°F.

14.21 Do Problem 14.7, if, in addition, the assembly is heated 250°C.

14.22 Do Problem 14.8, if, in addition, the assembly is heated 350°C.

14.23 The two aluminum rods of Fig. 14.8 are exactly 18 in. long and have a cross section of 1.5 in.^{2} The bronze rod has a cross section of 2.5 in.^{2} and is slightly short, leaving a gap of 0.008 in.
  (a) Find the change in temperature necessary to close the gap.
  (b) Find the stress in each member if the temperature drops another 150°F.
  (c) If after (b) 50,000 lb (compressive) are loaded on the structure, what will be the stress in each member?

14.24 The two aluminum rods of Fig. 14.8 are exactly 500 mm long and have a cross section of 900 mm^2. The bronze rod has a cross section of 1500 mm^2 and is slightly short, leaving a gap of 0.200 mm
  (a) Find the change in temperature necessary to close the gap.
  (b) Find the stress in each member if the temperature drops another 150°C.
  (c) If after (b) 200 kN (compressive) are loaded on the structure, what will be the stress in each member?

14.25 The steel cable in Fig. 14.10 is 10 ft long and 0.5 in.^{2} in area. The steel beam has a moment of inertia of 1200 in.^{4} If the loading is 300 lb/ft along the 16-ft span of the beam, find the load in the cable and the deflection of the end of the beam.

14.26 The aluminum cable in Fig. 14.10 is 6 ft long and 0.10 in.^{2} in area. The aluminum beam has a moment of inertia of 2400 in.^{4} If the load is 200

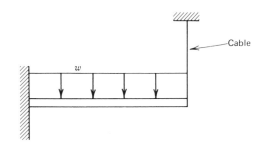

FIG. 14.10 Problems 14.25, 14.26, 14.27, 14.28.

lb/ft along the 20-ft span of the beam, find the load in the cable and the deflection of the end of the beam.

14.27 The aluminum cable in Fig. 14.10 is 2 m long and 60 mm² in area. The aluminum beam has a moment of inertia of $6.0E6$ mm⁴. If the load is 3 kN/m along the 7 m span of the beam, find the load in the cable and the deflection of the end of the beam.

14.28 The steel cable in Fig. 14.10 is 4 m long and 300 mm² in area. The aluminum beam has a moment of inertia of $4.0E6$ mm⁴. If the load is 5 kN/m along the 5 m-span of the beam, find the load in the cable and the deflection of the end of the beam.

14.29 In Fig. 14.11 the steel beam AC has a 12-ft span and I is 14.0 in.⁴ Beam BD (perpendicular to AC) has a span of 16 ft, I of 20.0 in.⁴ and is steel. Find the deflection of point B and the reaction at A. The load P is 8000 lb.

14.30 In Fig. 14.11, the steel beam AC has an 8-ft span and I is 10.0 in.⁴ Beam BD (perpendicular to AC) has a span of 12 ft, I of 15 in.⁴ and is aluminum. Find the deflection of point B and the reaction at C. The load P is 10,000 lb.

14.31 In Fig. 14.11, the aluminum beam AC has a 3-m span and I is $5.0E6$ mm⁴. Beam BD (perpendicular to AC) has a span of 4 m, I of $5.0E6$ mm and is aluminum. Find the deflection of point B and the reaction at D. The load P is 32 kN.

14.32 In Fig. 14.11, the aluminum beam AC has a 6-m span and I is $10E6$ mm⁴. Beam BD (perpendicular to AC) has a span of 6 m, I of $5.0E6$ mm⁴ and is aluminum. Find the deflection of point B and the reaction at A. The load P is 60 kN.

14.33 Find the reaction at each of the supports for the problems in Fig. 14.12.

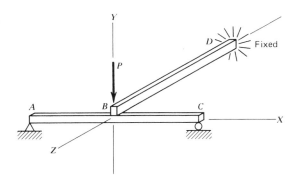

FIG. 14.11 Problems 14.29, 14.30, 14.31, 14.32.

368 STATICALLY INDETERMINATE MEMBERS

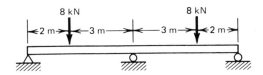

FIG. 14.12a   Problem 14.33a.

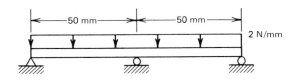

FIG. 14.12b   Problem 14.33b.

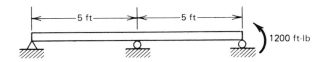

FIG. 14.12c   Problem 14.33c.

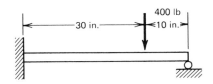

FIG. 14.12d   Problem 14.33d

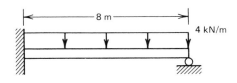

FIG. 14.12e   Problem 14.33e

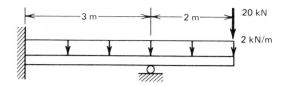

FIG. 14.12 f   Problem 14.33 f

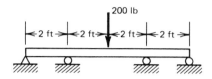

FIG. 14.12 g   Problem 14.33 g

Appendix Tables

A1.1 Metric-English conversion factors
A1.2 Derivatives of functions
A1.3 Integration of standard forms
A3.1 Typical physical properties of common materials
A4.1 Areas, centroids, and centroidal moments of inertia
A4.2 Standard beams
A4.3 Wide flange beams
A4.4 Channels
A4.5 Angles—equal legs
A4.6 Angles—unequal legs
A4.7 Pipe
A4.8 Screw threads
A4.9 Properties of structural lumber
A8.1 Beam deflection formulas

Table A1.1 Metric-English conversion factors

Length	Force	Angles [a]
$\dfrac{0.3048 \text{ m}^a}{\text{ft}}$	$\dfrac{4.448 \text{ N}}{\text{lb}_{\text{force}}}$	$\dfrac{2\pi \text{ rad}}{360°}$
$\dfrac{2.54 \text{ cm}^a}{\text{in.}}$	Stress	Power
Mass	$\dfrac{6895 \text{ Pa}}{\text{psi}}$	$\dfrac{745.7 \text{ W}}{\text{hp}}$
$\dfrac{0.4536 \text{ kg}}{\text{lb}_{\text{mass}}}$	Energy	$\dfrac{550 \text{ ft-lb/s}^a}{\text{hp}}$
	$\dfrac{1.356 \text{ J}}{\text{ft-lb}}$	

[a] Denotes conversion factors which are exact.
Note: Acceleration due to standard gravity: 9.807 m/s², 32.17 ft/s².

Table A1.2 Derivatives of functions

1. $\dfrac{d}{dx}(c) = 0$

2. $\dfrac{d}{dx}(x) = 1$

3. $\dfrac{d}{dx}(u+v-w) = \dfrac{du}{dx} + \dfrac{dv}{dx} - \dfrac{dw}{dx}$

4. $\dfrac{d}{dx}(cv) = \dfrac{c\,dv}{dx}$

5. $\dfrac{d}{dx}(uv) = \dfrac{u\,dv}{dx} + \dfrac{v\,du}{dx}$

6. $\dfrac{d}{dx}(v^n) = nv^{n-1}\dfrac{dv}{dx}$

7. $\dfrac{d}{dx}\left(\dfrac{u}{v}\right) = \dfrac{v\dfrac{du}{dx} - u\dfrac{dv}{dx}}{v^2}$

8. $\dfrac{d}{dx}(\ln v) = \dfrac{1}{v}\dfrac{dv}{dx}$

9. $\dfrac{d}{dx}(a^v) = a^v \ln a \dfrac{dv}{dx}$

10. $\dfrac{d}{dx}(\sin v) = \cos v \dfrac{dv}{dx}$

11. $\dfrac{d}{dx}(\cos v) = -\sin v \dfrac{dv}{dx}$

12. $\dfrac{d}{dx}(\tan v) = \sec^2 v \dfrac{dv}{dx}$

Table A1.3 Integration of standard forms

1. $\int (du + dv) = \int du + \int dv$ (u and v are variables)

2. $\int a\,dv = a\int dv$ (a is a constant)

3. $\int dx = x$

4. $\int x^n\,dx = \dfrac{x^{n+1}}{n+1}$ (n is a constant)

5. $\int \dfrac{dv}{v} = \ln v$

6. $\int e^x\,dx = e^x$

7. $\int \sin x\,dx = -\cos x$

8. $\int \cos x\,dx = \sin x$

Table A3.1 Typical physical properties of common materials

Material	Unit weight lb/in.³ (kN/m³)	Modulus of elasticity Mpsi (GPa)	Shear modulus of elasticity Mpsi (GPa)	Poisson's ratio	Coefficient of thermal expansion $\alpha \times E6$ /1°F (/1°C)	Ultimate strength kpsi (MPa)	Yield strength kpsi (MPa)
Aluminum	0.098 (26.6)	10.3 (71.0)	3.8 (26)	0.33	13 (23)	60 (414)	40 (276)
Bronze	0.295 (80.1)	16 (110)	6.0 (41)	0.35	10 (18)	40 (275)	10 (70)
Concrete	0.086 (23.3)	2 (14)			6.0 (10.8)	2 compression (14)	
Cast Iron	0.260 (70.6)	14.5 (100)	6.0 (41)	0.21	5.9 (10.6)	140 compression (965) 40 tension (476)	
Magnesium	0.065 (17.6)	6.5 (45)	2.4 (16)	0.35	14 (25)	44 (303)	31 (214)
Steel-carbon	0.282 (76.5)	30 (207)	11.5 (79.3)	0.29	6.5 (11.7)	50 (345)	36 (248)
Steel-nickel Alloy	0.280 (76.0)	30 (207)	11.5 (79.3)	0.29	6.5 (11.7)	200 (1380)	150 (1030)
Stainless Steel	0.280 (76.0)	27.6 (190)	10.6 (73.1)	0.30	6.2 (11.2)	110 (758)	40 (276)
Douglas Fir	0.017 (4.6)	1.6 (11.0)	0.6 (4.1)	0.33		7.4 (51)	
Southern Pine	0.021 (5.7)	1.6 (11.0)	0.6 (4.1)	0.33	3.0 (5.4)	8.4 (58)	
Nylon	0.039 (10.6)	1.8 (12.4)			80 (144)	7.4 (51)	6.0 (41)

APPENDIX TABLES 373

Table A4.1 Areas, centroids, and centroidal moments of inertia

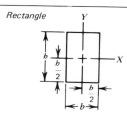

Rectangle

$A = bh$

$\bar{I}_x = \dfrac{bh^3}{12}$

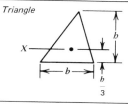

Triangle

$A = \dfrac{bh}{2}$

$\bar{I}_x = \dfrac{bh^3}{36}$

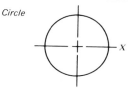

Circle

$A = \dfrac{\pi D^2}{4}$

$\bar{I}_x = \dfrac{\pi D^4}{64}$

$J_0 = \dfrac{\pi D^4}{32}$

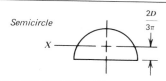

Semicircle

$A = \dfrac{\pi D^2}{8}$

$\bar{I}_x = 6.86E - 3 D^4$

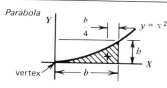

Parabola

$A = \tfrac{1}{3} bh$

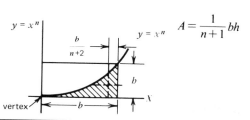

$A = \dfrac{1}{n+1} bh$

Table A4.2 Standard beams

S SHAPES — Properties for designing

Designation	Area A (In.²)	Depth d (In.)	Flange Width b_f (In.)	Flange Thickness t_f (In.)	Web Thickness t_w (In.)	Axis X-X I (In.⁴)	Axis X-X S (In.³)	Axis X-X r (In.)	Axis Y-Y I (In.⁴)	Axis Y-Y S (In.³)	Axis Y-Y r (In.)
S 24×120	35.3	24.00	8.048	1.102	0.798	3030	252	9.26	84.2	20.9	1.54
×105.9	31.1	24.00	7.875	1.102	0.625	2830	236	9.53	78.2	19.8	1.58
S 24×100	29.4	24.00	7.247	0.871	0.747	2390	199	9.01	47.8	13.2	1.27
× 90	26.5	24.00	7.124	0.871	0.624	2250	187	9.22	44.9	12.6	1.30
× 79.9	23.5	24.00	7.001	0.871	0.501	2110	175	9.47	42.3	12.1	1.34
S 20× 95	27.9	20.00	7.200	0.916	0.800	1610	161	7.60	49.7	13.8	1.33
× 85	25.0	20.00	7.053	0.916	0.653	1520	152	7.79	46.2	13.1	1.36
S 20× 75	22.1	20.00	6.391	0.789	0.641	1280	128	7.60	29.6	9.28	1.16
× 65.4	19.2	20.00	6.250	0.789	0.500	1180	118	7.84	27.4	8.77	1.19
S 18× 70	20.6	18.00	6.251	0.691	0.711	926	103	6.71	24.1	7.72	1.08
× 54.7	16.1	18.00	6.001	0.691	0.461	804	89.4	7.07	20.8	6.94	1.14
S 15× 50	14.7	15.00	5.640	0.622	0.550	486	64.8	5.75	15.7	5.57	1.03
× 42.9	12.6	15.00	5.501	0.622	0.411	447	59.6	5.95	14.4	5.23	1.07
S 12× 50	14.7	12.00	5.477	0.659	0.687	305	50.8	4.55	15.7	5.74	1.03
× 40.8	12.0	12.00	5.252	0.659	0.472	272	45.4	4.77	13.6	5.16	1.06
S 12× 35	10.3	12.00	5.078	0.544	0.428	229	38.2	4.72	9.87	3.89	0.980
× 31.8	9.35	12.00	5.000	0.544	0.350	218	36.4	4.83	9.36	3.74	1.00
S 10× 35	10.3	10.00	4.944	0.491	0.594	147	29.4	3.78	8.36	3.38	0.901
× 25.4	7.46	10.00	4.661	0.491	0.311	124	24.7	4.07	6.79	2.91	0.954
S 8× 23	6.77	8.00	4.171	0.425	0.441	64.9	16.2	3.10	4.31	2.07	0.798
× 18.4	5.41	8.00	4.001	0.425	0.271	57.6	14.4	3.26	3.73	1.86	0.831
S 7× 20	5.88	7.00	3.860	0.392	0.450	42.4	12.1	2.69	3.17	1.64	0.734
× 15.3	4.50	7.00	3.662	0.392	0.252	36.7	10.5	2.86	2.64	1.44	0.766
S 6× 17.25	5.07	6.00	3.565	0.359	0.465	26.3	8.77	2.28	2.31	1.30	0.675
× 12.5	3.67	6.00	3.332	0.359	0.232	22.1	7.37	2.45	1.82	1.09	0.705
S 5× 14.75	4.34	5.00	3.284	0.326	0.494	15.2	6.09	1.87	1.67	1.01	0.620
× 10	2.94	5.00	3.004	0.326	0.214	12.3	4.92	2.05	1.22	0.809	0.643
S 4× 9.5	2.79	4.00	2.796	0.293	0.326	6.79	3.39	1.56	0.903	0.646	0.569
× 7.7	2.26	4.00	2.663	0.293	0.193	6.08	3.04	1.64	0.764	0.574	0.581
S 3× 7.5	2.21	3.00	2.509	0.260	0.349	2.93	1.95	1.15	0.586	0.468	0.516
× 5.7	1.67	3.00	2.330	0.260	0.170	2.52	1.68	1.23	0.455	0.390	0.522

Courtesy American Institute of Steel Construction, Inc.

Table A4.3 Wide flange beams

**W SHAPES
Properties for designing**

Designation	Area A	Depth d	Flange		Web Thickness t_w	Elastic Properties					
			Width b_f	Thickness t_f		Axis X-X			Axis Y-Y		
						I	S	r	I	S	r
	In.²	In.	In.	In.	In.	In.⁴	In.³	In.	In.⁴	In.³	In.
W 36×300	88.3	36.72	16.655	1.680	0.945	20300	1110	15.2	1300	156	3.83
×280	82.4	36.50	16.595	1.570	0.885	18900	1030	15.1	1200	144	3.81
×260	76.5	36.24	16.551	1.440	0.841	17300	952	15.0	1090	132	3.77
×245	72.1	36.06	16.512	1.350	0.802	16100	894	15.0	1010	123	3.75
×230	67.7	35.88	16.471	1.260	0.761	15000	837	14.9	940	114	3.73
W 36×194	57.2	36.48	12.117	1.260	0.770	12100	665	14.6	375	61.9	2.56
×182	53.6	36.32	12.072	1.180	0.725	11300	622	14.5	347	57.5	2.55
×170	50.0	36.16	12.027	1.100	0.680	10500	580	14.5	320	53.2	2.53
×160	47.1	36.00	12.000	1.020	0.653	9760	542	14.4	295	49.1	2.50
×150	44.2	35.84	11.972	0.940	0.625	9030	504	14.3	270	45.0	2.47
×135	39.8	35.55	11.945	0.794	0.598	7820	440	14.0	226	37.9	2.39
W 33×240	70.6	33.50	15.865	1.400	0.830	13600	813	13.9	933	118	3.64
×220	64.8	33.25	15.810	1.275	0.775	12300	742	13.8	841	106	3.60
×200	58.9	33.00	15.750	1.150	0.715	11100	671	13.7	750	95.2	3.57
W 33×152	44.8	33.50	11.565	1.055	0.635	8160	487	13.5	273	47.2	2.47
×141	41.6	33.31	11.535	0.960	0.605	7460	448	13.4	246	42.7	2.43
×130	38.3	33.10	11.510	0.855	0.580	6710	406	13.2	218	37.9	2.38
×118	34.8	32.86	11.484	0.738	0.554	5900	359	13.0	187	32.5	2.32
W 30×210	61.9	30.38	15.105	1.315	0.775	9890	651	12.6	757	100	3.50
×190	56.0	30.12	15.040	1.185	0.710	8850	587	12.6	673	89.5	3.47
×172	50.7	29.88	14.985	1.065	0.655	7910	530	12.5	598	79.8	3.43
W 30×132	38.9	30.30	10.551	1.000	0.615	5760	380	12.2	196	37.2	2.25
×124	36.5	30.16	10.521	0.930	0.585	5360	355	12.1	181	34.4	2.23
×116	34.2	30.00	10.500	0.850	0.564	4930	329	12.0	164	31.3	2.19
×108	31.8	29.82	10.484	0.760	0.548	4470	300	11.9	146	27.9	2.15
× 99	29.1	29.64	10.458	0.670	0.522	4000	270	11.7	128	24.5	2.10

Courtesy American Institute of Steel Construction, Inc.

Table A4.3 Wide flange beams *(continued)*

W SHAPES
Properties for designing

Designation	Area A	Depth d	Flange Width b_f	Flange Thickness t_f	Web Thickness t_w	Axis X-X I	Axis X-X S	Axis X-X r	Axis Y-Y I	Axis Y-Y S	Axis Y-Y r
	In.²	In.	In.	In.	In.	In.⁴	In.³	In.	In.⁴	In.³	In.
W 14×136	40.0	14.75	14.740	1.063	0.660	1590	216	6.31	568	77.0	3.77
×127	37.3	14.62	14.690	0.998	0.610	1480	202	6.29	528	71.8	3.76
×119	35.0	14.50	14.650	0.938	0.570	1370	189	6.26	492	67.1	3.75
×111	32.7	14.37	14.620	0.873	0.540	1270	176	6.23	455	62.2	3.73
×103	30.3	14.25	14.575	0.813	0.495	1170	164	6.21	420	57.6	3.72
× 95	27.9	14.12	14.545	0.748	0.465	1060	151	6.17	384	52.8	3.71
× 87	25.6	14.00	14.500	0.688	0.420	967	138	6.15	350	48.2	3.70
W 14× 84	24.7	14.18	12.023	0.778	0.451	928	131	6.13	225	37.5	3.02
× 78	22.9	14.06	12.000	0.718	0.428	851	121	6.09	207	34.5	3.00
W 14× 74	21.8	14.19	10.072	0.783	0.450	797	112	6.05	133	26.5	2.48
× 68	20.0	14.06	10.040	0.718	0.418	724	103	6.02	121	24.1	2.46
× 61	17.9	13.91	10.000	0.643	0.378	641	92.2	5.98	107	21.5	2.45
W 14× 53	15.6	13.94	8.062	0.658	0.370	542	77.8	5.90	57.5	14.3	1.92
× 48	14.1	13.81	8.031	0.593	0.339	485	70.2	5.86	51.3	12.8	1.91
× 43	12.6	13.68	8.000	0.528	0.308	429	62.7	5.82	45.1	11.3	1.89
W 14× 38	11.2	14.12	6.776	0.513	0.313	386	54.7	5.88	26.6	7.86	1.54
× 34	10.0	14.00	6.750	0.453	0.287	340	48.6	5.83	23.3	6.89	1.52
× 30	8.83	13.86	6.733	0.383	0.270	290	41.9	5.74	19.5	5.80	1.49
W 14× 26	7.67	13.89	5.025	0.418	0.255	244	35.1	5.64	8.86	3.53	1.08
× 22	6.49	13.72	5.000	0.335	0.230	198	28.9	5.53	7.00	2.80	1.04

Courtesy American Institute of Steel Construction, Inc.

Table A4.3 Wide flange beams

W SHAPES
Properties for designing

Designation	Area A In.²	Depth d In.	Flange Width b_f In.	Flange Thickness t_f In.	Web Thickness t_w In.	Axis X-X I In.⁴	Axis X-X S In.³	Axis X-X r In.	Axis Y-Y I In.⁴	Axis Y-Y S In.³	Axis Y-Y r In.
W 36×300	88.3	36.72	16.655	1.680	0.945	20300	1110	15.2	1300	156	3.83
×280	82.4	36.50	16.595	1.570	0.885	18900	1030	15.1	1200	144	3.81
×260	76.5	36.24	16.551	1.440	0.841	17300	952	15.0	1090	132	3.77
×245	72.1	36.06	16.512	1.350	0.802	16100	894	15.0	1010	123	3.75
×230	67.7	35.88	16.471	1.260	0.761	15000	837	14.9	940	114	3.73
W 36×194	57.2	36.48	12.117	1.260	0.770	12100	665	14.6	375	61.9	2.56
×182	53.6	36.32	12.072	1.180	0.725	11300	622	14.5	347	57.5	2.55
×170	50.0	36.16	12.027	1.100	0.680	10500	580	14.5	320	53.2	2.53
×160	47.1	36.00	12.000	1.020	0.653	9760	542	14.4	295	49.1	2.50
×150	44.2	35.84	11.972	0.940	0.625	9030	504	14.3	270	45.0	2.47
×135	39.8	35.55	11.945	0.794	0.598	7820	440	14.0	226	37.9	2.39
W 33×240	70.6	33.50	15.865	1.400	0.830	13600	813	13.9	933	118	3.64
×220	64.8	33.25	15.810	1.275	0.775	12300	742	13.8	841	106	3.60
×200	58.9	33.00	15.750	1.150	0.715	11100	671	13.7	750	95.2	3.57
W 33×152	44.8	33.50	11.565	1.055	0.635	8160	487	13.5	273	47.2	2.47
×141	41.6	33.31	11.535	0.960	0.605	7460	448	13.4	246	42.7	2.43
×130	38.3	33.10	11.510	0.855	0.580	6710	406	13.2	218	37.9	2.38
×118	34.8	32.86	11.484	0.738	0.554	5900	359	13.0	187	32.5	2.32
W 30×210	61.9	30.38	15.105	1.315	0.775	9890	651	12.6	757	100	3.50
×190	56.0	30.12	15.040	1.185	0.710	8850	587	12.6	673	89.5	3.47
×172	50.7	29.88	14.985	1.065	0.655	7910	530	12.5	598	79.8	3.43
W 30×132	38.9	30.30	10.551	1.000	0.615	5760	380	12.2	196	37.2	2.25
×124	36.5	30.16	10.521	0.930	0.585	5360	355	12.1	181	34.4	2.23
×116	34.2	30.00	10.500	0.850	0.564	4930	329	12.0	164	31.3	2.19
×108	31.8	29.82	10.484	0.760	0.548	4470	300	11.9	146	27.9	2.15
× 99	29.1	29.64	10.458	0.670	0.522	4000	270	11.7	128	24.5	2.10

Courtesy American Institute of Steel Construction, Inc.

Table A4.3 Wide flange beams (continued)

W SHAPES
Properties for designing

Designation	Area A In.²	Depth d In.	Flange Width b_f In.	Flange Thickness t_f In.	Web Thickness t_w In.	Axis X-X I In.⁴	Axis X-X S In.³	Axis X-X r In.	Axis Y-Y I In.⁴	Axis Y-Y S In.³	Axis Y-Y r In.
W 27×177	52.2	27.31	14.090	1.190	0.725	6740	494	11.4	556	78.9	3.26
×160	47.1	27.08	14.023	1.075	0.658	6030	446	11.3	495	70.6	3.24
×145	42.7	26.88	13.965	0.975	0.600	5430	404	11.3	443	63.5	3.22
W 27×114	33.6	27.28	10.070	0.932	0.570	4090	300	11.0	159	31.6	2.18
×102	30.0	27.07	10.018	0.827	0.518	3610	267	11.0	139	27.7	2.15
× 94	27.7	26.91	9.990	0.747	0.490	3270	243	10.9	124	24.9	2.12
× 84	24.8	26.69	9.963	0.636	0.463	2830	212	10.7	105	21.1	2.06
W 24×160	47.1	24.72	14.091	1.135	0.656	5120	414	10.4	530	75.2	3.35
×145	42.7	24.49	14.043	1.020	0.608	4570	373	10.3	471	67.1	3.32
×130	38.3	24.25	14.000	0.900	0.565	4020	332	10.2	412	58.9	3.28
W 24×120	35.4	24.31	12.088	0.930	0.556	3650	300	10.2	274	45.4	2.78
×110	32.5	24.16	12.042	0.855	0.510	3330	276	10.1	249	41.4	2.77
×100	29.5	24.00	12.000	0.775	0.468	3000	250	10.1	223	37.2	2.75
W 24× 94	27.7	24.29	9.061	0.872	0.516	2690	221	9.86	108	23.9	1.98
× 84	24.7	24.09	9.015	0.772	0.470	2370	197	9.79	94.5	21.0	1.95
× 76	22.4	23.91	8.985	0.682	0.440	2100	176	9.69	82.6	18.4	1.92
× 68	20.0	23.71	8.961	0.582	0.416	1820	153	9.53	70.0	15.6	1.87
W 24× 61	18.0	23.72	7.023	0.591	0.419	1540	130	9.25	34.3	9.76	1.38
× 55	16.2	23.55	7.000	0.503	0.396	1340	114	9.10	28.9	8.25	1.34
W 21×142	41.8	21.46	13.132	1.095	0.659	3410	317	9.03	414	63.0	3.15
×127	37.4	21.24	13.061	0.985	0.588	3020	284	8.99	366	56.1	3.13
×112	33.0	21.00	13.000	0.865	0.527	2620	250	8.92	317	48.8	3.10
W 21× 96	28.3	21.14	9.038	0.935	0.575	2100	198	8.61	115	25.5	2.02
× 82	24.2	20.86	8.962	0.795	0.499	1760	169	8.53	95.6	21.3	1.99
W 21× 73	21.5	21.24	8.295	0.740	0.455	1600	151	8.64	70.6	17.0	1.81
× 68	20.0	21.13	8.270	0.685	0.430	1480	140	8.60	64.7	15.7	1.80
× 62	18.3	20.99	8.240	0.615	0.400	1330	127	8.54	57.5	13.9	1.77
× 55	16.2	20.80	8.215	0.522	0.375	1140	110	8.40	48.3	11.8	1.73
W 21× 49	14.4	20.82	6.520	0.532	0.368	971	93.3	8.21	24.7	7.57	1.31
× 44	13.0	20.66	6.500	0.451	0.348	843	81.6	8.07	20.7	6.38	1.27

Courtesy American Institute of Steel Construction, Inc.

Table A4.3 Wide flange beams

W SHAPES — Properties for designing

Designation	Area A (In.²)	Depth d (In.)	Flange Width b_f (In.)	Flange Thickness t_f (In.)	Web Thickness t_w (In.)	Axis X-X I (In.⁴)	Axis X-X S (In.³)	Axis X-X r (In.)	Axis Y-Y I (In.⁴)	Axis Y-Y S (In.³)	Axis Y-Y r (In.)
W 18×114	33.5	18.48	11.833	0.991	0.595	2040	220	7.79	274	46.3	2.86
×105	30.9	18.32	11.792	0.911	0.554	1850	202	7.75	249	42.3	2.84
× 96	28.2	18.16	11.750	0.831	0.512	1680	185	7.70	225	38.3	2.82
W 18× 85	25.0	18.32	8.838	0.911	0.526	1440	157	7.57	105	23.8	2.05
× 77	22.7	18.16	8.787	0.831	0.475	1290	142	7.54	94.1	21.4	2.04
× 70	20.6	18.00	8.750	0.751	0.438	1160	129	7.50	84.0	19.2	2.02
× 64	18.9	17.87	8.715	0.686	0.403	1050	118	7.46	75.8	17.4	2.00
W 18× 60	17.7	18.25	7.558	0.695	0.416	986	108	7.47	50.1	13.3	1.68
× 55	16.2	18.12	7.532	0.630	0.390	891	98.4	7.42	45.0	11.9	1.67
× 50	14.7	18.00	7.500	0.570	0.358	802	89.1	7.38	40.2	10.7	1.65
× 45	13.2	17.86	7.477	0.499	0.335	706	79.0	7.30	34.8	9.32	1.62
W 18× 40	11.8	17.90	6.018	0.524	0.316	612	68.4	7.21	19.1	6.34	1.27
× 35	10.3	17.71	6.000	0.429	0.298	513	57.9	7.05	15.5	5.16	1.23
W 16× 96	28.2	16.32	11.533	0.875	0.535	1360	166	6.93	224	38.8	2.82
× 88	25.9	16.16	11.502	0.795	0.504	1220	151	6.87	202	35.1	2.79
W 16× 78	23.0	16.32	8.586	0.875	0.529	1050	128	6.75	92.5	21.6	2.01
× 71	20.9	16.16	8.543	0.795	0.486	941	116	6.71	82.8	19.4	1.99
× 64	18.8	16.00	8.500	0.715	0.443	836	104	6.66	73.3	17.3	1.97
× 58	17.1	15.86	8.464	0.645	0.407	748	94.4	6.62	65.3	15.4	1.96
W 16× 50	14.7	16.25	7.073	0.628	0.380	657	80.8	6.68	37.1	10.5	1.59
× 45	13.3	16.12	7.039	0.563	0.346	584	72.5	6.64	32.8	9.32	1.57
× 40	11.8	16.00	7.000	0.503	0.307	517	64.6	6.62	28.8	8.23	1.56
× 36	10.6	15.85	6.992	0.428	0.299	447	56.5	6.50	24.4	6.99	1.52
W 16× 31	9.13	15.84	5.525	0.442	0.275	374	47.2	6.40	12.5	4.51	1.17
× 26	7.67	15.65	5.500	0.345	0.250	300	38.3	6.25	9.59	3.49	1.12

Courtesy American Institute of Steel Construction, Inc.

Table A4.3 Wide flange beams *(continued)*

W SHAPES
Properties for designing

Designation	Area A In.²	Depth d In.	Flange Width b_f In.	Flange Thickness t_f In.	Web Thickness t_w In.	Axis X-X I In.⁴	Axis X-X S In.³	Axis X-X r In.	Axis Y-Y I In.⁴	Axis Y-Y S In.³	Axis Y-Y r In.
W 14×730	215	22.44	17.889	4.910	3.069	14400	1280	8.18	4720	527	4.69
×665	196	21.67	17.646	4.522	2.826	12500	1150	7.99	4170	472	4.62
×605	178	20.94	17.418	4.157	2.598	10900	1040	7.81	3680	423	4.55
×550	162	20.26	17.206	3.818	2.386	9450	933	7.64	3260	378	4.49
×500	147	19.63	17.008	3.501	2.188	8250	840	7.49	2880	339	4.43
×455	134	19.05	16.828	3.213	2.008	7220	758	7.35	2560	304	4.37
W 14×426	125	18.69	16.695	3.033	1.875	6610	707	7.26	2360	283	4.34
×398	117	18.31	16.590	2.843	1.770	6010	657	7.17	2170	262	4.31
×370	109	17.94	16.475	2.658	1.655	5450	608	7.08	1990	241	4.27
×342	101	17.56	16.365	2.468	1.545	4910	559	6.99	1810	221	4.24
×314	92.3	17.19	16.235	2.283	1.415	4400	512	6.90	1630	201	4.20
×287	84.4	16.81	16.130	2.093	1.310	3910	465	6.81	1470	182	4.17
×264	77.6	16.50	16.025	1.938	1.205	3530	427	6.74	1330	166	4.14
×246	72.3	16.25	15.945	1.813	1.125	3230	397	6.68	1230	154	4.12
W 14×237	69.7	16.12	15.910	1.748	1.090	3080	382	6.65	1170	148	4.11
×228	67.1	16.00	15.865	1.688	1.045	2940	368	6.62	1120	142	4.10
×219	64.4	15.87	15.825	1.623	1.005	2800	353	6.59	1070	136	4.08
×211	62.1	15.75	15.800	1.563	0.980	2670	339	6.56	1030	130	4.07
×202	59.4	15.63	15.750	1.503	0.930	2540	325	6.54	980	124	4.06
×193	56.7	15.50	15.710	1.438	0.890	2400	310	6.51	930	118	4.05
×184	54.1	15.38	15.660	1.378	0.840	2270	296	6.49	883	113	4.04
×176	51.7	15.25	15.640	1.313	0.820	2150	282	6.45	838	107	4.02
×167	49.1	15.12	15.600	1.248	0.780	2020	267	6.42	790	101	4.01
×158	46.5	15.00	15.550	1.188	0.730	1900	253	6.40	745	95.8	4.00
×150	44.1	14.88	15.515	1.128	0.695	1790	240	6.37	703	90.6	3.99
×142	41.8	14.75	15.500	1.063	0.680	1670	227	6.32	660	85.2	3.97
W 14×320	94.1	16.81	16.710	2.093	1.890	4140	493	6.63	1640	196	4.17

Courtesy American Institute of Steel Construction, Inc.

Table A4.3 Wide flange beams

W SHAPES — Properties for designing

Designation	Area A (In.²)	Depth d (In.)	Flange		Web Thickness t_w (In.)	Elastic Properties					
			Width b_f (In.)	Thickness t_f (In.)		Axis X-X			Axis Y-Y		
						I (In.⁴)	S (In.³)	r (In.)	I (In.⁴)	S (In.³)	r (In.)
W 14×136	40.0	14.75	14.740	1.063	0.660	1590	216	6.31	568	77.0	3.77
×127	37.3	14.62	14.690	0.998	0.610	1480	202	6.29	528	71.8	3.76
×119	35.0	14.50	14.650	0.938	0.570	1370	189	6.26	492	67.1	3.75
×111	32.7	14.37	14.620	0.873	0.540	1270	176	6.23	455	62.2	3.73
×103	30.3	14.25	14.575	0.813	0.495	1170	164	6.21	420	57.6	3.72
× 95	27.9	14.12	14.545	0.748	0.465	1060	151	6.17	384	52.8	3.71
× 87	25.6	14.00	14.500	0.688	0.420	967	138	6.15	350	48.2	3.70
W 14× 84	24.7	14.18	12.023	0.778	0.451	928	131	6.13	225	37.5	3.02
× 78	22.9	14.06	12.000	0.718	0.428	851	121	6.09	207	34.5	3.00
W 14× 74	21.8	14.19	10.072	0.783	0.450	797	112	6.05	133	26.5	2.48
× 68	20.0	14.06	10.040	0.718	0.418	724	103	6.02	121	24.1	2.46
× 61	17.9	13.91	10.000	0.643	0.378	641	92.2	5.98	107	21.5	2.45
W 14× 53	15.6	13.94	8.062	0.658	0.370	542	77.8	5.90	57.5	14.3	1.92
× 48	14.1	13.81	8.031	0.593	0.339	485	70.2	5.86	51.3	12.8	1.91
× 43	12.6	13.68	8.000	0.528	0.308	429	62.7	5.82	45.1	11.3	1.89
W 14× 38	11.2	14.12	6.776	0.513	0.313	386	54.7	5.88	26.6	7.86	1.54
× 34	10.0	14.00	6.750	0.453	0.287	340	48.6	5.83	23.3	6.89	1.52
× 30	8.83	13.86	6.733	0.383	0.270	290	41.9	5.74	19.5	5.80	1.49
W 14× 26	7.67	13.89	5.025	0.418	0.255	244	35.1	5.64	8.86	3.53	1.08
× 22	6.49	13.72	5.000	0.335	0.230	198	28.9	5.53	7.00	2.80	1.04

Courtesy American Institute of Steel Construction, Inc.

Table A4.3 Wide flange beams *(continued)*

W SHAPES — Properties for designing

Designation	Area A (In.²)	Depth d (In.)	Flange Width b_f (In.)	Flange Thickness t_f (In.)	Web Thickness t_w (In.)	Axis X-X I (In.⁴)	Axis X-X S (In.³)	Axis X-X r (In.)	Axis Y-Y I (In.⁴)	Axis Y-Y S (In.³)	Axis Y-Y r (In.)
W 12×190	55.9	14.38	12.670	1.736	1.060	1890	263	5.82	590	93.1	3.25
×161	47.4	13.88	12.515	1.486	0.905	1540	222	5.70	486	77.7	3.20
×133	39.1	13.38	12.365	1.236	0.755	1220	183	5.59	390	63.1	3.16
×120	35.3	13.12	12.320	1.106	0.710	1070	163	5.51	345	56.0	3.13
×106	31.2	12.88	12.230	0.986	0.620	931	145	5.46	301	49.2	3.11
× 99	29.1	12.75	12.192	0.921	0.582	859	135	5.43	278	45.7	3.09
× 92	27.1	12.62	12.155	0.856	0.545	789	125	5.40	256	42.2	3.08
× 85	25.0	12.50	12.105	0.796	0.495	723	116	5.38	235	38.9	3.07
× 79	23.2	12.38	12.080	0.736	0.470	663	107	5.34	216	35.8	3.05
× 72	21.2	12.25	12.040	0.671	0.430	597	97.5	5.31	195	32.4	3.04
× 65	19.1	12.12	12.000	0.606	0.390	533	88.0	5.28	175	29.1	3.02
W 12× 58	17.1	12.19	10.014	0.641	0.359	476	78.1	5.28	107	21.4	2.51
× 53	15.6	12.06	10.000	0.576	0.345	426	70.7	5.23	96.1	19.2	2.48
W 12× 50	14.7	12.19	8.077	0.641	0.371	395	64.7	5.18	56.4	14.0	1.96
× 45	13.2	12.06	8.042	0.576	0.336	351	58.2	5.15	50.0	12.4	1.94
× 40	11.8	11.94	8.000	0.516	0.294	310	51.9	5.13	44.1	11.0	1.94
W 12× 36	10.6	12.24	6.565	0.540	0.305	281	46.0	5.15	25.5	7.77	1.55
× 31	9.13	12.09	6.525	0.465	0.265	239	39.5	5.12	21.6	6.61	1.54
× 27	7.95	11.96	6.497	0.400	0.237	204	34.2	5.07	18.3	5.63	1.52
W 12× 22	6.47	12.31	4.030	0.424	0.260	156	25.3	4.91	4.64	2.31	0.847
× 19	5.59	12.16	4.007	0.349	0.237	130	21.3	4.82	3.76	1.88	0.820
× 16.5	4.87	12.00	4.000	0.269	0.230	105	17.6	4.65	2.88	1.44	0.770
× 14	4.12	11.91	3.968	0.224	0.198	88.0	14.8	4.62	2.34	1.18	0.754

Courtesy American Institute of Steel Construction, Inc.

Table A4.3 Wide flange beams

W SHAPES
Properties for designing

Designation	Area A	Depth d	Flange Width b_f	Flange Thickness t_f	Web Thickness t_w	Axis X-X I	Axis X-X S	Axis X-X r	Axis Y-Y I	Axis Y-Y S	Axis Y-Y r
	In.²	In.	In.	In.	In.	In.⁴	In.³	In.	In.⁴	In.³	In.
W 10×112	32.9	11.38	10.415	1.248	0.755	719	126	4.67	235	45.2	2.67
×100	29.4	11.12	10.345	1.118	0.685	625	112	4.61	207	39.9	2.65
× 89	26.2	10.88	10.275	0.998	0.615	542	99.7	4.55	181	35.2	2.63
× 77	22.7	10.62	10.195	0.868	0.535	457	86.1	4.49	153	30.1	2.60
× 72	21.2	10.50	10.170	0.808	0.510	421	80.1	4.46	142	27.9	2.59
× 66	19.4	10.38	10.117	0.748	0.457	382	73.7	4.44	129	25.5	2.58
× 60	17.7	10.25	10.075	0.683	0.415	344	67.1	4.41	116	23.1	2.57
× 54	15.9	10.12	10.028	0.618	0.368	306	60.4	4.39	104	20.7	2.56
× 49	14.4	10.00	10.000	0.558	0.340	273	54.6	4.35	93.0	18.6	2.54
W 10× 45	13.2	10.12	8.022	0.618	0.350	249	49.1	4.33	53.2	13.3	2.00
× 39	11.5	9.94	7.990	0.528	0.318	210	42.2	4.27	44.9	11.2	1.98
× 33	9.71	9.75	7.964	0.433	0.292	171	35.0	4.20	36.5	9.16	1.94
W 10× 29	8.54	10.22	5.799	0.500	0.289	158	30.8	4.30	16.3	5.61	1.38
× 25	7.36	10.08	5.762	0.430	0.252	133	26.5	4.26	13.7	4.76	1.37
× 21	6.20	9.90	5.750	0.340	0.240	107	21.5	4.15	10.8	3.75	1.32
W 10× 19	5.61	10.25	4.020	0.394	0.250	96.3	18.8	4.14	4.28	2.13	0.874
× 17	4.99	10.12	4.010	0.329	0.240	81.9	16.2	4.05	3.55	1.77	0.844
× 15	4.41	10.00	4.000	0.269	0.230	68.9	13.8	3.95	2.88	1.44	0.809
× 11.5	3.39	9.87	3.950	0.204	0.180	52.0	10.5	3.92	2.10	1.06	0.787

Courtesy American Institute of Steel Construction, Inc.

Table A4.3 Wide flange beams

W SHAPES
Properties for designing

Designation	Area A (In.²)	Depth d (In.)	Flange Width b_f (In.)	Flange Thickness t_f (In.)	Web Thickness t_w (In.)	Axis X-X I (In.⁴)	Axis X-X S (In.³)	Axis X-X r (In.)	Axis Y-Y I (In.⁴)	Axis Y-Y S (In.³)	Axis Y-Y r (In.)
W 8×67	19.7	9.00	8.287	0.933	0.575	272	60.4	3.71	88.6	21.4	2.12
×58	17.1	8.75	8.222	0.808	0.510	227	52.0	3.65	74.9	18.2	2.10
×48	14.1	8.50	8.117	0.683	0.405	184	43.2	3.61	60.9	15.0	2.08
×40	11.8	8.25	8.077	0.558	0.365	146	35.5	3.53	49.0	12.1	2.04
×35	10.3	8.12	8.027	0.493	0.315	126	31.1	3.50	42.5	10.6	2.03
×31	9.12	8.00	8.000	0.433	0.288	110	27.4	3.47	37.0	9.24	2.01
W 8×28	8.23	8.06	6.540	0.463	0.285	97.8	24.3	3.45	21.6	6.61	1.62
×24	7.06	7.93	6.500	0.398	0.245	82.5	20.8	3.42	18.2	5.61	1.61
W 8×20	5.89	8.14	5.268	0.378	0.248	69.4	17.0	3.43	9.22	3.50	1.25
×17	5.01	8.00	5.250	0.308	0.230	56.6	14.1	3.36	7.44	2.83	1.22
W 8×15	4.43	8.12	4.015	0.314	0.245	48.1	11.8	3.29	3.40	1.69	0.876
×13	3.83	8.00	4.000	0.254	0.230	39.6	9.90	3.21	2.72	1.36	0.842
×10	2.96	7.90	3.940	0.204	0.170	30.8	7.80	3.23	2.08	1.06	0.839
W 6×25	7.35	6.37	6.080	0.456	0.320	53.3	16.7	2.69	17.1	5.62	1.53
×20	5.88	6.20	6.018	0.367	0.258	41.5	13.4	2.66	13.3	4.43	1.51
×15.5	4.56	6.00	5.995	0.269	0.235	30.1	10.0	2.57	9.67	3.23	1.46
W 6×16	4.72	6.25	4.030	0.404	0.260	31.7	10.2	2.59	4.42	2.19	0.967
×12	3.54	6.00	4.000	0.279	0.230	21.7	7.25	2.48	2.98	1.49	0.918
× 8.5	2.51	5.83	3.940	0.194	0.170	14.8	5.08	2.43	1.98	1.01	0.889
W 5×18.5	5.43	5.12	5.025	0.420	0.265	25.4	9.94	2.16	8.89	3.54	1.28
×16	4.70	5.00	5.000	0.360	0.240	21.3	8.53	2.13	7.51	3.00	1.26
W 4×13	3.82	4.16	4.060	0.345	0.280	11.3	5.45	1.72	3.76	1.85	0.991

Courtesy American Institute of Steel Construction, Inc.

Table A4.4 Channels

CHANNELS AMERICAN STANDARD
Properties for designing

Designation	Area A	Depth d	Flange Width b_f	Flange Average thickness t_f	Web thickness t_w	Axis X-X I	Axis X-X S	Axis X-X r	Axis Y-Y I	Axis Y-Y S	Axis Y-Y r	$\bar{x}$
	In.²	In.	In.	In.	In.	In.⁴	In.³	In.	In.⁴	In.³	In.	In.
C 15×50	14.7	15.00	3.716	0.650	0.716	404	53.8	5.24	11.0	3.78	0.867	0.799
×40	11.8	15.00	3.520	0.650	0.520	349	46.5	5.44	9.23	3.36	0.886	0.778
×33.9	9.96	15.00	3.400	0.650	0.400	315	42.0	5.62	8.13	3.11	0.904	0.787
C 12×30	8.82	12.00	3.170	0.501	0.510	162	27.0	4.29	5.14	2.06	0.763	0.674
×25	7.35	12.00	3.047	0.501	0.387	144	24.1	4.43	4.47	1.88	0.780	0.674
×20.7	6.09	12.00	2.942	0.501	0.282	129	21.5	4.61	3.88	1.73	0.799	0.698
C 10×30	8.82	10.00	3.033	0.436	0.673	103	20.7	3.42	3.94	1.65	0.669	0.649
×25	7.35	10.00	2.886	0.436	0.526	91.2	18.2	3.52	3.36	1.48	0.676	0.617
×20	5.88	10.00	2.739	0.436	0.379	78.9	15.8	3.66	2.81	1.32	0.691	0.606
×15.3	4.49	10.00	2.600	0.436	0.240	67.4	13.5	3.87	2.28	1.16	0.713	0.634
C 9×20	5.88	9.00	2.648	0.413	0.448	60.9	13.5	3.22	2.42	1.17	0.642	0.583
×15	4.41	9.00	2.485	0.413	0.285	51.0	11.3	3.40	1.93	1.01	0.661	0.586
×13.4	3.94	9.00	2.433	0.413	0.233	47.9	10.6	3.48	1.76	0.962	0.668	0.601
C 8×18.75	5.51	8.00	2.527	0.390	0.487	44.0	11.0	2.82	1.98	1.01	0.599	0.565
×13.75	4.04	8.00	2.343	0.390	0.303	36.1	9.03	2.99	1.53	0.853	0.615	0.553
×11.5	3.38	8.00	2.260	0.390	0.220	32.6	8.14	3.11	1.32	0.781	0.625	0.571
C 7×14.75	4.33	7.00	2.299	0.366	0.419	27.2	7.78	2.51	1.38	0.779	0.564	0.532
×12.25	3.60	7.00	2.194	0.366	0.314	24.2	6.93	2.60	1.17	0.702	0.571	0.525
× 9.8	2.87	7.00	2.090	0.366	0.210	21.3	6.08	2.72	0.968	0.625	0.581	0.541
C 6×13	3.83	6.00	2.157	0.343	0.437	17.4	5.80	2.13	1.05	0.642	0.525	0.514
×10.5	3.09	6.00	2.034	0.343	0.314	15.2	5.06	2.22	0.865	0.564	0.529	0.500
× 8.2	2.40	6.00	1.920	0.343	0.200	13.1	4.38	2.34	0.692	0.492	0.537	0.512
C 5× 9	2.64	5.00	1.885	0.320	0.325	8.90	3.56	1.83	0.632	0.449	0.489	0.478
× 6.7	1.97	5.00	1.750	0.320	0.190	7.49	3.00	1.95	0.478	0.378	0.493	0.484
C 4× 7.25	2.13	4.00	1.721	0.296	0.321	4.59	2.29	1.47	0.432	0.343	0.450	0.459
× 5.4	1.59	4.00	1.584	0.296	0.184	3.85	1.93	1.56	0.319	0.283	0.449	0.458
C 3× 6	1.76	3.00	1.596	0.273	0.356	2.07	1.38	1.08	0.305	0.268	0.416	0.455
× 5	1.47	3.00	1.498	0.273	0.258	1.85	1.24	1.12	0.247	0.233	0.410	0.438
× 4.1	1.21	3.00	1.410	0.273	0.170	1.66	1.10	1.17	0.197	0.202	0.404	0.437

Courtesy American Institute of Steel Construction, Inc.

Table A4.5 Angles—equal legs

ANGLES
Equal legs
Properties for designing

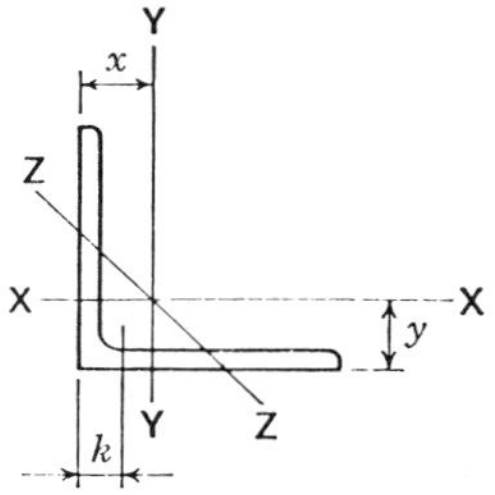

Size and Thickness	Weight per Foot	Area	AXIS X-X AND AXIS Y-Y				AXIS Z-Z
			I	S	r	x or y	r
In.	Lb.	In.²	In.⁴	In.³	In.	In.	In.
L 8 × 8 × 1⅛	56.9	16.7	98.0	17.5	2.42	2.41	1.56
1	51.0	15.0	89.0	15.8	2.44	2.37	1.56
⅞	45.0	13.2	79.6	14.0	2.45	2.32	1.57
¾	38.9	11.4	69.7	12.2	2.47	2.28	1.58
⅝	32.7	9.61	59.4	10.3	2.49	2.23	1.58
9/16	29.6	8.68	54.1	9.34	2.50	2.21	1.59
½	26.4	7.75	48.6	8.36	2.50	2.19	1.59
L 6 × 6 × 1	37.4	11.0	35.5	8.57	1.80	1.86	1.17
⅞	33.1	9.73	31.9	7.63	1.81	1.82	1.17
¾	28.7	8.44	28.2	6.66	1.83	1.78	1.17
⅝	24.2	7.11	24.2	5.66	1.84	1.73	1.18
9/16	21.9	6.43	22.1	5.14	1.85	1.71	1.18
½	19.6	5.75	19.9	4.61	1.86	1.68	1.18
7/16	17.2	5.06	17.7	4.08	1.87	1.66	1.19
⅜	14.9	4.36	15.4	3.53	1.88	1.64	1.19
5/16	12.4	3.65	13.0	2.97	1.89	1.62	1.20
L 5 × 5 × ⅞	27.2	7.98	17.8	5.17	1.49	1.57	.973
¾	23.6	6.94	15.7	4.53	1.51	1.52	.975
⅝	20.0	5.86	13.6	3.86	1.52	1.48	.978
½	16.2	4.75	11.3	3.16	1.54	1.43	.983
7/16	14.3	4.18	10.0	2.79	1.55	1.41	.986
⅜	12.3	3.61	8.74	2.42	1.56	1.39	.990
5/16	10.3	3.03	7.42	2.04	1.57	1.37	.994
L 4 × 4 × ¾	18.5	5.44	7.67	2.81	1.19	1.27	.778
⅝	15.7	4.61	6.66	2.40	1.20	1.23	.779
½	12.8	3.75	5.56	1.97	1.22	1.18	.782
7/16	11.3	3.31	4.97	1.75	1.23	1.16	.785
⅜	9.8	2.86	4.36	1.52	1.23	1.14	.788
5/16	8.2	2.40	3.71	1.29	1.24	1.12	.791
¼	6.6	1.94	3.04	1.05	1.25	1.09	.795

Courtesy American Institute of Steel Construction, Inc.

Table A4.5 Angles—equal legs *(continued)*

Size and Thickness	Weight per Foot	Area	AXIS X-X AND AXIS Y-Y				AXIS Z-Z
			I	S	r	x or y	r
In.	Lb.	In.²	In.⁴	In.³	In.	In.	In.
L 3½ × 3½ × ½	11.1	3.25	3.64	1.49	1.06	1.06	.683
⁷⁄₁₆	9.8	2.87	3.26	1.32	1.07	1.04	.684
⅜	8.5	2.48	2.87	1.15	1.07	1.01	.687
⁵⁄₁₆	7.2	2.09	2.45	.976	1.08	.990	.690
¼	5.8	1.69	2.01	.794	1.09	.968	.694
L 3 × 3 × ½	9.4	2.75	2.22	1.07	.898	.932	.584
⁷⁄₁₆	8.3	2.43	1.99	.954	.905	.910	.585
⅜	7.2	2.11	1.76	.833	.913	.888	.587
⁵⁄₁₆	6.1	1.78	1.51	.707	.922	.869	.589
¼	4.9	1.44	1.24	.577	.930	.842	.592
³⁄₁₆	3.71	1.09	.962	.441	.939	.820	.596
L 2½ × 2½ × ½	7.7	2.25	1.23	.724	.739	.806	.487
⅜	5.9	1.73	.984	.566	.753	.762	.487
⁵⁄₁₆	5.0	1.46	.849	.482	.761	.740	.489
¼	4.1	1.19	.703	.394	.769	.717	.491
³⁄₁₆	3.07	0.92	.547	.303	.778	.694	.495
L 2 × 2 × ⅜	4.7	1.36	.479	.351	.594	.636	.389
⁵⁄₁₆	3.92	1.15	.416	.300	.601	.614	.390
¼	3.19	.938	.348	.247	.609	.592	.391
³⁄₁₆	2.44	.715	.272	.190	.617	.569	.394
⅛	1.65	.484	.190	.131	.626	.546	.398
L 1¾ × 1¾ × ¼	2.77	.813	.227	.186	.529	.529	.341
³⁄₁₆	2.12	.621	.179	.144	.537	.506	.343
⅛	1.44	.422	.126	.099	.546	.484	.347
L 1½ × 1½ × ¼	2.34	.688	.139	.134	.449	.466	.292
³⁄₁₆	1.80	.527	.110	.104	.457	.444	.293
⁵⁄₃₂	1.52	.444	.094	.088	.461	.433	.295
⅛	1.23	.359	.078	.072	.465	.421	.296
L 1¼ × 1¼ × ¼	1.92	.563	.077	.091	.369	.403	.243
³⁄₁₆	1.48	.434	.061	.071	.377	.381	.244
⅛	1.01	.297	.044	.049	.385	.359	.246
L 1 × 1 × ¼	1.49	.438	.037	.056	.290	.339	.196
³⁄₁₆	1.16	.340	.030	.044	.297	.318	.195
⅛	.80	.234	.022	.031	.304	.296	.196

Courtesy American Institute of Steel Construction, Inc.

Table A4.6 Angles—unequal legs

ANGLES
Unequal legs
Properties for designing

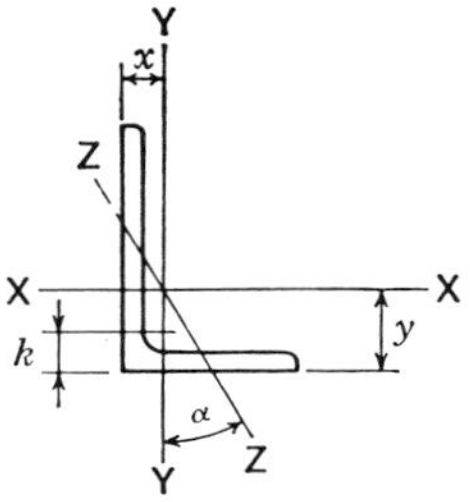

Size and Thickness	Weight per Foot	Area	AXIS X-X				AXIS Y-Y				AXIS Z-Z	
			I	S	r	y	I	S	r	x	r	Tan
In.	Lb.	In.²	In.⁴	In.³	In.	In.	In.⁴	In.³	In.	In.	In.	α
L 9 × 4 × 1	40.8	12.0	97.0	17.6	2.84	3.50	12.0	4.00	1.00	1.00	.834	.203
⅞	36.1	10.6	86.8	15.7	2.86	3.45	10.8	3.56	1.01	.953	.836	.208
¾	31.3	9.19	76.1	13.6	2.88	3.41	9.63	3.11	1.02	.906	.841	.212
⅝	26.3	7.73	64.9	11.5	2.90	3.36	8.32	2.65	1.04	.858	.847	.216
9⁄16	23.8	7.00	59.1	10.4	2.91	3.33	7.63	2.41	1.04	.834	.850	.218
½	21.3	6.25	53.2	9.34	2.92	3.31	6.92	2.17	1.05	.810	.854	.220
L 8 × 6 × 1	44.2	13.0	80.8	15.1	2.49	2.65	38.8	8.92	1.73	1.65	1.28	.543
⅞	39.1	11.5	72.3	13.4	2.51	2.61	34.9	7.94	1.74	1.61	1.28	.547
¾	33.8	9.94	63.4	11.7	2.53	2.56	30.7	6.92	1.76	1.56	1.29	.551
⅝	28.5	8.36	54.1	9.87	2.54	2.52	26.3	5.88	1.77	1.52	1.29	.554
9⁄16	25.7	7.56	49.3	8.95	2.55	2.50	24.0	5.34	1.78	1.50	1.30	.556
½	23.0	6.75	44.3	8.02	2.56	2.47	21.7	4.79	1.79	1.47	1.30	.558
7⁄16	20.2	5.93	39.2	7.07	2.57	2.45	19.3	4.23	1.80	1.45	1.31	.560
L 8 × 4 × 1	37.4	11.0	69.6	14.1	2.52	3.05	11.6	3.94	1.03	1.05	.846	.247
⅞	33.1	9.73	62.5	12.5	2.53	3.00	10.5	3.51	1.04	.999	.848	.253
¾	28.7	8.44	54.9	10.9	2.55	2.95	9.36	3.07	1.05	.953	.852	.258
⅝	24.2	7.11	46.9	9.21	2.57	2.91	8.10	2.62	1.07	.906	.857	.262
9⁄16	21.9	6.43	42.8	8.35	2.58	2.88	7.43	2.38	1.07	.882	.861	.265
½	19.6	5.75	38.5	7.49	2.59	2.86	6.74	2.15	1.08	.859	.865	.267
7⁄16	17.2	5.06	34.1	6.60	2.60	2.83	6.02	1.90	1.09	.835	.869	.269
L 7 × 4 × ⅞	30.2	8.86	42.9	9.65	2.20	2.55	10.2	3.46	1.07	1.05	.856	.318
¾	26.2	7.69	37.8	8.42	2.22	2.51	9.05	3.03	1.09	1.01	.860	.324
⅝	22.1	6.48	32.4	7.14	2.24	2.46	7.84	2.58	1.10	.963	.865	.329
9⁄16	20.0	5.87	29.6	6.48	2.24	2.44	7.19	2.35	1.11	.940	.868	.332
½	17.9	5.25	26.7	5.81	2.25	2.42	6.53	2.12	1.11	.917	.872	.335
7⁄16	15.8	4.62	23.7	5.13	2.26	2.39	5.83	1.88	1.12	.893	.876	.337
⅜	13.6	3.98	20.6	4.44	2.27	2.37	5.10	1.63	1.13	.870	.880	.340

Courtesy American Institute of Steel Construction, Inc.

Table A4.6 Angles—unequal legs *(continued)*

Size and Thickness		Weight per Foot	Area	AXIS X-X				AXIS Y-Y				AXIS Z-Z	
				I	S	r	y	I	S	r	x	r	Tan α
In.		Lb.	In.²	In.⁴	In.³	In.	In.	In.⁴	In.³	In.	In.	In.	
L 6 × 4 ×	7/8	27.2	7.98	27.7	7.15	1.86	2.12	9.75	3.39	1.11	1.12	.857	.421
	3/4	23.6	6.94	24.5	6.25	1.88	2.08	8.68	2.97	1.12	1.08	.860	.428
	5/8	20.0	5.86	21.1	5.31	1.90	2.03	7.52	2.54	1.13	1.03	.864	.435
	9/16	18.1	5.31	19.3	4.83	1.90	2.01	6.91	2.31	1.14	1.01	.866	.438
	1/2	16.2	4.75	17.4	4.33	1.91	1.99	6.27	2.08	1.15	.987	.870	.440
	7/16	14.3	4.18	15.5	3.83	1.92	1.96	5.60	1.85	1.16	.964	.873	.443
	3/8	12.3	3.61	13.5	3.32	1.93	1.94	4.90	1.60	1.17	.941	.877	.446
	5/16	10.3	3.03	11.4	2.79	1.94	1.92	4.18	1.35	1.17	.918	.882	.448
	1/4	8.3	2.44	9.27	2.26	1.95	1.89	3.41	1.10	1.18	.894	.887	.451
L 6 × 3½ ×	1/2	15.3	4.50	16.6	4.24	1.92	2.08	4.25	1.59	.972	.833	.759	.344
	3/8	11.7	3.42	12.9	3.24	1.94	2.04	3.34	1.23	.988	.787	.767	.350
	5/16	9.8	2.87	10.9	2.73	1.95	2.01	2.85	1.04	.996	.763	.772	.352
	1/4	7.9	2.31	8.86	2.21	1.96	1.99	2.34	0.847	1.01	.740	.777	.355
L 5 × 3½ ×	3/4	19.8	5.81	13.9	4.28	1.55	1.75	5.55	2.22	.977	.996	.748	.464
	5/8	16.8	4.92	12.0	3.65	1.56	1.70	4.83	1.90	.991	.951	.751	.472
	1/2	13.6	4.00	9.99	2.99	1.58	1.66	4.05	1.56	1.01	.906	.755	.479
	7/16	12.0	3.53	8.90	2.64	1.59	1.63	3.63	1.39	1.01	.883	.758	.482
	3/8	10.4	3.05	7.78	2.29	1.60	1.61	3.18	1.21	1.02	.861	.762	.486
	5/16	8.7	2.56	6.60	1.94	1.61	1.59	2.72	1.02	1.03	.838	.766	.489
	1/4	7.0	2.06	5.39	1.57	1.62	1.56	2.23	.830	1.04	.814	.770	.492
L 5 × 3 ×	1/2	12.8	3.75	9.45	2.91	1.59	1.75	2.58	1.15	.829	.750	.648	.357
	7/16	11.3	3.31	8.43	2.58	1.60	1.73	2.32	1.02	.837	.727	.651	.361
	3/8	9.8	2.86	7.37	2.24	1.61	1.70	2.04	.888	.845	.704	.654	.364
	5/16	8.2	2.40	6.26	1.89	1.61	1.68	1.75	.753	.853	.681	.658	.368
	1/4	6.6	1.94	5.11	1.53	1.62	1.66	1.44	.614	.861	.657	.663	.371

Courtesy American Institute of Steel Construction, Inc.

Table A4.6 Angles—unequal legs (continued)

ANGLES
Unequal legs
Properties for designing

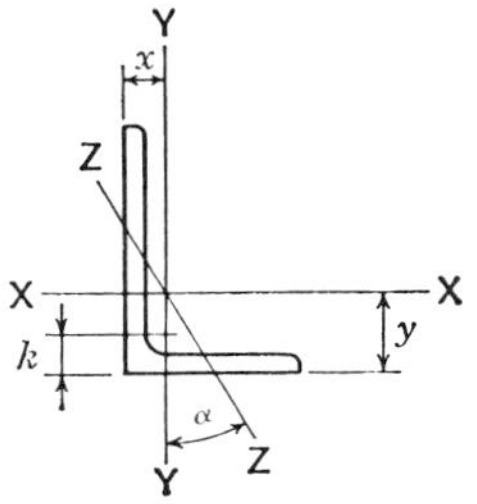

Size and Thickness	Weight per Foot	Area	AXIS X-X				AXIS Y-Y				AXIS Z-Z	
			I	S	r	y	I	S	r	x	r	Tan α
In.	Lb.	In.^{2}	In.^{4}	In.^{3}	In.	In.	In.^{4}	In.^{3}	In.	In.	In.	
L 4 × 3½ × ⅝	14.7	4.30	6.37	2.35	1.22	1.29	4.52	1.84	1.03	1.04	.719	.745
½	11.9	3.50	5.32	1.94	1.23	1.25	3.79	1.52	1.04	1.00	.722	.750
⁷⁄₁₆	10.6	3.09	4.76	1.72	1.24	1.23	3.40	1.35	1.05	.978	.724	.753
⅜	9.1	2.67	4.18	1.49	1.25	1.21	2.95	1.17	1.06	.955	.727	.755
⁵⁄₁₆	7.7	2.25	3.56	1.26	1.26	1.18	2.55	.994	1.07	.932	.730	.757
¼	6.2	1.81	2.91	1.03	1.27	1.16	2.09	.808	1.07	.909	.734	.759
L 4 × 3 × ⅝	13.6	3.98	6.03	2.30	1.23	1.37	2.87	1.35	.849	.871	.637	.534
½	11.1	3.25	5.05	1.89	1.25	1.33	2.42	1.12	.864	.827	.639	.543
⁷⁄₁₆	9.8	2.87	4.52	1.68	1.25	1.30	2.18	.992	.871	.804	.641	.547
⅜	8.5	2.48	3.96	1.46	1.26	1.28	1.92	.866	.879	.782	.644	.551
⁵⁄₁₆	7.2	2.09	3.38	1.23	1.27	1.26	1.65	.734	.887	.759	.647	.554
¼	5.8	1.69	2.77	1.00	1.28	1.24	1.36	.599	.896	.736	.651	.558
L 3½ × 3 × ½	10.2	3.00	3.45	1.45	1.07	1.13	2.33	1.10	.881	.875	.621	.714
⁷⁄₁₆	9.1	2.65	3.10	1.29	1.08	1.10	2.09	.975	.889	.853	.622	.718
⅜	7.9	2.30	2.72	1.13	1.09	1.08	1.85	.851	.897	.830	.625	.721
⁵⁄₁₆	6.6	1.93	2.33	.954	1.10	1.06	1.58	.722	.905	.808	.627	.724
¼	5.4	1.56	1.91	.776	1.11	1.04	1.30	.589	.914	.785	.631	.727
L 3½ × 2½ × ½	9.4	2.75	3.24	1.41	1.09	1.20	1.36	.760	.704	.705	.534	.486
⁷⁄₁₆	8.3	2.43	2.91	1.26	1.09	1.18	1.23	.677	.711	.682	.535	.491
⅜	7.2	2.11	2.56	1.09	1.10	1.16	1.09	.592	.719	.660	.537	.496
⁵⁄₁₆	6.1	1.78	2.19	.927	1.11	1.14	.939	.504	.727	.637	.540	.501
¼	4.9	1.44	1.80	.755	1.12	1.11	.777	.412	.735	.614	.544	.506
L 3 × 2½ × ½	8.5	2.50	2.08	1.04	.913	1.00	1.30	.744	.722	.750	.520	.667
⁷⁄₁₆	7.6	2.21	1.88	.928	.920	.978	1.18	.664	.729	.728	.521	.672
⅜	6.6	1.92	1.66	.810	.928	.956	1.04	.581	.736	.706	.522	.676
⁵⁄₁₆	5.6	1.62	1.42	.688	.937	.933	.898	.494	.744	.683	.525	.680
¼	4.5	1.31	1.17	.561	.945	.911	.743	.404	.753	.661	.528	.684
³⁄₁₆	3.39	.996	.907	.430	.954	.888	.577	.310	.761	.638	.533	.688

Courtesy American Institute of Steel Construction, Inc.

Table A4.6 Angles—unequal legs

Size and Thickness	Weight per Foot	Area	AXIS X-X				AXIS Y-Y				AXIS Z-Z	
			I	S	r	y	I	S	r	x	r	Tan α
In.	Lb.	In.^{2}	In.^{4}	In.^{3}	In.	In.	In.^{4}	In.^{3}	In.	In.	In.	
L 3 × 2 × ½	7.7	2.25	1.92	1.00	.924	1.08	.672	.474	.546	.583	.428	.414
⁷⁄₁₆	6.8	2.00	1.73	.894	.932	1.06	.609	.424	.553	.561	.429	.421
⅜	5.9	1.73	1.53	.781	.940	1.04	.543	.371	.559	.539	.430	.428
⁵⁄₁₆	5.0	1.46	1.32	.664	.948	1.02	.470	.317	.567	.516	.432	.435
¼	4.1	1.19	1.09	.542	.957	.993	.392	.260	.574	.493	.435	.440
³⁄₁₆	3.07	.902	.842	.415	.966	.970	.307	.200	.583	.470	.439	.446
L 2½ × 2 × ⅜	5.3	1.55	.912	.547	.768	.831	.514	.363	.577	.581	.420	.614
⁵⁄₁₆	4.5	1.31	.788	.466	.776	.809	.446	.310	.584	.559	.422	.620
¼	3.62	1.06	.654	.381	.784	.787	.372	.254	.592	.537	.424	.626
³⁄₁₆	2.75	.809	.509	.293	.793	.764	.291	.196	.600	.514	.427	.631
L 2½ × 1½ × ⁵⁄₁₆	3.92	1.15	.711	.444	.785	.898	.191	.174	.408	.398	.322	.349
¼	3.19	.938	.591	.364	.794	.875	.161	.143	.415	.375	.324	.357
³⁄₁₆	2.44	.715	.461	.279	.803	.852	.127	.111	.422	.352	.327	.364
L 2 × 1½ × ¼	2.77	.813	.316	.236	.623	.663	.151	.139	.432	.413	.320	.543
³⁄₁₆	2.12	.621	.248	.182	.632	.641	.120	.108	.440	.391	.322	.551
⅛	1.44	.422	.173	.125	.641	.618	.085	.075	.448	.368	.326	.558
L 2 × 1¼ × ¼	2.55	.750	.296	.229	.628	.708	.089	.097	.344	.333	.269	.378
³⁄₁₆	1.96	.574	.232	.177	.636	.686	.071	.075	.351	.311	.271	.387
⅛	1.33	.391	.163	.122	.645	.663	.050	.052	.359	.287	.274	.396
L 1¾ × 1¼ × ¼	2.34	.688	.202	.176	.543	.602	.085	.095	.352	.352	.267	.486
³⁄₁₆	1.80	.527	.160	.137	.551	.580	.068	.074	.359	.330	.269	.496
⅛	1.23	.359	.113	.094	.560	.557	.049	.051	.368	.307	.272	.506

Courtesy American Institute of Steel Construction, Inc.

Table A4.7 Pipe

PIPE
Dimensions and properties

Nominal Diameter In.	Outside Diameter In.	Inside Diameter In.	Wall Thickness In.	Weight per Foot Lbs. Plain Ends	A In.²	I In.⁴	S In.³	r In.
\multicolumn{9}{c}{Standard Weight}
½	.840	.622	.109	.85	.250	.017	.041	.261
¾	1.050	.824	.113	1.13	.333	.037	.071	.334
1	1.315	1.049	.133	1.68	.494	.087	.133	.421
1¼	1.660	1.380	.140	2.27	.669	.195	.235	.540
1½	1.900	1.610	.145	2.72	.799	.310	.326	.623
2	2.375	2.067	.154	3.65	1.07	.666	.561	.787
2½	2.875	2.469	.203	5.79	1.70	1.53	1.06	.947
3	3.500	3.068	.216	7.58	2.23	3.02	1.72	1.16
3½	4.000	3.548	.226	9.11	2.68	4.79	2.39	1.34
4	4.500	4.026	.237	10.79	3.17	7.23	3.21	1.51
5	5.563	5.047	.258	14.62	4.30	15.2	5.45	1.88
6	6.625	6.065	.280	18.97	5.58	28.1	8.50	2.25
8	8.625	7.981	.322	28.55	8.40	72.5	16.8	2.94
10	10.750	10.020	.365	40.48	11.9	161	29.9	3.67
12	12.750	12.000	.375	49.56	14.6	279	43.8	4.38
\multicolumn{9}{c}{Extra Strong}
½	.840	.546	.147	1.09	.320	.020	.048	.250
¾	1.050	.742	.154	1.47	.433	.045	.085	.321
1	1.315	.957	.179	2.17	.639	.106	.161	.407
1¼	1.660	1.278	.191	3.00	.881	.242	.291	.524
1½	1.900	1.500	.200	3.63	1.07	.391	.412	.605
2	2.375	1.939	.218	5.02	1.48	.868	.731	.766
2½	2.875	2.323	.276	7.66	2.25	1.92	1.34	.924
3	3.500	2.900	.300	10.25	3.02	3.89	2.23	1.14
3½	4.000	3.364	.318	12.50	3.68	6.28	3.14	1.31
4	4.500	3.826	.337	14.98	4.41	9.61	4.27	1.48
5	5.563	4.813	.375	20.78	6.11	20.7	7.43	1.84
6	6.625	5.761	.432	28.57	8.40	40.5	12.2	2.19
8	8.625	7.625	.500	43.39	12.8	106	24.5	2.88
10	10.750	9.750	.500	54.74	16.1	212	39.4	3.63
12	12.750	11.750	.500	65.42	19.2	362	56.7	4.33
\multicolumn{9}{c}{Double-Extra Strong}
2	2.375	1.503	.436	9.03	2.66	1.31	1.10	.703
2½	2.875	1.771	.552	13.69	4.03	2.87	2.00	.844
3	3.500	2.300	.600	18.58	5.47	5.99	3.42	1.05
4	4.500	3.152	.674	27.54	8.10	15.3	6.79	1.37
5	5.563	4.063	.750	38.55	11.3	33.6	12.1	1.72
6	6.625	4.897	.864	53.16	15.6	66.3	20.0	2.06
8	8.625	6.875	.875	72.42	21.3	162	37.6	2.76

The listed sections are available in conformance with ASTM Specification A53 Grade B or A501. Other sections are made to these specifications. Consult with pipe manufacturers or distributors for availability.

Courtesy American Institute of Steel Construction, Inc.

Table A4.8 Screw threads

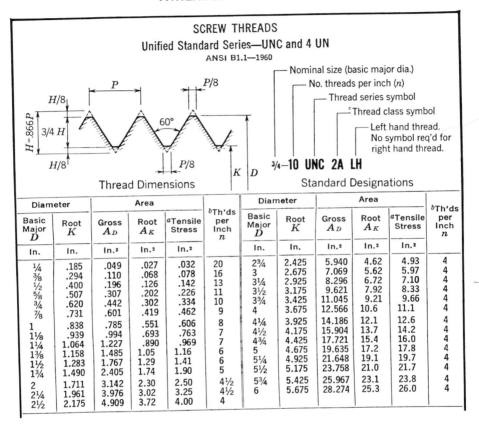

Courtesy American Institute of Steel Construction, Inc.

Table A4.9 Properties of structural lumber

Nominal size b(inches)d	Standard dressed size (S4S) b(inches)d	Area of Section A	Moment of inertia I	Section modulus S	Weight in pounds per linear foot of piece when weight of wood per cubic foot equals:					
					25 lb.	30 lb.	35 lb.	40 lb.	45 lb.	50 lb.
1 × 3	3/4 × 2-1/2	1.875	0.977	0.781	0.326	0.391	0.456	0.521	0.586	0.651
1 × 4	3/4 × 3-1/2	2.625	2.680	1.531	0.456	0.547	0.638	0.729	0.820	0.911
1 × 6	3/4 × 5-1/2	4.125	10.398	3.781	0.716	0.859	1.003	1.146	1.289	1.432
1 × 8	3/4 × 7-1/4	5.438	23.817	6.570	0.944	1.133	1.322	1.510	1.699	1.888
1 × 10	3/4 × 9-1/4	6.938	49.466	10.695	1.204	1.445	1.686	1.927	2.168	2.409
1 × 12	3/4 × 11-1/4	8.438	88.989	15.820	1.465	1.758	2.051	2.344	2.637	2.930
2 × 3	1-1/2 × 2-1/2	3.750	1.953	1.563	0.651	0.781	0.911	1.042	1.172	1.302
2 × 4	1-1/2 × 3-1/2	5.250	5.359	3.063	0.911	1.094	1.276	1.458	1.641	1.823
2 × 5	1-1/2 × 4-1/2	6.750	11.391	5.063	1.172	1.406	1.641	1.875	2.109	2.344
2 × 6	1-1/2 × 5-1/2	8.250	20.797	7.563	1.432	1.719	2.005	2.292	2.578	2.865
2 × 8	1-1/2 × 7-1/4	10.875	47.635	13.141	1.888	2.266	2.643	3.021	3.398	3.776
2 × 10	1-1/2 × 9-1/4	13.875	98.932	21.391	2.409	2.891	3.372	3.854	4.336	4.818
2 × 12	1-1/2 × 11-1/4	16.875	177.979	31.641	2.930	3.516	4.102	4.688	5.273	5.859
2 × 14	1-1/2 × 13-1/4	19.875	290.775	43.891	3.451	4.141	4.831	5.521	6.211	6.901
3 × 1	2-1/2 × 3/4	1.875	0.088	0.234	0.326	0.391	0.456	0.521	0.586	0.651
3 × 2	2-1/2 × 1-1/2	3.750	0.703	0.938	0.651	0.781	0.911	1.042	1.172	1.302
3 × 4	2-1/2 × 3-1/2	8.750	8.932	5.104	1.519	1.823	2.127	2.431	2.734	3.038
3 × 5	2-1/2 × 4-1/2	11.250	18.984	8.438	1.953	2.344	2.734	3.125	3.516	3.906
3 × 6	2-1/2 × 5-1/2	13.750	34.661	12.604	2.387	2.865	3.342	3.819	4.297	4.774
3 × 8	2-1/2 × 7-1/4	18.125	79.391	21.901	3.147	3.776	4.405	5.035	5.664	6.293
3 × 10	2-1/2 × 9-1/4	23.125	164.886	35.651	4.015	4.818	5.621	6.424	7.227	8.030
3 × 12	2-1/2 × 11-1/4	28.125	296.631	52.734	4.883	5.859	6.836	7.813	8.789	9.766
3 × 14	2-1/2 × 13-1/4	33.125	484.625	73.151	5.751	6.901	8.051	9.201	10.352	11.502
3 × 16	2-1/2 × 15-1/4	38.125	738.870	96.901	6.619	7.943	9.266	10.590	11.914	13.238
4 × 1	3-1/2 × 3/4	2.625	0.123	0.328	0.456	0.547	0.638	0.729	0.820	0.911
4 × 2	3-1/2 × 1-1/2	5.250	0.984	1.313	0.911	1.094	1.276	1.458	1.641	1.823
4 × 3	3-1/2 × 2-1/2	8.750	4.557	3.646	1.519	1.823	2.127	2.431	2.734	3.038
4 × 4	3-1/2 × 3-1/2	12.250	12.505	7.146	2.127	2.552	2.977	3.403	3.828	4.253
4 × 5	3-1/2 × 4-1/2	15.750	26.578	11.813	2.734	3.281	3.828	4.375	4.922	5.469
4 × 6	3-1/2 × 5-1/2	19.250	48.526	17.646	3.342	4.010	4.679	5.347	6.016	6.684
4 × 8	3-1/2 × 7-1/4	25.375	111.148	30.661	4.405	5.286	6.168	7.049	7.930	8.811
4 × 10	3-1/2 × 9-1/4	32.375	230.840	49.911	5.621	6.745	7.869	8.933	10.117	11.241
4 × 12	3-1/2 × 11-1/4	39.375	415.283	73.828	6.836	8.203	9.570	10.938	12.305	13.672
4 × 14	3-1/2 × 13-1/4	46.375	678.475	102.411	8.047	9.657	11.266	12.877	14.485	16.094
4 × 16	3-1/2 × 15-1/4	53.375	1034.418	135.66	9.267	11.121	12.975	14.828	16.682	18.536
5 × 2	4-1/2 × 1-1/2	6.750	1.266	1.688	1.172	1.406	1.641	1.875	2.109	2.344
5 × 3	4-1/2 × 2-1/2	11.250	5.859	4.688	1.953	2.344	2.734	3.125	3.516	3.906
5 × 4	4-1/2 × 3-1/2	15.750	16.078	9.188	2.734	3.281	3.828	4.375	4.922	5.469
5 × 5	4-1/2 × 4-1/2	20.250	34.172	15.188	3.516	4.219	4.922	5.675	6.328	7.031
6 × 1	5-1/2 × 3/4	4.125	0.193	0.516	0.716	0.859	1.003	1.146	1.289	1.432
6 × 2	5-1/2 × 1-1/2	8.250	1.547	2.063	1.432	1.719	2.005	2.292	2.578	2.865
6 × 3	5-1/2 × 2-1/2	13.750	7.161	5.729	2.387	2.865	3.342	3.819	4.297	4.774
6 × 4	5-1/2 × 3-1/2	19.250	19.651	11.229	3.342	4.010	4.679	5.347	6.016	6.684
6 × 6	5-1/2 × 5-1/2	30.250	76.255	27.729	5.252	6.302	7.352	8.403	9.453	10.503
6 × 8	5-1/2 × 7-1/2	41.250	193.359	51.563	7.161	8.594	10.026	11.458	12.891	14.323
6 × 10	5-1/2 × 9-1/2	52.250	392.963	82.729	9.071	10.885	12.700	14.514	16.328	18.142
6 × 12	5-1/2 × 11-1/2	63.250	697.068	121.229	10.981	13.177	15.373	17.569	19.766	21.962
6 × 14	5-1/2 × 13-1/2	74.250	1127.672	167.063	12.891	15.469	18.047	20.625	23.203	25.781
6 × 16	5-1/2 × 15-1/2	85.250	1706.776	220.229	14.800	17.760	20.720	23.681	26.641	29.601
6 × 18	5-1/2 × 17-1/2	96.250	2456.380	280.729	16.710	20.052	23.394	26.736	30.078	33.420
6 × 20	5-1/2 × 19-1/2	107.250	3398.484	348.563	18.620	22.344	26.068	29.792	33.516	37.240
6 × 22	5-1/2 × 21-1/2	118.250	4555.086	423.729	20.530	24.635	28.741	32.847	36.953	41.059
6 × 24	5-1/2 × 23-1/2	129.250	5948.191	506.229	22.439	26.927	31.415	35.903	40.391	44.878
8 × 1	7-1/4 × 3/4	5.438	0.255	0.680	0.944	1.133	1.322	1.510	1.699	1.888
8 × 2	7-1/4 × 1-1/2	10.875	2.039	2.719	1.888	2.266	2.643	3.021	3.398	3.776
8 × 3	7-1/4 × 2-1/2	18.125	9.440	7.552	3.147	3.776	4.405	5.035	5.664	6.293
8 × 4	7-1/4 × 3-1/2	25.375	25.904	14.803	4.405	5.286	6.168	7.049	7.930	8.811
8 × 6	7-1/2 × 5-1/2	41.250	103.984	37.813	7.161	8.594	10.026	11.458	12.891	14.323
8 × 8	7-1/2 × 7-1/2	56.250	263.672	70.313	9.766	11.719	13.672	15.625	17.578	19.531
8 × 10	7-1/2 × 9-1/2	71.250	535.859	112.813	12.370	14.844	17.318	19.792	22.266	24.740
8 × 12	7-1/2 × 11-1/2	86.250	950.547	165.313	14.974	17.969	20.964	23.958	26.953	29.948
8 × 14	7-1/2 × 13-1/2	101.250	1537.734	227.813	17.578	21.094	24.609	28.125	31.641	35.156
8 × 16	7-1/2 × 15-1/2	116.250	2327.422	300.313	20.182	24.219	28.255	32.292	36.328	40.365
8 × 18	7-1/2 × 17-1/2	131.250	3349.609	382.813	22.786	27.344	31.901	36.458	41.016	45.573
8 × 20	7-1/2 × 19-1/2	146.250	4634.297	475.313	25.391	30.469	35.547	40.625	45.703	50.781
8 × 22	7-1/2 × 21-1/2	161.250	6211.484	577.813	27.995	33.594	39.193	44.792	50.391	55.990
8 × 24	7-1/2 × 23-1/2	176.250	8111.172	690.313	30.599	36.719	42.839	48.958	55.078	61.198

Courtesy National Forest Products Association.

Table A4.9 Properties of structural lumber *(continued)*

Nominal size b(inches)d	Standard dressed size (S4S) b(inches)d	Area of Section A	Moment of inertia I	Section modulus S	Weight in pounds per linear foot of piece when weight of wood per cubic foot equals					
					25 lb.	30 lb.	35 lb.	40 lb.	45 lb.	50 lb.
10 × 1	9-1/4 × 3/4	6.938	0.325	0.867	1.204	1.445	1.686	1.927	2.168	2.409
10 × 2	9-1/4 × 1-1/2	13.875	2.602	3.469	2.409	2.891	3.372	3.854	4.336	4.818
10 × 3	9-1/4 × 2-1/2	23.125	12.044	9.635	4.015	4.818	5.621	6.424	7.227	8.030
10 × 4	9-1/4 × 3-1/2	32.375	33.049	18.885	5.621	6.745	7.869	8.993	10.117	11.241
10 × 6	9-1/2 × 5-1/2	52.250	131.714	47.896	9.071	10.885	12.700	14.514	16.328	18.142
10 × 8	9-1/2 × 7-1/2	71.250	333.984	89.063	12.370	14.844	17.318	19.792	22.266	24.740
10 × 10	9-1/2 × 9-1/2	90.250	678.755	142.896	15.668	18.802	21.936	25.069	28.203	31.337
10 × 12	9-1/2 × 11-1/2	109.250	1204.026	209.396	18.967	22.760	26.554	30.347	34.141	37.934
10 × 14	9-1/2 × 13-1/2	128.250	1947.797	288.563	22.266	26.719	31.172	35.625	40.078	44.531
10 × 16	9-1/2 × 15-1/2	147.250	2948.068	380.396	25.564	30.677	35.790	40.903	46.016	51.128
10 × 18	9-1/2 × 17-1/2	166.250	4242.836	484.896	28.863	34.635	40.408	46.181	51.953	57.726
10 × 20	9-1/2 × 19-1/2	185.250	5870.109	602.063	32.161	38.594	45.026	51.458	57.891	64.323
10 × 22	9-1/2 × 21-1/2	204.250	7867.879	731.896	35.460	42.552	49.644	56.736	63.828	70.920
10 × 24	9-1/2 × 23-1/2	223.250	10274.148	874.396	38.759	46.510	54.262	62.014	69.766	77.517
12 × 1	11-1/4 × 3/4	8.438	0.396	1.055	1.465	1.758	2.051	2.344	2.637	2.930
12 × 2	11-1/4 × 1-1/2	16.875	3.164	4.219	2.930	3.516	4.102	4.688	5.273	5.859
12 × 3	11-1/4 × 2-1/2	28.125	14.648	11.719	4.883	5.859	6.836	7.813	8.789	9.766
12 × 4	11-1/4 × 3-1/2	39.375	40.195	22.969	6.836	8.203	9.570	10.938	12.305	13.672
12 × 6	11-1/2 × 5-1/2	63.250	159.443	57.979	10.981	13.177	15.373	17.569	19.766	21.962
12 × 8	11-1/2 × 7-1/2	86.250	404.297	107.813	14.974	17.969	20.964	23.958	26.953	29.948
12 × 10	11-1/2 × 9-1/2	109.250	821.651	172.979	18.967	22.760	26.554	30.347	34.141	37.934
12 × 12	11-1/2 × 11-1/2	132.250	1457.505	253.479	22.960	27.552	32.144	36.736	41.328	45.920
12 × 14	11-1/2 × 13-1/2	155.250	2357.859	349.313	26.953	32.344	37.734	43.125	48.516	53.906
12 × 16	11-1/2 × 15-1/2	178.250	3568.713	460.479	30.946	37.135	43.325	49.514	55.703	61.892
12 × 18	11-1/2 × 17-1/2	201.250	5136.066	586.979	34.939	41.927	48.915	55.903	62.891	69.878
12 × 20	11-1/2 × 19-1/2	224.250	7105.922	728.813	38.932	46.719	54.505	62.292	70.078	77.865
12 × 22	11-1/2 × 21-1/2	247.250	9524.273	885.979	42.925	51.510	60.095	68.681	77.266	85.851
12 × 24	11-1/2 × 23-1/2	270.250	12437.129	1058.479	46.918	56.302	65.686	75.069	84.453	93.837
14 × 2	13-1/4 × 1-1/2	19.875	3.727	4.969	3.451	4.141	4.831	5.521	6.211	6.901
14 × 3	13-1/4 × 2-1/2	33.125	17.253	13.802	5.751	6.901	8.051	9.201	10.352	11.502
14 × 4	13-1/4 × 3-1/2	46.375	47.34	27.052	8.047	9.657	11.266	12.877	14.485	16.094
14 × 6	13-1/2 × 5-1/2	74.250	187.172	68.063	12.891	15.469	18.047	20.625	23.203	25.781
14 × 8	13-1/2 × 7-1/2	101.250	474.609	126.563	17.578	21.094	24.609	28.125	31.641	35.156
14 × 10	13-1/2 × 9-1/2	128.250	964.547	203.063	22.266	26.719	31.172	35.625	40.078	44.531
14 × 12	13-1/2 × 11-1/2	155.250	1710.984	297.563	26.953	32.344	37.734	43.125	48.516	53.906
14 × 16	13-1/2 × 15-1/2	209.250	4189.359	540.563	36.328	43.594	50.859	58.125	65.391	72.656
14 × 18	13-1/2 × 17-1/2	236.250	6029.297	689.063	41.016	49.219	57.422	65.625	73.828	82.031
14 × 20	13-1/2 × 19-1/2	263.250	8341.734	855.563	45.703	54.844	63.984	73.125	82.266	91.406
14 × 22	13-1/2 × 21-1/2	290.250	11180.672	1040.063	50.391	60.469	70.547	80.625	90.703	100.781
14 × 24	13-1/2 × 23-1/2	317.250	14600.109	1242.563	55.078	66.094	77.109	88.125	99.141	110.156
16 × 3	15-1/4 × 2-1/2	38.125	19.857	15.885	6.619	7.944	9.267	10.592	11.915	13.240
16 × 4	15-1/4 × 3-1/2	53.375	54.487	31.135	9.267	11.121	12.975	14.828	16.682	18.536
16 × 6	15-1/2 × 5-1/2	85.250	214.901	78.146	14.800	17.760	20.720	23.681	26.641	29.601
16 × 8	15-1/2 × 7-1/2	116.250	544.922	145.313	20.182	24.219	28.255	32.292	36.328	40.365
16 × 10	15-1/2 × 9-1/2	147.250	1107.443	233.146	25.564	30.677	35.790	40.903	46.016	51.128
16 × 12	15-1/2 × 11-1/2	178.250	1964.463	341.646	30.946	37.135	43.325	49.514	55.703	61.892
16 × 14	15-1/2 × 13-1/2	209.250	3177.984	470.813	36.328	43.594	50.859	58.125	65.391	72.656
16 × 16	15-1/2 × 15-1/2	240.250	4810.004	620.646	41.710	50.052	58.394	66.736	75.078	83.420
16 × 18	15-1/2 × 17-1/2	271.250	6922.523	791.146	47.092	56.510	65.929	75.347	84.766	94.184
16 × 20	15-1/2 × 19-1/2	302.250	9577.547	982.313	52.474	62.969	73.464	83.958	94.453	104.948
16 × 22	15-1/2 × 21-1/2	333.250	12837.066	1194.146	57.856	69.427	80.998	92.569	104.141	115.712
16 × 24	15-1/2 × 23-1/2	364.250	16763.086	1426.646	63.238	75.885	88.533	101.181	113.828	126.476
18 × 6	17-1/2 × 5-1/2	96.250	242.630	88.229	16.710	20.052	23.394	26.736	30.078	33.420
18 × 8	17-1/2 × 7-1/2	131.250	615.234	164.063	22.786	27.344	31.901	36.458	41.016	45.573
18 × 10	17-1/2 × 9-1/2	166.250	1250.338	263.229	28.863	34.635	40.408	46.181	51.953	57.726
18 × 12	17-1/2 × 11-1/2	201.250	2217.943	385.729	34.939	41.927	48.915	55.903	62.891	69.878
18 × 14	17-1/2 × 13-1/2	236.250	3588.047	531.563	41.016	49.219	57.422	65.625	73.828	82.031
18 × 16	17-1/2 × 15-1/2	271.250	5430.668	700.729	47.092	56.510	65.929	75.347	84.766	94.184
18 × 18	17-1/2 × 17-1/2	306.250	7815.754	893.229	53.168	63.802	74.436	85.069	95.703	106.337
18 × 20	17-1/2 × 19-1/2	341.250	10813.359	1109.063	59.245	71.094	82.943	94.792	106.641	118.490
18 × 22	17-1/2 × 21-1/2	376.250	14493.461	1348.229	65.321	78.385	91.450	104.514	117.578	130.642
18 × 24	17-1/2 × 23-1/2	411.250	18926.066	1610.729	71.398	85.677	99.957	114.236	128.516	142.795

Courtesy National Forest Products Association.

Table A4.9 Properties of structural lumber (continued)

Nominal size b(inches)d	Standard dressed size (S4S) b(inches)d	Area of Section A	Moment of inertia I	Section modulus S	Weight in pounds per linear foot of piece when weight of wood per cubic foot equals:					
					25 lb.	30 lb.	35 lb.	40 lb.	45 lb.	50 lb.
20 × 6	19-1/2 × 5-1/2	107.250	270.359	98.313	18.620	22.344	26.068	29.792	33.516	37.240
20 × 8	19-1/2 × 7-1/2	146.250	685.547	182.813	25.391	30.469	35.547	40.625	45.703	50.781
20 × 10	19-1/2 × 9-1/2	185.250	1393.234	293.313	32.161	38.594	45.026	51.458	57.891	64.323
20 × 12	19-1/2 × 11-1/2	224.250	2471.422	429.813	38.932	46.719	54.505	62.292	70.078	77.865
20 × 14	19-1/2 × 13-1/2	263.250	3998.109	592.313	45.703	54.844	63.984	73.125	82.266	91.406
20 × 16	19-1/2 × 15-1/2	302.250	6051.297	780.813	52.474	62.969	73.464	83.958	94.453	104.948
20 × 18	19-1/2 × 17-1/2	341.250	8708.984	995.313	59.245	71.094	82.943	94.792	106.641	118.490
20 × 20	19-1/2 × 19-1/2	380.250	12049.172	1235.813	66.016	79.219	92.422	105.625	118.828	132.031
20 × 22	19-1/2 × 21-1/2	419.250	16149.859	1502.313	72.786	87.344	101.901	116.458	131.016	145.573
20 × 24	19-1/2 × 23-1/2	458.250	21089.047	1794.813	79.557	95.469	111.380	127.292	243.203	159.115
22 × 6	21-1/2 × 5-1/2	118.250	298.088	108.396	20.530	24.635	28.741	32.847	36.953	41.059
22 × 8	21-1/2 × 7-1/2	161.250	755.859	201.563	27.995	33.594	39.193	44.792	50.391	55.990
22 × 10	21-1/2 × 9-1/2	204.250	1536.130	323.396	35.460	42.552	49.644	56.736	63.828	70.920
22 × 12	21-1/2 × 11-1/2	247.250	2724.901	473.896	42.925	51.510	60.095	68.681	77.266	85.851
22 × 14	21-1/2 × 13-1/2	290.250	4408.172	653.063	50.391	60.469	70.547	80.625	90.703	100.781
22 × 16	21-1/2 × 15-1/2	333.250	6671.941	860.896	57.856	69.427	80.998	92.569	104.141	115.712
22 × 18	21-1/2 × 17-1/2	376.250	9602.211	1097.396	65.321	78.385	91.450	104.514	117.578	130.642
22 × 20	21-1/2 × 19-1/2	419.250	13284.984	1362.563	72.786	87.344	101.901	116.458	131.016	145.573
22 × 22	21-1/2 × 21-1/2	462.250	17806.254	1656.396	80.252	96.302	112.352	128.403	144.453	160.503
22 × 24	21-1/2 × 23-1/2	505.250	23252.023	1978.896	87.717	105.260	122.804	140.347	157.891	175.434
24 × 6	23-1/2 × 5-1/2	129.250	325.818	118.479	22.439	26.927	31.415	35.903	40.391	44.878
24 × 8	23-1/2 × 7-1/2	176.250	826.172	220.313	30.599	36.719	42.839	48.958	55.078	61.198
24 × 10	23-1/2 × 9-1/2	223.250	1679.026	353.479	38.759	46.510	54.262	62.014	69.766	77.517
24 × 12	23-1/2 × 11-1/2	270.250	2978.380	517.979	46.918	56.302	65.686	75.069	84.453	93.837
24 × 14	23-1/2 × 13-1/2	317.250	4818.234	713.813	55.078	66.094	77.109	88.125	99.141	110.156
24 × 16	23-1/2 × 15-1/2	364.250	7292.586	940.979	63.238	75.885	88.533	101.181	113.828	126.476
24 × 18	23-1/2 × 17-1/2	411.250	10495.441	1199.479	71.398	85.677	99.957	114.236	128.516	142.795
24 × 20	23-1/2 × 19-1/2	458.250	14520.797	1489.313	79.557	95.469	111.380	127.292	143.203	159.115
24 × 22	23-1/2 × 21-1/2	505.250	19462.648	1810.479	87.717	105.260	122.804	140.347	157.891	175.434
24 × 24	24-1/2 × 23-1/2	552.250	25415.004	2162.979	95.877	115.052	134.227	153.403	172.578	191.753

Courtesy National Forest Products Association.

Table A8.1 Beam deflection formulas

Loading	Deflection

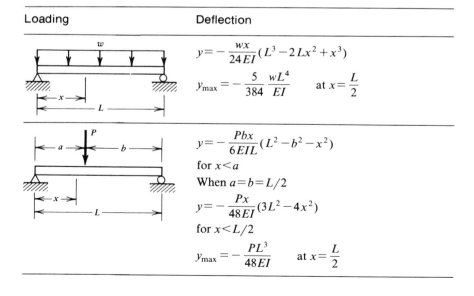

$$y = -\frac{wx}{24EI}(L^3 - 2Lx^2 + x^3)$$

$$y_{max} = -\frac{5}{384}\frac{wL^4}{EI} \quad \text{at } x = \frac{L}{2}$$

$$y = -\frac{Pbx}{6EIL}(L^2 - b^2 - x^2)$$

for $x < a$

When $a = b = L/2$

$$y = -\frac{Px}{48EI}(3L^2 - 4x^2)$$

for $x < L/2$

$$y_{max} = -\frac{PL^3}{48EI} \quad \text{at } x = \frac{L}{2}$$

Table A8.1 Beam deflection formulas

Loading	Deflection
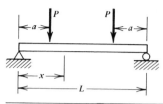	$y = -\dfrac{Pa^2}{6EI}(3L-4a)$ at $x=a$ $y_{max} = -\dfrac{Pa}{24EI}(3L^2 - 4a^2)$ at $x = \dfrac{L}{2}$
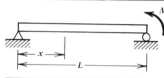	$y = -\dfrac{Mx}{6EIL}(L^2 - x^2)$ $y_{max} = -\dfrac{ML^2}{9\sqrt{3}\,EI}$ at $x = \dfrac{L}{\sqrt{3}}$
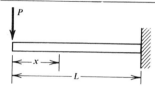	$y = -\dfrac{P}{6EI}(2L^3 - 3L^2 x + x^3)$ $y_{max} = -\dfrac{PL^3}{3EI}$ at $x=0$
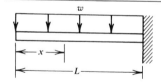	$y = -\dfrac{w}{24EI}(x^4 - 4L^3 x + 3L^4)$ $y_{max} = -\dfrac{wL^4}{8EI}$ at $x=0$
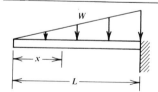	W is total load $y = -\dfrac{W}{60EIL^2}(x^5 - 5L^4 x + 4L^5)$ $y_{max} = -\dfrac{WL^3}{15EI}$ at $x=0$
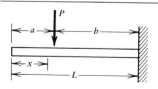	$y = -\dfrac{Pb^2}{6EI}(3L - 3x - b)$ for $x < a$ $y = -\dfrac{P(L-x)^2}{6EI}(3b - L + x)$ for $x > a$ $y_{max} = -\dfrac{Pb^2}{6EI}(3L - b)$ at $x=0$

Answers to Problems

Answers are generally omitted for every third problem.

CHAPTER 1

1.1   (a) in.
     (b) lb
     (c) in.
     (d) lb/in.^{2}
     (e) in.^{4}

1.2   (a) and (d) valid

1.4   (a) 866 700; 867 000; 870 000; 900 000
     (b) 942.1; 942; 940; 900
     (c) 0.000 123 4; 0.000 123; 0.000 12, 0.000 1
     (d) 2.000; 2.00; 2.0; 2
     (e) 0.02000; 0.0200; 0.020; 0.02
     (f) 3.142; 3.14; 3.1; 3

1.5   (a) 508 in. lb
     (b) 8.68 lb/in.^{2}
     (c) 4.09E9 ft-lb.
     (d) 134 000 lb/in.^{2}
     (e) 6.81 rad/s

1.7   (a) 207 kPa
     (b) 6.09 m^3
     (c) 24.1 km/hr
     (d) 3280 N/m
     (e) 45.2 W

1.8   (a) $18x$
     (b) $16x+2$
     (c) $4\cos 2x$
     (d) $3/x$
     (e) $\frac{x}{2}\cos\frac{x}{2}+\sin\frac{x}{2}$

1.10 (a) $\frac{x^4}{2}+C$
     (b) $\frac{x^3}{9}+x^2+4x+C$
     (c) $-\frac{1}{2}\cos 2x+C$
     (d) $\frac{2\sqrt{3}}{3}x^{3/2}+C$
     (e) $\frac{4}{3}\sin 3\theta+C$

1.11 (a) 648
     (b) 26.2
     (c) 0.117
     (d) 22.9
     (e) 0.943

1.13   60 m/s, 150 m

1.14   129 ft/s

ANSWERS TO PROBLEMS

CHAPTER 2

2.1		P_x	P_y	Units
	(a)	86.6	50.0	lb
	(b)	50.0	86.6	lb
	(c)	25.9	96.6	lb
	(d)	47.8	14.8	kN
	(e)	41.2	28.2	kN
	(f)	27.0	42.0	kN

2.2	(a)	−85.5	235	kips
	(b)	−161	192	kips
	(c)	−246	43.4	kips
	(d)	−703	1660	kips
	(e)	−1290	1250	kips
	(f)	−1680	645	kips

2.4	(a)	17.7	17.7	kN
	(b)	17.7	−17.7	kN
	(c)	−17.7	17.7	kN
	(d)	−17.7	−17.7	kN
	(e)	−17.7	−17.7	kN
	(f)	−17.7	17.7	kN

2.5		R	θ (degrees)
	(a)	50.0 lb	53.1
	(b)	50.0 lb	−53.1
	(c)	130 lb	247
	(d)	130 lb	157
	(e)	130 N	67.4
	(f)	130 N	203
	(g)	36.0 kips	326
	(h)	36.0 kips	33.7
	(i)	721 kN	56.3
	(j)	721 kN	146
	(k)	721 kN	236
	(l)	721 kN	326

2.7	(a)	140 lb	60.4
	(b)	140 lb	44.6
	(c)	1130 lb	90.0
	(d)	56.7 kips	177
	(e)	608 N	−54.7
	(f)	26.9 kN	0.405 rad
	(g)	24.4 kN	1.95 rad

2.8		R	θ (degrees)
	(a)	140 lb	60.4
	(b)	140 lb	44.6
	(c)	1130 lb	90.0
	(d)	56.7 kips	177
	(e)	608 N	−54.7
	(f)	26.9 kN	0.405 rad
	(g)	24.4 kN	1.95 rad

2.10	(a)	57.0 N·m
	(b)	84.0 N·m
	(c)	68.0 N·m
	(d)	86.0 N·m
	(e)	64.0 N·m
	(f)	71.0 N·m
	(g)	51.0 N·m
	(h)	1.00 N·m

2.11	(a)	56.6 lb @ −45.0°
		72.1 lb @ 33.7°
	(b)	84.9 lb @ 135°
		72.1 lb @ 33.7°
	(c)	28.3 lb @ 135°
		144 lb @ 33.7°
	(d)	14.1 lb @ −45°
		288 lb @ 33.7°

2.13	(a)	32.0 lb, 48.0 lb
	(b)	45.0 lb, 75.0 lb
	(c)	66.7 N, 133 N
	(d)	13.3 kN, 26.7 kN

2.14	(a)	−53.3 lb, 133 lb
	(b)	−72.0 lb, 192 lb
	(c)	−100 N, 300 N
	(d)	−20.0 kN, 60.0 kN

CHAPTER 3

3.1	10.0 psi, 3.33 psi
3.2	0.562 psi, 1.12 psi
3.4	127 psi
3.5	127 MPa

ANSWERS TO PROBLEMS 401

CHAPTER 3 cont'd

3.7 7.98 mm

3.8 (a) 204 ksi
     (b) 10.2 ksi

3.10 92.9 psi

3.11 600 kPa

3.13 AB: 278 MPa, CE: 278 MPa
      BC: 222 MPa, CF: 500 MPa
      CD: 555 MPa, EF: 3560 MPa
      AE: 3780 MPa, FD: 3560 MPa
      BE: 165 MPa

3.14 AB: 1.97E-6, 1.18E-4 in.
      AE: 2.68E-5, 1.29E-3 in.
      BE: 1.18E-6, 4.26E-5 in.
      BC: 1.88E-6, 7.57E-5 in.
      EC: 1.97E-6, 1.18E-4 in.

      EF: 2.52E-5, 1.21E-3 in.
      CD: 3.15E-5, 1.89E-3 in.
      CF: 1.77E-5, 6.38E-4 in.
      FD: 2.52E-5, 1.21E-3 in.

3.16 6190 lb

3.17 30.2 kN

3.19 407 MPa, 5.74E-3, 2.30 mm

3.20 50.9 ksi, 2.50E-4, 204E 3 ksi

3.22 619 MPa, 0.100, 6.19 GPa

3.23 27.2 ksi, 2.78E-4, 9.79E 7 psi

3.25 603 MPa, 6100 μ, 98.8 GPa
      0.320

3.26 3.00E-3, 160 ksi, 53.3E 3 ksi
      $-4.54E$-4

3.28 9.75E 6 N/m

3.29 87.0 ksi, 124 ksi, 15.8%

3.31 0.426 mm, 0.544 mm^2

3.32 2.34E-2 in., 6.03E 4 lb

3.34 0.0186 in., 29,800 lb

3.35 34.4 kips

3.37 $-231°$F, $+115°$F, $+231°$F

3.38 10,700 lb

3.40 4.51E 6 psi

3.43 5.79 mm

3.44 0.199 in.

3.46 9.29E-5, 5.80E-6 in.

3.49 0.217 in.

3.50 243 psi, 8.10E-6, 1.13E-3 in.

3.52 2020 lb, 1.65E-4 in.

3.53 0.259 in.

3.55 30.3 mm

3.56 16.7 kN, 51.3 MPa, 722E-6,
      2.17 mm

3.58 0.124 in.

3.59 134 mm

CHAPTER 4

4.1 4.00 ft

4.2 1.00 ft

4.4 5.39 in.

CHAPTER 4 cont'd

4.5	162 ft^4		5.8	245 N·m
4.7	162 ft^4		5.10	118 MPa
4.8	13.5 ft^4		5.11	74.8 hp
4.10	398 in.^{4}		5.13	2.52 in.
4.11	(a) 5.44 in., 1300 in.^{4}		5.14	66.3 mm
4.13	17.2 in.^{4}, 1.77 in.		5.16	145 N·m
4.14	17.2 in.^{4}		5.17	14.7 ksi
4.16	200 in.^{4}, 2.89 in.		5.19	51.1 hp
4.17	236 in.^{4}, 4.00 in.		5.20	36.0 kW
4.19	318 in.^{4}, 4.65 in.		5.22	91.7 mm
4.20	4.72 in., 767 in.^{4}, 3.03 in.		5.23	4520 in.-lb 6.82 ksi
4.22	9.17 kips, 20.8 kips		5.25	3630 in.-lb 6820 psi
4.23	12.4 kips, 19.6 kips		5.26	193 N·m 18.2 MPa
4.25	941 N, 859 N		5.28	742 N·m 2.20E-2 rad
4.26	3.29 in., 264 in.^{4}		5.29	8290 in.-lb 7.70E-2 rad
4.28	3.25 in., 125 in.^{4}, 1.85 in.		5.31	79.3 hp 127 ksi 6.64E-2 rad

CHAPTER 5

5.1	158 ft-lb		5.32	6.54 kW 102 MPa 7.71E-2 rad
5.2	215 N·m			
5.4	118 W		5.34	1780 N·m 2.78E-2 rad 21.4 MPa
5.5	2.19 lb			
5.7	2360 in.-lb			

ANSWERS TO PROBLEMS 403

CHAPTER 5 cont'd

5.35 79.3 hp
      30.7 ksi
      0.160 RAD

5.37 1300 ft-lb
      5.70 ksi
      7.18E-2 rad

5.38 1780 N·m
      7.45E-2 rad
      36.3 MPa

5.30 20400 in.-lb

5.41 No
      39.2 lb. max.

5.43 $\tau_0 = 0$
      $\tau_{.5} = 1260$ psi
      $\tau_{1.0} = 2520$ psi
      $\tau_{1.5} = 3770$ psi

5.44 $\tau_0 = 0$
      $\tau_{10} = 15.4$ MPa
      $\tau_{20} = 30.8$ MPa
      $\tau_{30} = 46.0$ MPa

5.46 75.0 MPa

5.47 1.89E-4 rad

5.49 513 lb, 1310 psi

5.50 0.116 rad

CHAPTER 6

6.1 $V_2 = 8.00$ kN, $M_2 = 16.0$ kN·m
     $V_8 = 0$, $M_8 = 24.0$ kN·m

6.2 $V_2 = 4.00$ kip, $M_2 = 20.0$ kip-ft
     $V_8 = 0$, $M_8 = 0$

6.4 $V_2 = 0.400$ kip, $M_2 = 4.40$ kip-ft
     $V_{8L} = 5.60$ kips, $M_{8L} = 19.8$ kip-ft

6.5 $V_2 = 134$ lb, $M_2 = 578$ ft-lb
     $V_8 = -133$ lb, $M_8 = 136$ ft-lb

6.7 $V_2 = 800$ lb, $M_2 = -1070$ ft-lb
     $V_6 = 0$, $M_6 = 0$

6.8 $V_2 = -60.0$ N, $M_2 = -120$ N·m
     $V_6 = -20.0$ N, $M_6 = -320$ N·m

6.10

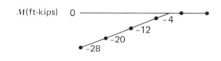

6.11

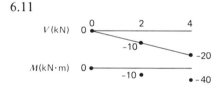

6.13

6.14

CHAPTER 6 cont'd

6.16

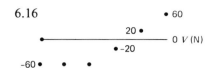

6.17 0 to 3
$V = 8$
$M = 8x$
3 to 9
$V = 0$
$M = 24$
9 to 12
$V = -8$
$M = 96 - 8x$

6.19 0 to 4
$V = -5x$
$M = -2.5x^2$

6.20 0 to 5
$V = 4.4 - 2x$
$M = 4.4x - x^2$
5 to 8
$V = -5.6$
$M = -5.6x + 25$
8 to 10
$V = 10$
$M = 10x - 99.8$

6.22 0 to 3
$V = -1$
$M = -x$
3 to 10
$V = -1$
$M = 10 - x$

6.23 0 to 6
$V = 1800 - 300x - 50x^2$
$M = -3600 + 1800x - 200x^2 - 16.7x^3$
6 to 8
$V = 0 = M$

6.25

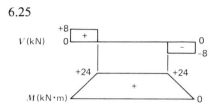

6.26

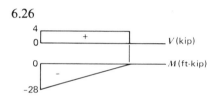

6.28

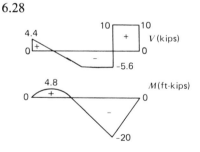

6.29

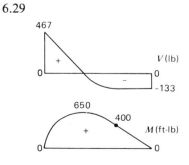

ANSWERS TO PROBLEMS 405

CHAPTER 6 cont'd

6.31
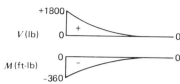

6.32
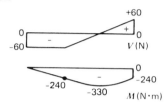

6.34
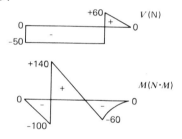

6.35
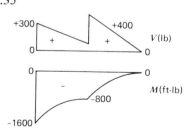

CHAPTER 7

7.1 28.0 ft-kips, 7.50 ksi T

7.2 56.0 ft-kips, 5.81 ksi C

7.4 4.36 ksi

7.5 13.2 ksi

7.7 $1.79T, 1.49T, 0.609T, 0, 0.568C,$ 1.16 ksi C

7.8 $830C, 577C, 325C, 0, 181T, 433T,$ 686 psi T

7.10 2.22 MPa

7.11 (a) 576 MPa   (i) 72.0 MPa
      (b) 288 MPa   (j) 36.0 MPa
      (c) 177 MPa   (k) 22.1 MPa
      (d) 477 MPa   (l) 59.6 MPa
      (e) 541 MPa   (m) 793 kPa
      (f) 270 MPa   (n) 396 kPa
      (g) 166 MPa   (o) 243 kPa
      (h) 448 MPa   (p) 657 kPa

7.13 (a) 101 mm
      (b) 120 mm
      (c) 66.9 mm
      (d) 11.2 mm

7.14 (a) 63.6 mm wide
      (b) 75.4 mm wide
      (c) 42.4 mm wide
      (d) 7.07 mm wide

7.16 8.30 ksi C

7.17 8.54 ksi

7.19 39.5 ksi

7.20 442 psi

7.22 $W14 \times 30$

7.23 $W10 \times 15$

7.25 $W21 \times 49$

7.26 $W4 \times 13$

CHAPTER 7 cont'd

7.28 18.5 MPa

7.29 $\sigma_{\text{flat}} = 2\sigma_{\text{edge}}$

CHAPTER 8

8.1 $\dfrac{1}{\text{ft}} = \dfrac{1}{\text{ft}}$

8.2 8330 in.-lb, 8330 psi

8.4 146 in.-lb, 28.1 ksi

8.5 20.3 N·m, 207 MPa

8.7 1.25 m

8.8 (a) $M = 200 - 300x + x^3$
$w = 6x$
(b) $M = 384x - 0.750x^4$
$w = -9x^2$
(c) $M = -12x^2 + 108x - 360$
$w = -24$

8.10 $w = 0$
$V(0) = -200\,N$
$M(0) = 0$
$\theta(1.5) = 0$
$y(1.5) = 0$

8.11 $w = w$
$V(0) = 0$
$M(0) = 0$
$\theta(L) = 0$
$y(L) = 0$

8.13 $M = C_1 x + C_2$

$EIy = \dfrac{C_1 x^3}{6} + \dfrac{C_2 x^2}{2} + C_3 x + C_4$

8.14 $M = -\dfrac{wx^2}{2} + C_1 x + C_2$
$EIy = \dfrac{-wx^4}{24} + \dfrac{C_1 x^3}{6} + \dfrac{C_2 x}{2} + C_3 x + C_4$

8.16 $V = -200$
$M = -200x$
$EI\theta = -100x^2 + 225$
$EIy = -33.3x^3 + 225x - 225$

8.17 $V = -wx$
$M = \dfrac{-wx^2}{2}$
$EI\theta = \dfrac{w}{6}(x^3 - L^3)$
$EIy = \dfrac{-w}{24}(x^4 - 4L^3 x + 192L^4)$

8.19 $\dfrac{-225 \text{ N·m}^3}{EI}$

8.20 $\dfrac{wL^4}{8EI}$

8.22
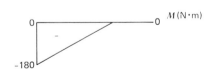

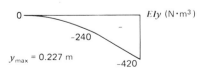
$y_{\max} = 0.227$ m

CHAPTER 8 cont'd

8.23

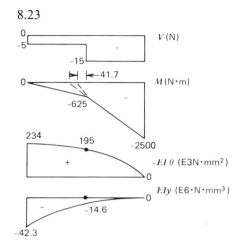

y_{max} = 17.8 mm

8.25 $\dfrac{-25{,}000}{EI}$ lb-ft^3

8.26 $\dfrac{-225}{EI}$ N·m^3

8.28 -0.227 m

8.29 -17.8 mm

8.31 $y(0) = -225/EI$
$y(0.3) = -158/EI$
$y(0.6) = -97.2/EI$
$y(0.9) = -46.8/EI$
$y(1.2) = -12.6/EI$
$y(1.5) = 0$

8.32 $y(0) = 0$
$y(.5) = 11.2$ mm
$y(1.0) = 40.6$ mm
$y(1.5) = 82.1$ mm
$y(2.0) = 130$ mm
$y(2.5) = 178$ mm
$y(3.0) = 228$ mm

8.34 0.548 m

8.35 6.66 mm

8.37 (a) 9.67E-3 in.
(b) 6.66 mm

8.38 (a) $-2.04E$-4 rad, $-5.99E$-3 in.

8.40 $EIy = 2.5x^4 - 250x^2 - 5000x$

8.41 $EIy = -16.7x^4 - 1330x^3 + 751{,}000$

8.43 0.0446 in.

8.44 max $y = -1200$ N·m^3 at $x = 0$

CHAPTER 9

9.1 0, 51.3, 87.9, 110, 117, 110,
87.9, 51.3, 0 psi

9.2 0, 12.0, 18.0, 18.0, 12.0, 0 MPa

9.4 150, 140, 112, 66.0, 0 MPa

9.5 234, 230, 220, 202, 176, 143,
102, 54.9, 0 psi

9.7 284 psi, 7500 psi

9.8 341 psi, 12 900 psi

9.10 203 psi on right 5 ft,
8300 psi at 10 K load

9.16 (a) 2.40 MPa
(b) 2.40 MPa
(c) 4.67 MPa
(d) 4.24 MPa
(e) 4.50 MPa
(f) 4.50 MPa
(g) 8.76 MPa
(h) 7.95 MPa
(i) 300 kPa

CHAPTER 9 cont'd

(j) 300 kPa
(k) 584 kPa
(l) 530 kPa
(m–p) 0

9.17 (a) 2.40 MPa, 576 MPa
(b) 2.40 MPa, 288 MPa
(c) 4.67 MPa, 177 MPa
(d) 4.24 MPa, 477 MPa
(e) 6.00 MPa, 962 MPa
(f) 6.00 MPa, 480 MPa
(g) 11.7 MPa, 295 MPa
(h) 10.6 MPa, 796 MPa
(i) 300 kPa, 168 MPa
(j) 300 kPa, 83.9 MPa
(k) 584 kPa, 51.5 MPa
(l) 530 kPa, 139 MPa
(m) 180 kPa, 793 kPa
(n) 180 kPa, 396 kPa
(o) 351 kPa, 243 kPa
(p) 318 kPa, 657 kPa

9.19 267 N/mm, 30.0 mm

9.20 528 lb/in.

CHAPTER 10

10.1 1890 psi T, 2040 psi C

10.2 990 psi T

10.4 144 MPa T

10.5 156 MPa C

10.7 5330 psi C

10.8 6670 psi T

10.10 21.9 ksi T

10.11 34.0 ksi T

10.13 293 MPa C

10.14 87.5 psi C

10.16 2.59 MPa C

10.17 3.32 MPa T

10.19 3.73 MPa, 168 MPa, 172 MPa

10.20 73.6 lb

10.22 (a) 11,300 psi
(b) 10,900 psi
(c) 10,400 psi
(d) 10,900 psi

10.23 (a) 90.5 psi
(b) 724 psi
(c) 1540 psi
(d) 724 psi

10.25 (a) 0.60 in.

10.26 (a) 260 MPa
(b) 239 MPa
(c) 218 MPa
(d) 239 MPa

10.28 133 mm

10.29 97 mm

10.31 19.7 MPa

10.32 0.334 in.

10.34 16, 287 ksi

10.35 15, 263 MPa

10.37 196 N

ANSWERS TO PROBLEMS 409

CHAPTER 11

11.1 (a) 280 psi, 75.0 psi
      (b) 225 psi, 130 psi
      (c) 150 psi, 150 psi

11.2 (a) −1.46 GPa, −1.46 GPa
      (b) −1.46 GPa, 1.46 GPa
      (c) 0, 4.00 GPa

11.4 (a) 476 kPa, −91.6 kPa
      (b) 282 kPa, −440 kPa

11.5 (a) 11.5 ksi, −8.28 ksi
      (b) 9.36 ksi, 2.32 ksi

11.7 (a) 280 psi, −75.0 psi
      (b) 225 psi, −130 psi
      (c) 20.1 psi, 75.0 psi
      (d) 75.0 psi, 130 psi
      (e) 20.1 psi, −75.0 psi

11.8 (a) −1.46 GPa, −1.46 GPa
      (b) −1.46 GPa, 1.46 GPa
      (c) 9.46 GPa, 1.46 GPa
      (d) 9.46 GPa, −1.46 GPa
      (e) 5.46 GPa, 5.46 GPa

11.10 (a) 476 kPa, −91.6 kPa
      (b) 282 kPa, −440 kPa
      (c) −676 kPa, 129 kPa
      (d) −482 kPa, 440 kPa
      (e) −290 kPa, −551 kPa

11.11 (a) 11.5 ksi, −8.28 ksi
      (b) 9.36 ksi, 2.32 ksi
      (c) 38.5 ksi, 8.28 ksi
      (d) 40.6 ksi, −2.32 ksi
      (e) 19.2 ksi, 14.7 ksi

11.13 (a) 0, 90°
      (b) 22.5°, 112°
      (c) 13.3°, 103°
      (d) 15.5°, 105°
      (e) 35.8°, 126°
      (f) 39.3°, 129°

11.14 (a) 0, 300 psi
      (b) 9.66, −1.66 ksi
      (c) 10.9, −6.94 ksi
      (d) −683, 483 kPa
      (e) 40.8, 9.19 ksi
      (f) −50.5, 0.495 ksi

11.16

θ	a psi	b GPa	c ksi	d kPa	e ksi	f MPa
0	0	8.00	10.0	−600	30.0	−30.0
15	20.1	9.64	10.9	−683	36.8	−41.8
30	75.0	9.64	9.46	−610	40.5	−49.2
45	150	8.00	6.00	−400	40.0	−50.0
60	225	5.64	1.46	−110	35.5	−44.2
75	280	2.54	−2.93	183	28.2	−33.2
90	300	0.	−6.00	400	20.0	−20.0
105	280	−1.46	−6.92	483	13.2	−8.17
120	225	−1.46	−5.46	410	9.51	−0.85
135	150	0.	−2.00	200	10.0	0.
150	75.0	2.54	2.54	−90.2	14.5	−5.85
165	20.1	5.64	6.92	−383	21.8	−16.8
180	0	8.00	10.0	−600	30.0	−30.0
max	300	9.66	10.9	483	40.8	0.50
min	0	−1.66	−6.94	−683	9.19	−50.5

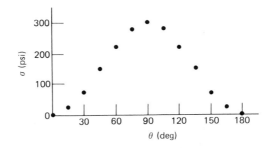

11.16 cont'd

11.17
(a) 45°, 135°
(b) −22.5°, 67.5°
(c) −31.7°, 58.3°
(d) −29.5°, 60.5°
(e) −9.22°, 80.8°
(f) −5.65°, 84.3°

11.19
(a) 150, 150 psi
(b) 5.66, 4.00 GPa
(c) 8.94, 2.00 ksi
(d) 583, −100 kPa
(e) 15.8, 25.0 ksi
(f) 25.5, −25.0 MPa

11.20

				σ			
θ	a psi	b GPa	c ksi	d kPa	e ksi	f MPa
0	0	4.00	4.00	−300	15.0	−25.0
15	75	1.46	− .54	− 10	10.5	−19.2
30	130	−1.46	−4.93	283	3.2	− 8.2
45	150	−4.00	−8.00	500	− 5.0	5.0
60	130	−5.46	−8.93	583	−11.8	16.8
75	75	−5.46	−7.46	510	−15.5	24.2
90	0	−4.00	−4.00	300	−15.0	25.0
105	− 75	−1.46	.54	10	−10.5	19.2
120	−130	1.46	4.93	−283	− 3.2	8.2
135	−150	4.00	8.00	−500	5.0	− 5.0
150	−130	5.46	8.93	−583	11.8	−16.8
165	− 75	5.46	7.46	−510	15.5	−24.2
180	0	4.00	4.00	−300	15.0	−25.0
max	150	−5.66	8.94	583	−15.8	25.5

11.21

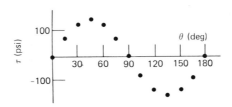

CHAPTER 11 cont'd

11.22

(a)

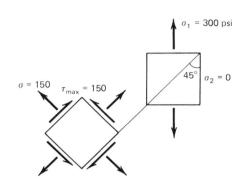

(b)

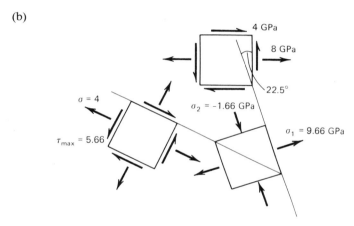

(c)

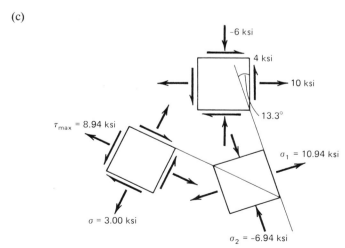

CHAPTER 11 cont'd

11.22 cont'd

(d)

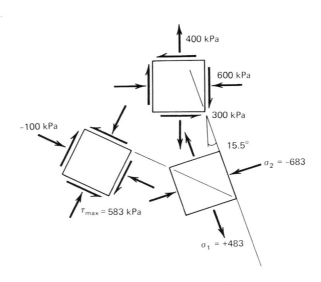

(e)

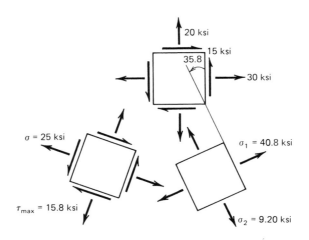

CHAPTER 11 cont'd

11.22 cont'd

(f)

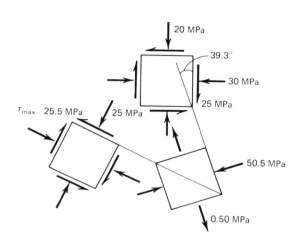

11.23 (a) 223, 129 psi
       (b) 5.45, 5.50 GPa
       (c) 9.41, 5.02 ksi
       (d) −100, 583 psi
       (e) 40.2, 5.1 ksi
       (f) 0, 5.2 MPa

11.25 72.0, 144, 0 MPa

11.26 78.0, 0, −156 MPa

11.28 3.34, 6.67, 0 ksi

11.29 (a) 11.3, 11.3, −11.3 ksi
       (b) 10.9, 10.9, −10.9 ksi
       (c) 10.4, 10.4, −10.4 ksi
       (d) 10.9, 10.9, −10.9 ksi

11.31 (a) 260, 260, −260 MPa
       (b) 239, 239, −239 MPa
       (c) 218, 218, −218 MPa
       (d) 239, 239, −239 MPa

11.32 (a) 22.6, 22.6, −22.6 MPa
       (b) 59.7, 59.7, −59.7 MPa
       (c) 96.8, 96.8, −96.8 MPa
       (d) 59.7, 59.7, −59.7 MPa

CHAPTER 12

12.1 (a) 249, 125
      (b) 143, 44.3
      (c) 180, 46.5
      (d) 124
      (e) 192
      (f) 185, 121
      (g) 193, 152, 92.9

12.2 (a) 277, 139
      (b) 139, 46.2
      (c) 200
      (d) 160

12.4 (a) 7.07 kN
      (b) 169 kN
      (c) 257 kN
      (d) 176 kN

12.5 (a) 8160 lb
      (b) 510 lb
      (c) 4080 lb

CHAPTER 12 cont'd

12.7   (a)   122 kips
       (b)   7.60 kips
       (c)   60.8 kips

12.8   (a)   676 kN
       (b)   42.2 kN
       (c)   338 kN

12.10   (a)   1030 kN
       (b)   64.2 kN
       (c)   514 kN

12.11   (a)   172 kips
       (b)   10.7 kips
       (c)   85.8 kips

12.13   (a)   227 kips
       (b)   14.2 kips
       (c)   114 kips

12.16   Lower value will govern
       (a)   103, 343 kips
       (b)   201, 103 000 kips
       (c)   540, 521 kips
       (d)   600, 179 kips

12.17   Lower value will govern
       (a)   500, 1850 kN
       (b)   1.34, 421 MN
       (c)   1.49, 144 MN

12.19   2.5 in. standard

12.20   70.2 mm O.D.

12.22 AB, BE, BC, CD
       (a)   0.787 in.
       (b)   0.175 in.
       (c)   0.514 in.
       (d)   0.759 in.
       (e)   0.138 in.
       (f)   0.124 in.
       (g)   0.406 in.
       (h)   0
       (i)   0.124 in.

12.23 AB: 1.05 in.
       BC: 1.05 in.

12.25 AB: 35.9 mm
       CD: 52.0 mm
       BC: 23.3 mm

CHAPTER 13

13.1   (a)   28.8 ksi
       (b)   14.4 ksi
       (d)   7.20 ksi

13.2   (a)   200 MPa
       (b)   100 MPa
       (d)   50.0 MPa

13.4   5.40 MPa

13.5   15.0 ksi

13.7 $D = 7.01$ ft, $t = 0.525$ in.

13.8 $D = 1.56$ m, $t = 9.77$ mm

13.10   (a)   2.26 MPa
       (b)   21.2 MPa

13.11   (a)   159 psi
       (b)   2120 psi

13.13   39.0 kN

13.14   (a)   10.2 ksi
       (b)   4.00 ksi
       (c)   5.33 ksi
       (d)   8.00 ksi
       (e)   10.7 ksi

13.16   (a)   3.40 ksi
       (b)   2.00 ksi
       (c)   2.67 ksi

CHAPTER 13 cont'd

      (d) 2.67 ksi
      (e) 3.56 ksi

13.17 (a) 88.4 MPa
      (b) 29.8 MPa
      (c) 35.7 MPa
      (d) 69.4 MPa
      (e) 83.3 MPa

13.19 (a) 29.4 MPa
      (b) 18.9 MPa
      (c) 22.7 MPa
      (d) 23.1 MPa
      (e) 27.8 MPa

13.20 4.91 kips

13.22 Shear: 5.09 ksi
      Bearing: 8.00 ksi
      Tensile: 4.00 ksi

13.34 Shear: 40.7 MPa
      Bearing: 66.7 MPa
      Tensile: 33.3 MPa

13.25 102 kN

13.26 0.109 in.

13.28 4.49 mm

13.29 754 psi

13.31 6 in. by 0.101 in.

13.23 5 mm by 84.9 mm

13.34 2710 psi

13.35 11.5 MPa

13.37 10.25 MPa

13.38 0.0608 in.

CHAPTER 14

14.1 (a) 1660, 345 lb
     (b) 58.5, 878 psi
     (c) 1.76E-3 in.

14.2 (a) 696, 1300 lb
     (b) 27.7, 415 psi
     (c) 8.30E-4 in.

14.4 (a) 7.02, 13.0 kN
     (b) 0.447, 6.61 MPa
     (c) 47.9 μm

14.5 (a) 870, 1260 lb
     (b) 4430, 2850 psi
     (c) 277, 277μ
     (d) 13.3E-3 in.

14.7 (a) 3.48, 5.04 kN
     (b) 44.3, 28.6 MPa
     (c) 402, 402μ
     (d) 0.483 mm

14.8 (a) 9.24, 11.5 kN
     (b) 36.3, 23.4 MPa
     (c) 330, 330μ
     (d) 0.661 mm

14.10 (a) 13.7 kips
      (b) Aluminum: 10.9 kips each,
          Bronze: 28.2 kips
      (c) 17.979 in.

14.11 (a) 22.8 kips
      (b) Aluminum: 6.93 kips each,
          Bronze: 16.1 kips
      (c) 29.973 in.

14.13 (a) 79.9 kN
      (b) Aluminum: 588 kN each,
          Bronze: 1820 kN
      (c) 795.28 mm

14.14 (a) 1150 lb and 0.0336 in.,
          1150 lb and 0.0538 in.

CHAPTER 14 cont'd

(b) 1400 lb and 0.0652 in., 1400 lb and 0.0979 in.
(c) 10.0 kN and 2.82 mm, 10.0 kN and 8.45 mm
(d) 16.0 kN and 5.63 mm, 16.0 kN and 8.45 mm

14.16 47.0 MPa C, 0.602 MPa T

14.17 5850 psi C, 10.5 psi T

14.19 Bronze: 320 lb C, 1620 psi C, 3.80E-3, 0.182 in.
Aluminum: 3640 lb T, 8230 psi T, 3.80E-3, 0.182 in.

14.20 Bronze: 5630 lb T, 12,700 psi T, 4.80E-3, 0.173 in.
Aluminum: 3620 lb C, 4150 psi C, 4.80E-3, 0.173 in.

14.22 Bronze: 28.0 kN T, 110 MPa T 7.30E-3, 14.6 mm
Aluminum: 26.1 kN C, 53.2 MPa C, 7.30E-3, 14.6 mm

14.23
(a) $-148°F$
(b) Aluminum: 2610 psi T, Bronze: 3140 psi C
(c) Aluminum: 4650 psi C, Bronze 14,400 psi C

14.25 1600 lb, 0.0128 in.

14.26 1450 lb, 0.101 in.

14.28 9.37 kN, 0.604 mm

14.29 1.14 in., 3850 lb

14.31 49.4 mm, 822 N

14.32 369 mm, 29.1 kN

Index

Angular twist, 126, 129
Angular velocity, 113
Axial load diagram, 141
Axial loads, 46, 141, 147, 239, 352
Axial strain, 53, 63

Beams, 143
   deflection, 189, 195, 199, 202, 204, 206, 358
   design, 180
   diagrams, 152, 177, 182, 201, 220
   graph, 151
   slope, 194
   statically indeterminate, 145, 358
   types, 145
Bending moment, 112, 145, 148, 155, 175, 190, 192, 195, 239, 306
Bending moment diagram, 157, 177, 182, 202, 206
Bolted connections, 332
Boundary conditions, 195, 200, 307
Brittle materials, 61, 245, 296, 341
Buckling, 124, 296, 304

Center of gravity, 85, 88
Centroid, 87, 174, 225, 238
   of composite areas, 89
   by integration, 88
   of structural shapes, 91, 101
   tabular method, 99, 101
Coefficient of thermal expansion, 66
Coiled springs, 260
Column design, 319, 321
Columns, 304, 316

Combined stresses, 238, 271
Concrete structures, 245
Concurrent forces, 30
Connections, 51, 332
Couplings, 132
Critical load, 307
Critical stress, 309, 316
Curvature, 190, 192, 193

Deflection, 59, 84, 129, 189, 203, 353
   angular, 126, 129
   of beams, 189, 195, 199, 202, 204, 206, 358
Deformation, 54
Derivative, 12, 194, 279
Dimensional algebra, 3
Dimensional analysis, 3
Distributed load, 85
Ductile material, 57, 296, 341

Elastic limit, 57
Endurance limit, 344
Equilibrium, 30
Equivalent length, 312
Equivalent load, 85
Euler's equation, 305, 307, 317

Factor of safety, 70, 320
Fatigue, 342
Flexure formula, 173, 175, 180
Free body diagram, 27, 29, 34, 35, 36, 37, 51, 52, 71, 117, 141, 143, 146, 150, 153, 155, 158, 163, 177, 182, 196, 201, 241, 243, 248, 250, 254, 258, 260, 273, 274, 305, 334, 352, 357, 358, 360

417

Graphical integration, 199

Hollow shafts, 123
Hooke's Law, 55, 69, 117, 129, 171, 191
Horsepower, 113

Integration, 14, 16, 88, 93, 104, 119, 156, 173, 195, 199, 221, 306
Intermediate columns, 317
Internal forces, 46, 140, 145
Internal shear, 145, 147
International Systems of Units, 11

J. B. Johnson Formula, 321

Kern, 245

Lateral strain, 64
Long columns, 316

Margin of safety, 70
Masonry structures, 245
Metric units, 3, 10
Modulus of elasticity, 56, 69, 171, 191, 307
Modulus of rigidity, 69
Mohr's Circle, 287, 294, 331
Moment of area, 88, 222, 230
Moment of area method, 206
Moment of force, 26, 86, 112, 113, 122, 175
Moment of inertia, 92, 102, 175, 178, 191, 222, 230, 307, 308
   of composite areas, 97
   by integration, 93
   polar, 103, 119, 124, 129, 132
   of structural shapes, 101
   tabular method, 99, 101
Moore fatigue test, 343

Neutral axis, 169, 174, 222
Newton's Laws, 22, 30, 141, 182, 254, 272
Normal strain, 53, 170

Parallel axis theorem, 94, 100, 103
Parallelogram law, 23
Pascal, 48
Poisson's ratio, 63, 69
Power, 113
Power transmission, 111
Pressure vessels, 327, 330
Proportional limit, 56, 117

Radius of curvature, 192
Radius of gyration, 102, 103, 309
Riveted connections, 332

Secant formula, 321
Second moment of area, 92
Section modulus, 180
Shaft design, 120
Shear diagram, 152, 157, 220
Shear flow, 230
Shear load, 145, 147, 152, 155, 222, 230, 252, 262
Shear modulus of elasticity, 69, 129
Shear strain, 68, 127
Short columns, 316
SI 3, 10
Significant figures, 6
Slenderness ratio, 309, 316
Slope, 194
S-N diagram, 343
Spring constant, 59
Spring index, 262
Springs, 144, 260
Statically indeterminate members, 351
Strain:
   axial, 53, 63
   lateral, 64
   normal, 53, 170
   shear, 68, 127
   thermal, 66
   total, 54
   unit, 54
Strength:
   bearing, 333
   fatigue, 344
   offset yield, 62
   shear, 333
   tensile, 58, 333
   ultimate, 58, 70
   yield, 57, 62, 316
Stress:
   allowable, 69
   bearing, 48, 334
   bending, 171, 239, 241, 249, 339
   combined, 242
   concentration, 338
   critical, 309, 316
   design, 70
   hoop, 329
   longitudinal, 328
   longitudinal shear, 121, 223
   normal, 46, 171, 238, 239, 241, 249, 272, 277, 284, 289, 338
   plane, 271, 277
   principal, 281, 289, 294, 331
   shear, 49, 68, 117, 121, 219, 222, 226, 227, 229, 249, 252, 262, 272, 277, 284, 289, 294, 334, 336, 339

tensile, 335, 336
  thermal, 68
Stress concentration factors, 340
Stress-strain curve, 57, 62
Superposition, 204, 242, 249, 252, 262, 359
Support reactions, 29

Tensile test, 56
Tensors, 271
Thermal expansion, 66
Thermal strain, 66
Three force members, 33
Torque, 113, 117, 253
Torsion, 118, 294
Total strain, 54

Twist, 126
Two force members, 33

Unit conversion, 5, 48, 114, 199
Units, 9, 11, 48, 115, 129
Unit strain, 54

Vectors, 25, 149, 271

Wahl correction factor, 263
Weldments, 335
Work, 112

Yield point, 57
Young's modulus, 56